Chemical Ecology in Forests

Chemical Ecology in Forests

Chunjian Zhao
Zhi-Chao Xia
Chunying Li
Jingle Zhu

Basel • Beijing • Wuhan • Barcelona • Belgrade • Novi Sad • Cluj • Manchester

Editors

Chunjian Zhao
Key Laboratory of Forest
Plant Ecology
Ministry of Education
Northeast Forestry University
Harbin
China

Zhi-Chao Xia
School of Forestry and
Landscape Architecture
Anhui Agricultural
University
Hefei
China

Chunying Li
College of Chemistry,
Chemical Engineering and
Resource Utilization
Northeast Forestry University
Harbin
China

Jingle Zhu
Research Institute of
Non-timber Forestry
Chinese Academy of Forestry
Zhengzhou
China

Editorial Office
MDPI AG
Grosspeteranlage 5
4052 Basel, Switzerland

This is a reprint of articles from the Special Issue published online in the open access journal *Forests* (ISSN 1999-4907) (available at: www.mdpi.com/journal/forests/special_issues/forests_chemical).

For citation purposes, cite each article independently as indicated on the article page online and as indicated below:

Lastname, A.A.; Lastname, B.B. Article Title. *Journal Name* **Year**, *Volume Number*, Page Range.

ISBN 978-3-7258-2210-2 (Hbk)
ISBN 978-3-7258-2209-6 (PDF)
doi.org/10.3390/books978-3-7258-2209-6

© 2024 by the authors. Articles in this book are Open Access and distributed under the Creative Commons Attribution (CC BY) license. The book as a whole is distributed by MDPI under the terms and conditions of the Creative Commons Attribution-NonCommercial-NoDerivs (CC BY-NC-ND) license.

Contents

About the Editors vii

Preface ix

Chunjian Zhao, Zhi-Chao Xia, Chunying Li and Jingle Zhu
Chemical Ecology in Forests
Reprinted from: *Forests* 2024, *15*, 1571, doi:10.3390/f15091571 1

You Xu, Xin Chen, Le Ding and Chui-Hua Kong
Allelopathy and Allelochemicals in Grasslands and Forests
Reprinted from: *Forests* 2023, *14*, 562, doi:10.3390/f14030562 5

Qun Zhao, Zheng Wang, Gaiping Wang, Fuliang Cao, Xiaoming Yang and Huiqin Zhao et al.
Effects of UVA on Flavonol Accumulation in *Ginkgo biloba*
Reprinted from: *Forests* 2024, *15*, 909, doi:10.3390/f15060909 27

Lijun Zhong, Hongxing Dong, Zhijun Deng, Jitao Li, Li Xu and Jiaolin Mou et al.
Physiological Mechanisms of *Bretschneidera sinensis* Hemsl. Seed Dormancy Release and Germination
Reprinted from: *Forests* 2023, *14*, 2430, doi:10.3390/f14122430 44

Josiane Costa Maciel, Tayna Sousa Duque, Aline Cristina Carvalho, Brenda Thaís Barbalho Alencar, Evander Alves Ferreira and José Cola Zanuncio et al.
Development of Commercial Eucalyptus Clone in Soil with Indaziflam Herbicide Residues
Reprinted from: *Forests* 2023, *14*, 1923, doi:10.3390/f14091923 57

Hong Jiang, Shanchun Yan, Zhaojun Meng, Shen Zhao, Dun Jiang and Peng Li
Effects of the Larch–Ashtree Mixed Forest on Contents of Secondary Metabolites in *Larix olgensis*
Reprinted from: *Forests* 2023, *14*, 871, doi:10.3390/f14050871 71

Qiqiang Guo, Huie Li, Xueguang Sun, Zhengfeng An and Guijie Ding
Patterns of Needle Nutrient Resorption and Ecological Stoichiometry Homeostasis along a Chronosequence of *Pinus massoniana* Plantations
Reprinted from: *Forests* 2023, *14*, 607, doi:10.3390/f14030607 82

Rongrong Chen, Jingle Zhu, Jiabing Zhao, Xinru Shi, Wenshi Shi and Yue Zhao et al.
Relationship between Leaf Scorch Occurrence and Nutrient Elements and Their Effects on Fruit Qualities in Chinese Chestnut Orchards
Reprinted from: *Forests* 2022, *14*, 71, doi:10.3390/f14010071 95

Chunjian Zhao, Sen Shi, Naveed Ahmad, Yinxiang Gao, Chunguo Xu and Jiajing Guan et al.
Promotion Effects of *Taxus chinensis* var. *mairei* on *Camptotheca acuminata* Seedling Growth in Interplanting Mode
Reprinted from: *Forests* 2022, *13*, 2119, doi:10.3390/f13122119 109

Yiming Zhong, Ang Zhang, Xiaowei Qin, Huan Yu, Xunzhi Ji and Shuzhen He et al.
Effects of Intercropping *Pandanus amaryllifolius* on Soil Properties and Microbial Community Composition in *Areca Catechu* Plantations
Reprinted from: *Forests* 2022, *13*, 1814, doi:10.3390/f13111814 124

Jiaying Liu, Yawei Wei, Haitao Du, Wenxu Zhu, Yongbin Zhou and You Yin
Effects of Intercropping between *Morus alba* and Nitrogen Fixing Species on Soil Microbial Community Structure and Diversity
Reprinted from: *Forests* **2022**, *13*, 1345, doi:10.3390/f13091345 . . . . **144**

About the Editors

Chunjian Zhao

Chunjian Zhao is a professor and doctoral supervisor at the College of Chemistry, Chemical Engineering and Resource Utilization, Northeast Forestry University. He currently holds the position of first-level subject leader in pharmacy, director of the Engineering Research Center of Forest Bio-Preparation, Ministry of Education, as well as deputy director of the Key Laboratory of Forest Plant Ecology, Ministry of Education. Additionally, he serves as Vice Chairman of the Botanical Society and Executive Director of the Ecological Society in Heilongjiang Province. His primary research focuses on targeted induction, separation, and purification techniques; quality control methods; and chemical inter-relationships among bioactive substances found in medicinal plant resources.

Zhi-Chao Xia

Prof. Zhi-Chao Xia is the head of the Forest and Soil Ecology Team at the College of Forestry and Landscape Architecture, Anhui Agricultural University. His research primarily focuses on subtropical forests, specifically conducting independent studies on the interactions between microorganisms and trees, rhizospheric microbial processes and regulation, as well as the environmental adaptation mechanisms of trees in hydraulic structures.

Chunying Li

Dr. Chunying Li is an associate professor and supervisor of doctoral students at the College of Chemistry, Chemical Engineering, and Resource Utilization, Northeast Forestry University. She also serves as the director of the Heilongjiang Botanical Society. Her research primarily focuses on plant interspecies relationships, plant resource science, medicinal plant active ingredient screening, and functional ingredient evaluation.

Jingle Zhu

Dr. Jingle Zhu is a researcher and director of the Transformation and Industrial Development Division of the Economic Forest Research Institute of the Chinese Academy of Forestry. He is mainly engaged in the cultivation and breeding research of woody grain species such as chestnut and science popularization. Dr. Zhu presided over the National Forestry and Grassland Administration's science popularization project "Exploration and Popularization of the Charm of Forest Food on National Science Popularization Day", and completed six science popularization works such as "Wall Map of Major forest food types in China". His research is primarily focused on non-timber forest cultivation.

Preface

In nature, forest ecosystems contain the wonderful secrets of life. This reprint seeks to explore a pivotal aspect of this enigmatic domain: chemical ecology. As a significant branch of ecology, chemical ecology reveals how organisms use chemicals to communicate, compete, defend, and reproduce. Using detailed scientific data and case studies in the literature, this reprint explains how chemical signals generate, transmit, receive, and respond in forest ecosystems.

Chemical ecology provides us with a unique insight into the interactions of bioorganisms that are hidden deep in the forest and invisible to the naked eye. This reprint aims to present interesting findings in this field to a general audience through scientific language and inspire more people to examine the chemical ecology of forests.

This reprint is suitable for all readers interested in ecology, biochemistry, and environmental science who are seeking inspiration and useful knowledge in this field. We hope to bridge theory and practice in chemical ecology.

This reprint is co-authored by a group of experts and scholars with extensive research experience and profound academic achievements in the field of chemical ecology. They are enthusiastic and have a rigorous approach to translating complex scientific issues into easily comprehensible language, aiming to captivate readers with the allure of science while fostering a deep understanding of the indispensable role played by chemical ecology in forest ecosystems. Throughout the writing process, our colleagues generously shared their research findings, and we would like to express our gratitude to them.

Chunjian Zhao, Zhi-Chao Xia, Chunying Li, and Jingle Zhu
Editors

Editorial

Chemical Ecology in Forests

Chunjian Zhao [1,2,3,4,*], Zhi-Chao Xia [5], Chunying Li [1,2,3,4] and Jingle Zhu [6]

1. Key Laboratory of Forest Plant Ecology, Ministry of Education, Northeast Forestry University, Harbin 150040, China; lcy@nefu.edu.cn
2. College of Chemistry, Chemical Engineering and Resource Utilization, Northeast Forestry University, Harbin 150040, China
3. Engineering Research Center of Forest Bio-Preparation, Ministry of Education, Northeast Forestry University, Harbin 150040, China
4. Heilongjiang Provincial Key Laboratory of Ecological Utilization of Forestry-Based Active Substances, Harbin 150040, China
5. School of Forestry & Landscape Architecture, Anhui Agricultural University, Hefei 230036, China; zhichaoxia@hznu.edu.cn
6. Research Institute of Non-Timber Forestry, Chinese Academy of Forestry, Zhengzhou 450003, China; zhujingle@caf.ac.cn
* Correspondence: zcj@nefu.edu.cn

Citation: Zhao, C.; Xia, Z.-C.; Li, C.; Zhu, J. Chemical Ecology in Forests. *Forests* 2024, *15*, 1571. https://doi.org/10.3390/f15091571

Received: 30 July 2024
Accepted: 12 August 2024
Published: 7 September 2024

Copyright: © 2024 by the authors. Licensee MDPI, Basel, Switzerland. This article is an open access article distributed under the terms and conditions of the Creative Commons Attribution (CC BY) license (https://creativecommons.org/licenses/by/4.0/).

There is a competitive and coordinated relationship among organisms which depends on their chemical connections. Chemical relationships are an important way for organisms to interact with each other. Various levels of organisms and those without a nutritional relationship are linked by chemicals, forming a vast network of chemical information. It can be said that the relationship between organisms is actually a chemical relationship. Forest ecosystems, which have a complex structure and diverse chemical relationships, encompass the majority of woody plants and animals worldwide. Chemical ecology is an important aspect of the exploration of forest ecosystems.

Plants can sense and recognize coexisting species of the same or different species, thereby adjusting their growth, reproduction, and defense strategies. Throughout its life cycle, a plant can interact directly or indirectly with its plant neighbors, thereby influencing the plant population and community structure. Agroforestry is an effective method to improve plant productivity by rationally regulating the interspecific relationships between plants. Zhao et al. conducted a comparative analysis of the growth indicators of *Camptotheca acuminata* (*C. acuminata*) cultivated in monoculture and intercropping systems, revealing a significantly higher growth rate for *C. acuminata* in the intercropping system compared to that in the monoculture system [1], and the soil properties in the mixed planting system were significantly improved. The authors believe that this positive effect is probably due to plant allelopathy. Using mass spectrometry, it was found that in *C. acuminata* rhizosphere soil exists taxanes, a class of allelochemicals unique to *Taxus chinensis* var. *mairei*. Allelopathy may be one of the important factors in promoting the growth of *C. acuminata* seedlings by interplanting *Taxus chinensis* var. *mairei*, which is an important chemical link in the interactions between plant species [1].

Plant allelopathy is a natural ecological phenomenon. When plants are ingested by animals and infected by microorganisms, plants often respond by synthesizing and releasing allelochemicals. Plants adjust their biomass distribution by recognizing information from neighboring species to decide whether to adopt chemical defense strategies. Hong et al. established a mixed forest of *Larix olgensis* and *Fraxinus mandshurica*, and found that intercropping significantly increased the content of secondary metabolites (phenolic compounds) in *Larix olgensis*, thereby enhancing its chemical defense ability [2]. The allelomic effect of *Fraxinus mandshurica* on *Larix olgensis* was found to be related to the mixing ratio, and the chemical defense ability of high-proportionally mixed forests was effective [2]. Secondary metabolites are the result of long-term evolutionary interactions between plants and their living environment, affecting the color, odor, and taste of the plant. Phenolic compounds can inhibit the digestion and utilization of food by herbivorous insects, thus affecting insect activity. Phenolic compounds are an important indicator for plants to resist pests, and they play a crucial role in the process of plant resistance to pests [2].

Allelochemicals can regulate forest biodiversity, productivity, and sustainability. A comprehensive understanding of allelochemicals can offer novel insights into the sustainable development of forest ecosystems. Xu et al. described allelopathy in forest ecosystems from three levels: forest, plantation, and understory vegetation [3]. Meanwhile, the author also summarized the main categories of allelochemicals in forest ecosystems and proposed that the identification of allelochemicals requires accurate information on the quantity, quality, and temporal and spatial dynamics of allelochemicals, otherwise it will be difficult to accurately understand the functional significance of allelopathic plant–plant interactions in forests [3]. The authors emphasize that allelochemicals can change the consequences of underground ecological interactions, and suitable mixed tree species can enhance their growth through underground chemical interactions [3]. Allelochemicals and signaling chemicals work synergistically to affect the coexistence, diversity, and community structure of forest plants. The proper use of kinship identification between plants can even help forest regeneration.

In addition to interactions between plants, plants also produce endogenous chemical signals to regulate their own growth. Plant hormones are active substances produced by plant cells in response to specific environmental signals and can regulate plant physiological responses. Zhong et al. comprehensively analyzed the mechanisms for seed dormancy release and germination in *Bretschneidera sinensis* Hemsl, revealing that the ratio of GA3 (gibberellin A3)/ABA (abscisic acid) in seeds plays a pivotal role in determining seed dormancy release and germination, and also revealing that seed germination requires the interaction of hormones [4]. Additionally, the author points out that seeds can break their dormancy and stimulate germination by increasing the level of soluble sugars, which provide energy for seed dormancy to germination [4]. The increased soluble sugar levels can also promote the removal of ROS (reactive oxygen species), protecting the seeds from oxidative stress [4].

Chemical herbicides are often used in plantation cultivation, and the application of herbicides can be toxic to plants, so chemical controls must be used carefully. Herbicides have a lower selectivity for eucalyptus plantations, which may cause losses in early tree development and lead to a loss in productivity. Indaziflam herbicide is one of the herbicides that is relatively safe for crops. Little is known about the tolerance of indaziflam herbicide in *Eucalyptus* plantations. Maciel et al. evaluated the persistent effects of indaziflam herbicide and its impact on plant growth [5]. The results show that the content of chlorophyll a and b, the rate of electron transport, height, and stem mass of plants in soil contaminated by indaziflam herbicide residues were all lower [5]. Indaziflam herbicide was applied to eucalyptus plants, and it was found that indaziflam herbicide could be leached to a depth of 30 cm in the soil [5]. Residues of indaziflam herbicide in the soil inhibit the growth of *Eucalyptus Clone*.

The mechanism by which organisms perceive information chemicals is an important subject of study in chemoecology. Plants can respond to their surroundings by producing chemical signaling substances, and they can also share these chemical signals with other plants. This "communication", dominated by chemicals, can change the microenvironment for plant growth, regulate nutrient supply, and even affect plant yield.

Microorganisms are the executors and drivers of energy flow in soil, and plants can influence the rhizosphere soil microbial community through root exudates or litter. In turn, changes in soil biological characteristics can affect the host plant and its coexisting plants. Zhong et al. established an intercropping field between *Areca catechu* L. and *Pandanus amaryllifolius* Roxb [6]. It was found that intercropping had positive effects on soil microbial homeostasis in plantations by comparing them with monoculture [6]. What is more interesting is that the authors found a special correlation between the soil's physical and chemical properties, enzyme activity, and microorganisms [6]. Urease and phosphatase are the key factors that regulate the abundance of the soil's microbial community. Compared with fungi, the authors suggest that bacterial communities are more sensitive to interplant relationships and that bacteria are more responsive to changes in soil environmental factors [6].

The composition of soil microbial communities affects the availability of soil nutrients. To determine the correlation between soil nutrients and microbial diversity after the introduction of other plant species, Liu et al. conducted a mixed planting experiment with the legume species *Lespedeza bicolor* Turcz. and the mulberry species *Morus alba* [7]. It was found that intercropping significantly increased the contents of C, N, and P in soil. Nitrogen-fixing plants increase the productivity of plants by increasing the availability of soil nutrients, especially nitrogen, and provide essential base metabolites for more microbial growth [7]. *Actinobacteria* has a high soil abundance due to the soil's nutrient and organic matter content. Due to the harsh environment, *Proteobacteria*, which has certain resistance in the face of extreme environment, dominates the soil [7]. By conducting 16S rRNA and ITS sequencing on soil microorganisms, the authors found that there was a significant change in diversity within the bacterial community compared to the fungal community [7]. Therefore, soil microbial communities serve as an important link between aboveground plant communities and underground ecological processes, regulating the material cycling process in forest ecosystems and the flow of energy in the soil.

Due to biological and abiotic influences, the accumulation, release, and transformation of soil nutrients often change. No matter how environmental factors change, the ecological stoichiometric value of a plant species usually remains relatively constant, which is called stoichiometric homeostasis. In order to further understand the adaptability of trees under nutrient changes, Guo et al. explored the changing rules of nutrient uptake and ecological stoichiometric homeostasis in *Pinus massoniana* plantation [8]. The authors found that there was a synergistic effect between the leaf litter and soil, and that the ecological stoichiometry and nutrient uptake of different aged trees were variable [8]. The results showed that P content decreased first and then increased with the increase in plantation age, which was different from the conventional rule of increasing accumulation of nutrients alongside an increase in the time sequence [8]. The absorption efficiency of N and P first increased and then decreased during the growth of the *Pinus massoniana* plantation [8]. The increase in nutrient element absorption promoted the growth of *Pinus massoniana*. The author believes that introducing suitable tree species and planting them with *Pinus massoniana* can achieve more effective artificial forest cultivation, which is an effective strategy to alleviate nutrient limitations [8].

The pharmacological effects of plants are derived from compounds produced by the secondary metabolism in plants. External environmental factors can regulate the production of a plant's active substances. Photosynthesis is the process by which plants produce nutrients, which is crucial for the production of secondary metabolites. Zhao et al. treated *Ginkgo biloba* with UVA to explore the molecular mechanism of the influence of light on the synthesis of flavonols, the plant's active ingredient, thereby improving the quality of *Ginkgo biloba* [9]. The results showed significant differences in flavonol content and enzyme activity in the phenylpropane pathway in plants under different intensities of UVA [9]. Moderate UVA intensity can promote enzyme activity related to the flavonoid synthesis pathway and flavonol accumulation in *Ginkgo biloba*, while excessive UVA plays an inhibitory role [9]. The authors indicate that the enhancement of the flavonoid content's medicinal value in *Ginkgo biloba* can be achieved by stimulating the expression of related enzyme genes (MYB (Gb_02997), bHLH (Gb_05320), bZIP (Gb_00122), and NAC (Gb_13200, Gb_37720)) [9].

Environmental factors can also trigger plant diseases and reduce plant quality. In Yanshan chestnut garden, there were symptoms of scorching between the leaf's margin and vein. Different from the previous reports of leaf burn disease, no pathogenic bacteria were detected in the infected plants. In order to explore the main factors leading to *Castanea mollissima* leaf scorching and the effect of leaf scorching disease on the characteristics of the nuts, Chen et al. analyzed and compared the differences in the leaf, root, and soil nutrients, nut phenotypes, and antioxidant enzyme activities between healthy and leaf-scorched trees [10]. Leaf scorching has a significant impact on the morphology and traits of *Castanea mollissima*' nuts [10]. The correlation analysis results show that B, Zn, Mg, and Fe have a significant impact on the health of leaves [10]. The soil AK, K Fe, B, and Cu have a significant impact on the leaf's B concentration. The author believes that *Castanea mollissima* leaf scorching may be caused by the high content of B in leaves and the lack of Mg, which is related to the change in the balance of AK, B, Mg, Cu, and Fe in the soil [10]. The decrease in Mg is most likely caused by the soil's AK [10].

This Special Issue of the *Forests* journal, "Forest Chemical Ecology", covers the study of the chemical connection between organisms and their mechanisms. This Special Issue covers discussions on pesticide pollution, pest resistance, and the intrinsic causes of interspecies relationships, and provides guidance on pest control, biodiversity conservation, and the rational utilization of biological resources in forest ecosystems. The research reports contained in this Special Issue provide important insights for realizing the sustainable development of forest ecological systems.

Conflicts of Interest: The author declares no conflicts of interest.

References

1. Zhao, C.J.; Shi, S.; Ahmad, N.; Gao, Y.X.; Xu, C.G.; Guan, J.J.; Fu, X.D.; Li, C.Y. Promotion Effects of *Taxus chinensis* var. *mairei* on *Camptotheca acuminata* Seedling Growth in Interplanting Mode. *Forests* **2022**, *13*, 2119. [CrossRef]
2. Jiang, H.; Yan, S.C.; Meng, Z.J.; Zhao, S.; Jiang, D.; Li, P. Effects of the Larch-Ashtree Mixed Forest on Contents of Secondary Metabolites in *Larix olgensis*. *Forests* **2023**, *14*, 871. [CrossRef]
3. Xu, Y.; Chen, X.; Ding, L.; Kong, C.H. Allelopathy and Allelochemicals in Grasslands and Forests. *Forests* **2023**, *14*, 562. [CrossRef]
4. Zhong, L.J.; Dong, H.X.; Deng, Z.J.; Li, J.T.; Xu, L.; Mou, J.L.; Deng, S.M.; Valbuena, L. Physiological Mechanisms of *Bretschneidera sinensis* Hemsl. Seed Dormancy Release and Germination. *Forests* **2023**, *14*, 2430. [CrossRef]
5. Maciel, J.C.; Duque, T.S.; Carvalho, A.C.; Alencar, B.T.B.; Ferreira, E.A.; Zanuncio, J.C.; Castro, B.M.; da Silva, F.D.; Silva, D.V.; dos Santos, J.B. Development of Commercial Eucalyptus Clone in Soil with Indaziflam Herbicide Residues. *Forests* **2023**, *14*, 1923. [CrossRef]
6. Zhong, Y.M.; Zhang, A.; Qin, X.W.; Yu, H.; Ji, X.Z.; He, S.Z.; Zong, Y.; Wang, J.; Tang, J.X. Effects of Intercropping *Pandanus amaryllifolius* on Soil Properties and Microbial Community Composition in *Areca catechu* Plantations. *Forests* **2022**, *13*, 1814. [CrossRef]
7. Liu, J.Y.; Wei, Y.W.; Du, H.T.; Zhu, W.X.; Zhou, Y.B.; Yin, Y. Effects of Intercropping between *Morus alba* and Nitrogen Fixing Species on Soil Microbial Community Structure and Diversity. *Forests* **2022**, *13*, 1345. [CrossRef]
8. Guo, Q.Q.; Li, H.E.; Sun, X.G.; An, Z.F.; Ding, G.J. Patterns of Needle Nutrient Resorption and Ecological Stoichiometry Homeostasis Along a Chronosequence of *Pinus massoniana* Plantations. *Forests* **2023**, *14*, 607. [CrossRef]
9. Zhao, Q.; Wang, Z.; Wang, G.P.; Cao, F.L.; Yang, X.M.; Zhao, H.Q.; Zhai, J.T. Effects of Uva on Flavonol Accumulation in *Ginkgo biloba*. *Forests* **2024**, *15*, 909. [CrossRef]
10. Chen, R.R.; Zhu, J.L.; Zhao, J.B.; Shi, X.R.; Shi, W.S.; Zhao, Y.; Yan, J.W.; Pei, L.; Jia, Y.X.; Wu, Y.Y.; et al. Relationship between Leaf Scorch Occurrence and Nutrient Elements and Their Effects on Fruit Qualities in Chinese Chestnut Orchards. *Forests* **2023**, *14*, 71. [CrossRef]

Disclaimer/Publisher's Note: The statements, opinions and data contained in all publications are solely those of the individual author(s) and contributor(s) and not of MDPI and/or the editor(s). MDPI and/or the editor(s) disclaim responsibility for any injury to people or property resulting from any ideas, methods, instructions or products referred to in the content.

Review

Allelopathy and Allelochemicals in Grasslands and Forests

You Xu [1], Xin Chen [2], Le Ding [2] and Chui-Hua Kong [2,*]

[1] Institute of Ecological Conservation and Restoration, Chinese Academy of Forestry, Beijing 100091, China
[2] College of Resources and Environmental Sciences, China Agricultural University, Beijing 100193, China
* Correspondence: kongch@cau.edu.cn

Abstract: Plants can produce and release allelochemicals to interfere with the establishment and growth of conspecific and interspecific plants. Such allelopathy is an important mediator among plant species in natural and managed ecosystems. This review focuses on allelopathy and allelochemicals in grasslands and forests. Allelopathy drives plant invasion, exacerbates grassland degradation and contributes to natural forest regeneration. Furthermore, autotoxicity (intraspecific allelopathy) frequently occurs in pastures and tree plantations. Various specialized metabolites, including phenolics, terpenoids and nitrogen-containing compounds from herbaceous and woody species are responsible for allelopathy in grasslands and forests. Terpenoids with a diversity of metabolites are qualitative allelochemicals occurring in annual grasslands, while phenolics with a few specialized metabolites are quantitative allelochemicals occurring in perennial forests. Importantly, allelochemicals mediate below-ground ecological interactions and plant–soil feedback, subsequently affecting the biodiversity, productivity and sustainability of grasslands and forests. Interestingly, allelopathic plants can discriminate the identity of neighbors via signaling chemicals, adjusting the production of allelochemicals. Therefore, allelochemicals and signaling chemicals synergistically interact to regulate interspecific and intraspecific interactions in grasslands and forests. Allelopathy and allelochemicals in grasslands and forests have provided fascinating insights into plant–plant interactions and their consequences for biodiversity, productivity and sustainability, contributing to our understanding of terrestrial ecosystems and global changes.

Keywords: allelopathic interference; autotoxicity; below-ground chemical interactions; plant neighbor detection; plant–soil feedback; qualitative and quantitative allelochemicals

Citation: Xu, Y.; Chen, X.; Ding, L.; Kong, C.-H. Allelopathy and Allelochemicals in Grasslands and Forests. *Forests* **2023**, *14*, 562. https://doi.org/10.3390/f14030562

Academic Editor: Mark D. Coleman

Received: 30 December 2022
Revised: 5 March 2023
Accepted: 10 March 2023
Published: 13 March 2023

Copyright: © 2023 by the authors. Licensee MDPI, Basel, Switzerland. This article is an open access article distributed under the terms and conditions of the Creative Commons Attribution (CC BY) license (https://creativecommons.org/licenses/by/4.0/).

1. Introduction

Grasslands and forests are integral components of the global ecosystem, totally covering about 70% of the earth's terrestrial area. Both function as the crucial global pool of biodiversity to supply a wide range of species, and their productivity and sustainability modulate global changes [1–3]. Importantly, grasslands and forests play substantial roles in diverse ecological services to generate tremendous benefits for humans, such as water conservation, sand fixation, carbon sequestration, oxygen release and global biogeochemical cycles [4,5]. Understanding the biodiversity, productivity and sustainability of grasslands and forests and their underlying mechanisms has been of great interest to ecologists for decades.

The biodiversity, productivity and sustainability of grasslands and forests are the net outcomes of various biotic versus abiotic feedbacks between plants and their environment. These can arise through a variety of mechanisms such as resource partitioning, niche divergence, plant–soil and other species-specific interactions [6–8], but the central driver must be interspecific and intraspecific plant–plant interactions that can be neutral (consummation and recognition), positive (facilitation and kin selection) and negative (competition and allelopathy) to allow local coexistence. The interactions, either beneficial, harmful or commensal, eventually contribute to the biodiversity, productivity and sustainability of grasslands and forests. While most studies have focused on resource competition, environmental factors and global changes, relatively little is known about the importance of allelopathy in grassland and forest ecological processes [9].

A plant may interfere with the growth and establishment of neighboring plants through competition, allelopathy or both. Differing from competition for resources, allelopathy is an interference mechanism in which living or dead plants release allelochemicals exerting an effect (mostly negative) on co-occurring plants [10,11], even within a species (i.e., autotoxicity or intraspecific allelopathy). Four ecological processes, volatilization, leaching, litter decomposition and root exudation, can bring allelochemicals into air or soil. When allelochemicals contact or approach the associated plants, they directly demonstrate allelopathic action by disturbing the systems of photosynthesis, respiration, and metabolism, or indirectly affect target species by altering environmental conditions, particularly for soil physicochemical properties and microbial communities [12–14]. In fact, allelopathy originates from interspecific and intraspecific plant–plant interactions in grasslands and forests. The first classical case is black walnut (*Juglans nigra*), which produced and released a 1,4-naphthoquinone (juglone) to interfere with the growth of understory plants thousands of years ago [15]. The allelopathic interference of shrubs in grass through the release of volatile terpenes into southern California coastal grassland was reported in the 1960s [16]. Subsequently, an increasing number of studies have shown that many ecological events occurring in grasslands and forests are associated with allelopathy and certain allelochemicals [9–11,17,18].

Allelopathy in grasslands and forests is key for understanding terrestrial ecosystems and global changes. In recent decades, numerous control experiments and field investigations have been conducted to estimate the functional consequences of allelopathy for plant communities in natural and managed grasslands and forests. However, a comprehensive allelopathy, particularly for allelochemical-mediated below-ground and above-ground interactions in grasslands and forests, is rare. Understanding allelopathy with allelochemicals and their consequences for biodiversity, productivity and sustainability in grasslands and forests can provide new insight into terrestrial ecosystems and global changes. Hence, capturing recent advances and applications in allelopathy and allelochemicals is becoming valuable in advancing interdisciplinary research in grasslands and forests.

2. Allelopathy in Grasslands
2.1. Allelopathy Drives Plant Invasion in Grasslands

The occurrence of invasive plants threatens the structure and function of grassland ecosystems, especially in biodiversity and stability [17]. Several plant species have been confirmed to invade grasslands with an allelopathic mechanism. Spotted knapweed (*Centaurea stoebe*), native to Europe and introduced into North America, is an example of an invasive plant in western American grasslands. Spotted knapweed can take advantage of root-secreted allelochemicals against local grassland species and alter nutrition availability and underground microbial community composition [18,19]. However, the allelopathy of spotted knapweed is conditional, and there is discrepancy between geographical sites. Spotted knapweed does not exhibit allelopathic invasion in eastern American grasslands [20]. Additionally, sufficient light or infection with fungal endophytes can enhance the allelopathic invasion of spotted knapweed in American grasslands [20,21].

Allelopathic invasion of spotted knapweed in American grasslands results in the novel weapons hypothesis (NWH) that the success of plant invasion can be attributed to the allelochemicals of invaders [22]. Generally, allelochemicals of invasive species have little effect on their original neighbors due to long-term mutual adaptation, but as they are novel to the species of the invaded habitat, they exert a strongly allelopathic interference on the native species [22]. Much evidence has demonstrated that allelochemicals appear to confer a competitive advantage to the invasive plants [23–26]. However, some studies did not fully support the NWH, and questioned the necessity of secondary metabolites for nonnative species to ensure invasive success [27–29]. Another hypothesis, the biochemical recognition hypothesis (BRH), postulates that plant seeds can adaptively detect phytochemicals released from potential competitors and respond by extending their period of dormancy until better establishment conditions occur [30]. Leachates from spotted knapweed reduced the germination rate of grassland species. Importantly, they had no effect on seeding biomass, implying that the allelochemicals in the leachates are non-phytotoxic and do not impede plant growth [30].

Although both the NWH and BRH focus on plant-derived chemicals and predict similar results that phytochemicals released from invasive plants inhibit the emergence of native plants, their fundamental mechanisms are distinct. This can be explained either a negative exposure to toxic chemicals by NWH or a positive recognition of facilitative chemicals by BRH [22,30]. Nevertheless, whether the success of invasive plants is attributed to allelochemicals has been debated. Actually, allelopathy is pervasive in invasive plants [31]. Interestingly, allelopathy of native grassland communities seems to increase their resistance to invasion by introduced plants [32], but there was no evidence that native plant communities' tolerance to allelopathy contributes to the degree of invasiveness of introduced plants. A more vital linkage between allelopathic traits and invasive performance needs to be explored in further studies.

2.2. Allelopathy Exacerbates Grassland Degradation

Grassland degradation is a phenomenon in which grass struggles to grow or hardly survives, which usually leads to an irreversible reduction in grassland productivity and biodiversity [33]. Many factors have been regarded as the drivers of grassland degradation, of which the main factors are natural climate change and human disturbance [34,35]. One early sign of degraded grassland is that the originally dominant species are gradually replaced by other adaptable plants, such as toxic weeds with allelopathic traits [36–38]. Toxic weeds in degraded grassland are adapted to extremely harsh environmental conditions and exhibit high aggression toward surrounding plants, even poisoning livestock or humans [39,40].

In the process of grassland degradation, toxic weeds not only vigorously compete with forage plants for water and nutrition resources, but also produce a wide range of secondary metabolites to exert allelopathic effects on the establishment of the co-occurring plants, subsequently reducing species richness and exacerbating grassland degradation [41–43]. Several studies have shown that extracts of toxic weeds, regardless of plant tissues or growing soil, can reduce the seed germination rate and seedling biomass of the receiving plants [38,44,45]. However, the allelopathic effects have distinct differences among the extract concentration, extract source and tested species [44]. Many phytotoxic compounds, such as coumarins, flavonoids and terpenoids, have been isolated and identified from toxic weeds. These potential allelochemicals could jeopardize the photosynthesis, respiration, and metabolic system of plants [46–48].

Stellera chamaejasme and *Artemisia frigida* are representatives of toxic weeds and generally serve as bioindicators to characterize the degree of grassland degradation. *S. chamaejasme* is a common toxic weed in the degraded grasslands of northern China, which can restrict the growth of co-occurring plants via root exudates [38,49]. *A. frigida*, a perennial dicotyledonous semi-shrub species, has a wide distribution range in the global temperate grasslands, covering Eurasian steppes and northern mixed-grass prairies. Differing from the mainly allelopathic pathway of *S. chamaejasme*, *A. frigida* can significantly decrease seed germination and seedling growth by emitting volatile organic compounds (VOCs) as allelochemicals [50,51]. This environmental disturbance may severely influence the composition and abundance of VOCs emitted from *A. frigida*. Artificial damage can induce *A. frigida* to release more categories and greater concentrations of VOCs [51]. In particular, grazing activity can enhance the allelopathic effect on the growth of other grassland species, suggesting that allelopathy may interact with over-grazing grassland to accelerate the grassland deterioration by frequently simulating *A. frigida* [52].

Overall, allelopathy is one of the critical factors driving grassland degradation. Comprehensively understanding of how allelochemicals from toxic weeds mediate intraspecific and interspecific plant–plant interactions would be useful for rehabilitating degraded grassland.

2.3. Allelopathy in Pasture Management

A pasture is a piece of grassland that mainly grows forage grass for livestock. Its quantity and quality are closely related to grassland ecosystem health and animal husbandry development. Hence, the management of pasture, whether natural or managed, is essential to ensure adequate forage grass and to support livestock production.

Allelopathy-based interspecific and intraspecific interactions have ecological consequences for the productivity and biodiversity of a pasture. Particularly in a managed pasture, pasture weeds can immensely decrease forage yield and quality, negatively affecting livestock production. Fortunately, some forage species can take full advantage of allelopathy and allelochemicals to retard the emergence and growth of co-occurring weeds, from which they will obtain growing benefits [53,54]. For example, rye (*Secale cereale*) is a cool-season forage species with high frost and drought resistance; it is generally planted in infertile or acid soils due to its strong adaptability. Rye can produce and release benzoxazinoids to selectively inhibit broadleaf weeds, modifying the spectrum of weed species in the pasture [55,56]. Therefore, some fine forage cultivars with allelopathic traits can be used for weed control. In particular, natural allelochemicals released from allelopathic forage cultivars may act as biological herbicides to a large extent, lowering the consumption of chemical herbicides and the cost of pasture management [57,58]. Many studies have shown that the application of allelopathic forage cultivars can effectively control pasture weeds and increase pasture productivity [55,57,59]. Notably, allelopathic forage species such as rye not only suppressed the pasture weeds but also succeeding forage species. To avoid failure in rotation systems, it is warranted to select resistant succeeding forage species [60].

Autotoxicity (intraspecific allelopathy) is ubiquitous in pastures. Autotoxicity in pasture has been well verified in alfalfa (*Medicago sativa*) [61–63]. Alfalfa is a major forage legume used as a high-quality livestock feed and cultivated in pastures throughout the world. Several phytotoxic phenolics, saponins and medicarpin in alfalfa can remarkably suppress their own seed germination. To attenuate the autotoxicity, the most obvious solution is to develop a new autotoxicity-tolerant alfalfa cultivar. A recent study has picked out the most autotoxicity-tolerant alfalfa from 22 cultivars based on a technique for order of preference by similarity to ideal solution analysis [64], which provides a theoretical basis for the breeding of autotoxicity-tolerant alfalfa cultivars. However, a long-term and large-scale field verification is needed to assess the tolerance of different alfalfa cultivars to autotoxicity.

A mixture of diverse forage species is considered as another option to experimentally prove effectiveness in improving forage productivity [65,66]. Directly, some highly allelopathy-tolerant forage seeds can be used as a subsequent alternative for restoring sparse natural grassland caused by allelopathy [67]. Additionally, the pattern of mixing species also has another benefit for pastures. The mixture of rye with berseem clover (*Trifolium alexandrinum*) may promote rye pathogen-resistant capabilities [68]. In the coexistence system of *Artemisia adamsii* with *Stipa krylovii*, volatiles emitted by *A. adamsii* can strengthen photosynthesis of *S. krylovii* by enhancing stomatal conductance even with water deficiency [69]. When grown with the P-mobilizing species *Filifolium sibiricum*, *Leymus chinensis* exhibited greater shoot and root P content [70]. These positive interactions are prevalent in pastures and mostly attributed to plant–plant chemical communication.

3. Allelopathy in Forests

3.1. Allelopathy in Natural Forests

Natural forests usually possess plant diversity and stable productivity. The role of allelopathy and the mechanisms underpinning it remain poorly resolved in species-rich forests, but allelopathy does contribute to natural forest regeneration. Forest regeneration is commonly considered as a critical ecological process that sustains resource reproduction through the establishment of saplings and the replacement of dead trees; it has profound implications for the perpetuation of tree species in the temporal and spatial dimensions. However, long-term exposure to allelochemicals from woody species may create a barrier effect on the understory-regenerated saplings, resulting in forest regeneration failure. In particular, endangered and rare plant species are inherently difficult to generate due to their scarce propagules and low adaptability. Allelopathy additively reduces the likelihood of the sapling establishment and probably leads to locally rare species' extinction. *Cinnamomum migao* and *Metasequoia glyptostroboides* are two endangered woody species. Their regeneration is extremely restrained, and the natural population would be gradually diminished over time without active management. Generally, most natural populations only occasionally have 1~2 saplings in their understories [71,72]. Recent studies found that leaf extracts or litters of *C. migao* and *M. glyptostroboides* dramatically impeded their seedling growth by impairing the lipid structure of the cell membrane, suggesting that autotoxicity might aggravate the obstruction of the natural forest regeneration among some endangered tree species [72,73].

Apart from autotoxicity or self-inhibition, allelochemical-mediated interspecific interactions also hinder natural forest regeneration and impact the plant community's composition. In the context of forests dominated by two tree species, dominant tree species may chemically inhibit the sapling regeneration of the others. For example, *Kandelia obovate* and *Aegiceras corniculatum* are two dominant species in mangrove forests. Leaf litter leachates of *K. obovate* are detrimental to the propagule germination and sapling growth of *A. corniculatum*, ultimately modulating the natural regeneration of the whole mangrove forest [74]. In the later successional forests of maple-beech codominance (*Acer saccharum* and *Fagus grandifolia*), the abundance of beech progressively increases as maple decreases with the years. This result, in part, can be explained by the allelopathic advantage of beech leading to the regeneration failure of maple [75,76].

Monopolistic herbaceous plants grown in the floor layer may inhibit natural forest regeneration. For example, the natural regeneration of sessile oak (*Quercus petraea*) is often hampered by the dense moor grass (*Molinia caerulea*) understory [77]. When watered with root exudates of moor grass, a significant decrease in oak biomass occurred, suggesting the allelopathic interference of moor grass in oak growth. Even though this negative impact was lower than that of resource depletion, it demonstrated the crucial contribution of herbaceous allelopathy to natural forest regeneration [78].

Based on the understanding of the allelopathic mechanisms underlying natural forest regeneration, some appropriate methods of forest management are proposed to alleviate the adverse effects of allelopathy and promote long-term natural regeneration. One of the most direct and efficient ways is to reduce the frequency of allelopathic interactions by removing litter, or eradicating the allelopathic species. Prevention of saplings from potential allelochemicals facilitates the sustainability of forest health [78,79]. In addition, attempts to enhance the diversity of the shrub layer and floor layer may be an alternative way to promote natural forest regeneration [80].

3.2. Allelopathy in Tree Plantations

A tree plantation is an artificial forest for the large-scale production of wood; usually, easily established and fast-growing tree species are selected as a monoculture forest. The productivity and sustainability of tree plantations intimately links the economic and ecological benefits of forestry. However, successive rotations of some forestry species may cause a replanting problem or soil disease, resulting in a decline in productivity and the loss of biodiversity in plantations [81,82]. Although the underlying mechanism for this issue is still being disentangled, a growing amount of evidence has shown that allelochemicals enriched in soil are mainly responsible for this problem [83,84].

Eucalyptus is one of the most widely planted forestry genera on the planet, but it has suffered from autotoxicity for a long time. Most studies have demonstrated that allelochemicals of *Eucalyptus* penetrate into the soil through the decomposition of litter and leachates, exerting an allelopathic effect on understory plants, thus limiting the regeneration of native vegetation [85,86]. However, Zhang et al. (2016) argued that the poor establishment of indigenous vegetation on plantations mainly arose from *Eucalyptus* roots rather than *Eucalyptus* litter. Retention of understory litter is more likely to facilitate the performance of native species [87]. Whatever the case is, a consensus is that allelopathy is more crucial than resource competition in the replanting problem of *Eucalyptus* plantations [88]. Chinese fir (*Cunninghamia lanceolata*) is another tree plantation severely disrupted by autotoxicity. Regeneration failure and poor establishment have remained critical problems in monocultural plantations of this species [89]. However, root exudates contribute more to soil allelochemicals than the litter in Chinese fir plantations. Root-secreted allelochemicals, therefore, are considered a primary source leading to the decline in the plantation of Chinese fir [90].

The mixture of multiple tree species is an effective way to improve the self-inhibition and soil deterioration caused by allelopathy and allelochemicals in plantations [91–93]. In *Eucalyptus* plantations, *Albizia lebbeck*, an introduced N-fixing species, has been regarded as a 'good partner' to *Eucalyptus*. Mixed-species plantations of *Eucalyptus* with *A. lebbeck* increase productivity and maintain soil fertility compared with pure *Eucalyptus* stands [91]. Similarly, the establishment and productivity of autotoxic Manchurian walnut (*Juglans mandshurica*) can be improved in the presence of larch (*Larix gmelini*). Larch root exudates and soil in mixed-species plantations greatly stimulated the growth of Manchurian walnut seedlings and rapidly degraded the allelochemical juglone [92]. The growth and regeneration of Chinese fir is improved in *Michelia macclurei* and Chinese fir mixed-species plantations. One of the explanations for this beneficial promotion is that there may be interspecific facilitation mediated by the root exudates from *M. macclurei*, which not only attenuate the release of allelochemicals from Chinese fir roots but also induce a microbial shift to accelerate the decomposition rates of allelochemicals [93]. These studies illustrate the importance of mixed-species stands in plantations. However, most successful mixtures were empirically established from traditional practices, or were assessed from haphazard experimental combinations. We lack effective strategies for a priori selection of mixtures to achieve relevant benefits. Therefore, understanding intraspecific or interspecific allelopathy will be a key step in screening appropriate combinations of tree species to design plantations.

3.3. Tree-Understory Vegetation Allelopathic Interactions

The canopy position and soil occupancy of dominant forest trees remarkably reduce light and soil nutrient availability for understory vegetation. Even so, some shrub and herbaceous species in understory vegetation can adapt to these diverse conditions and coexist with trees. Apart from competition for resources, the allelopathy of the trees is an interference mechanism for the growth of understory vegetation [94,95]. The allelopathic trait of some trees is highly associated with forest abundance and biodiversity, particularly for woody invasive species. The presence of allelopathic tree species in forests can reduce the abundance of understory vegetation, ultimately becoming dense monospecific stands and extending to the whole forests [96–98]. In this process, allelochemicals may act as a meditator [99].

For the allelopathic effect of trees on understory plants, leaf litter and leachates have long been considered the main source [100,101]. Leaf litter and leachates from trees falling into the ground may prevent the colonization and development of understory vegetation [102,103]. This suppression is mainly attributed to their physical and chemical effects [104,105]. However, allelochemicals from leaf litter and leachates also have a measurable effect on understory vegetation [106–108]. Through the decomposition of leaf litter, allelochemicals can be gradually liberated into the soil and come into effect by altering soil pH, nutrient availability, the nitrogen cycle and microbial community structures [109,110]. Especially intriguing is leaf litter and leachates that may modify plant coexistence in the grass layer. For example, spotted knapweed and *Bromus tectorum* exhibit strong competition with each other, while leaf litter and leachates of *Pinus ponderosa* can mitigate the competitive effect of spotted knapweed on *B. tectorum*. In other words, the presence of

P. ponderosa shifted competitive outcomes through physical and allelopathic effects, thereby indirectly facilitating *B. tectorum* by more strongly inhibiting spotted knapweed [111].

In some cases, leaf litter and leachates cannot solely show allelopathic potential. It must unite other biotic or abiotic factors to jointly impact the ecological process [102,112]. *Prosopis juliflora* is one of the world's most aggressive invasive species, the leaf litter of which causes the increase of total phenolics in soil and toxifies understory vegetation [113]. When incubated with similar levels of leaf leachate from *P. juliflora*, the content of allelochemicals varies in different soil textures. Sandy soil accumulates higher levels of phenolics than sandy loam soil due to the greater absorption of inactive phenolics fettered in sandy loam soil [112]. In addition, the allelopathic effect of *P. juliflora* is also limited by soil moisture because their water-soluble allelochemicals in the soil are more likely to be washed away by rain. Therefore, *P. juliflora* could not manifest their allelopathic potential in humid soil. Only in dry environments, *P. juliflora* can create a depressive impact on understory plants [114].

Dense understory species with highly allelopathic potential, in turn, may directly slow the growth of trees and indirectly cause trouble by dissolving the fungal hyphae of trees. Garlic mustard (*Alliaria petiolata*) is a typical understory invasive species that may suppress fungal mutualists via allelochemicals, leading to significant declines in a series of physiological and metabolic functions [115–117]. Nevertheless, arbuscular mycorrhizal fungi (AMF) strains can be quickly selected by the allelopathic stress from garlic mustard. After the initial decline in AMF abundance, resistant AMF strains gradually displace sensitive AMF strains and the abundance rises again after the long-term invasion of garlic mustard [118,119]. Moreover, as an invader, the novelty of allelochemicals to resident species, regardless of the plant or microorganism, diminishes over time. Ultimately, garlic mustard may enter a new coevolutionary relationship with native competitors and slowly be integrated into the native community [120,121].

4. Allelochemicals in Grasslands and Forests
4.1. Category of Allelochemicals

All the occurrences of allelopathic phenomena can be attributed to a certain or a set of allelochemicals. Allelochemicals and their properties largely determine the allelopathic effectiveness. In the past decades, numerous old and new allelochemicals have been detected and identified from herbaceous and woody species in grasslands and forests. These allelochemicals involve a diversity of plant secondary metabolites, but are mainly divided into three categories of phenolics, terpenoids and nitrogen-containing compounds.

Phenolics have a wide distribution in plants and represent a diverse group of compounds with an aromatic ring possessing at least one hydroxyl group and possibly other substituents, including simple phenolic acids, coumarins, flavonoids and quinones. In forests, a tremendous amount of lignin from litter is decomposed into a variety of phenolic acids. These lignin-derived phenolic acids are main allelochemicals in forest soil, leading to a decline in forest species' abundance and biodiversity. For example, the soil of the *Eucalyptus urograndis* plantation contains high levels of hydroxybenzoic, vanillic, coumaric and ferulic acids (Figure 1), resulting in autotoxicity of *E. urograndis* [85]. However, the allelopathic effect of phenolic acids is concentration-dependent. In particular, individual phenolic acids are insufficient to effectively suppress the growth of co-occurring plants, but their mixtures exhibit phytotoxic effects [122].

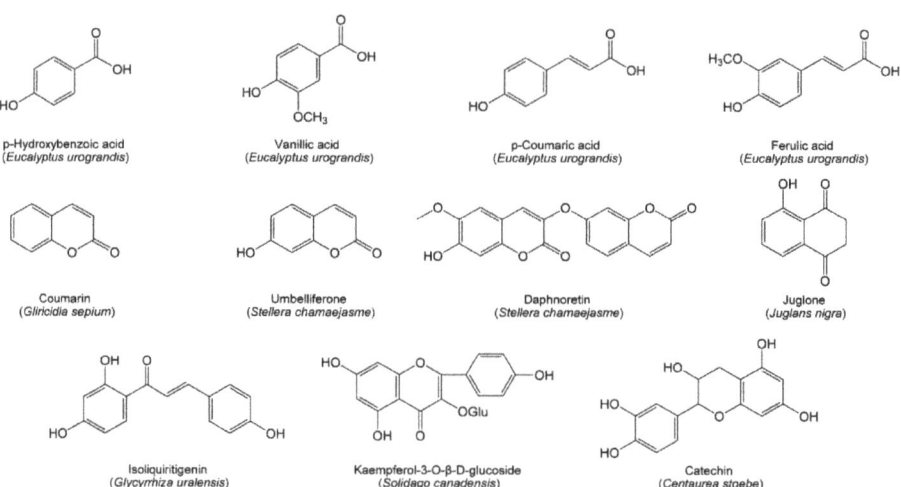

Figure 1. Phenolic allelochemicals from herbaceous and woody species in grasslands and forests.

Many coumarins possess phytotoxicity and act as potential allelochemicals in grasslands and forests. Coumarin exacted from the leaf of *Gliricidia sepium* was identified as an allelochemical to inhibit the growth of plants [123]. Umbelliferone and daphnoretin (Figure 1) are two coumarin allelochemicals in *Stellera chamaejasme* [38]. Umbelliferone can inhibit plant growth by inducing membrane lipid peroxidation and retarding cell division, while daphnoretin inhibits plant growth by arresting the mitosis process [124].

Flavonoids generally perform a broad range of ecological functions. Several flavonoids have proved to be allelochemicals in grasslands and forests. Isoliquiritigenin (Figure 1) is a flavonoid allelochemical in *Glycyrrhiza uralensis*. It is able to trigger a chain of reactions in plant cells, including the overproduction of reactive oxygen species, lipid peroxidation, and a decline in cell vitality and chlorophyll content, ultimately resulting in seedling growth inhibition [125]. Another flavonoid, kaempferol-3-O-β-D-glucoside (Figure 1), is an allelochemical of *Solidago canadensis* [126]. Catechin (Figure 1) is a controversial flavonoid allelochemical secreted by spotted knapweed. Many studies have found high catechin concentrations in spotted knapweed soils [127,128] and proposed that catechin acts as a novel allelochemical of spotted knapweed, which contributes to growth limitation of the native plants [18]. However, several studies pointed out that catechin was hardly present in the bulk soils of spotted knapweed, and possessed low phytotoxicity to a variety of plant species [27,28].

Quinones are the classical allelochemicals in forests. Juglone (Figure 1) is an exclusive allelochemical of *Juglandaceae* family and represents one of the best-known members of quinones [129]. Initially, juglone is stored in leaves, barks and roots in the form of non-toxic naphthol O-glycoside. When released from plant living tissues to the environment, it is hydrolyzed into a less phytotoxic naphthol, and subsequently oxidized into phytotoxic juglone. The allelopathic mechanisms of juglone are associate with the disruption of leaf photosynthesis, transpiration, respiration and stomatal conductance. Additionally, juglone has high stability in soil. The toxicity of juglone can maintain for up to a year in spite of the removal of the walnut trees [130].

Terpenoids, including monoterpenes, sesquiterpenes, diterpenes, triterpenes and steroids, are a class of compounds derived from the 5-carbon isoprene. Monoterpenes and their derivatives possess strong volatility and may interact with neighboring plants in their gaseous phase. Volatile allelochemicals emitted by donor plants generally impact surrounding plants through two main pathways, either forming 'terpene clouds' of directly impacted target plants [16], or leaching into the soil of indirectly affected target plants. The volatiles of *A. frigida* contain a copious quantity of terpenoids, among which monoterpene camphor is a key allelochemical affecting the neighboring species [36]. Another monoterpene, (−)-α-thujone, emitted from *Thuja occidentalis* (Figure 2), can display phytotoxic activities against seed germination and seedling growth of *Taraxacum mongolicum* and *Arabidopsis thaliana* [131]. β-Caryophyllene (Figure 2), a sesquiterpene within the needle litter of *Pinus halepensis*, exerts a deleterious effect on the germination and growth of herbaceous target species [132]. Dihydromikanolide (Figure 2) is another sesquiterpene allelochemical from *Mikania micrantha*, which contributes to promoting soil bacterial diversity but reduces fungal diversity [133]. Two diterpenes, ent-kaurene and phyllocladane (Figure 2), isolated from senescent needles of *Araucaria angustifolia* can act as allelochemicals to inhibit the germination and seedling growth of neighboring plants [134]. Similarly, diterpene allelochemicals, 7-oxodehydroabietic acid and 15-hydroxy-7-oxodehydroabietate (Figure 2), were found in the understory soil of *Pinus densiflora*. Both allelochemicals may be the underlying cause of sparse understory vegetation within the *P. densiflora* canopy [135]. Besides, some pentacyclic triterpenoids may function as allelochemicals, such as betulinic, oleanolic and ursolic acids (Figure 2) within the litter of *Alstonia scholaris*, limiting the growth of co-occurring species by inhibiting seed germination, radicle growth and the functioning of photosystem II [136].

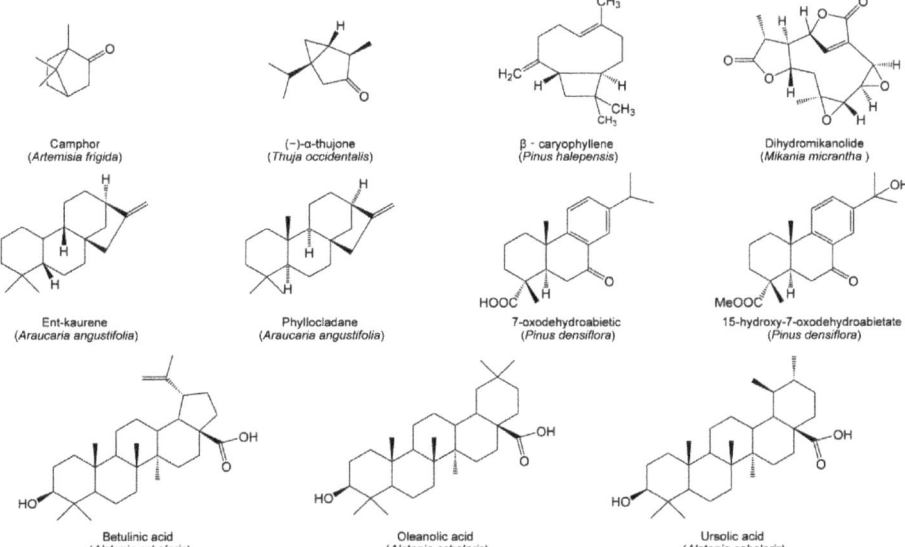

Figure 2. Allelopathic terpenoids from herbaceous and woody species in grasslands and forests.

Nitrogen-containing compounds mainly include alkaloids, non-protein amino acids, benzoxazinoids and cyanogenic glycosides. Compared with phenolics and terpenoids, nitrogen-containing allelochemicals are relatively unknown. However, several specialized nitrogen-containing metabolites have been identified as allelochemicals that have significant ecological implications for grasslands and forests. Hexadecahydro-1-azachrysen-8-yl ester (Figure 3), identified as a potential alkaloid allelochemical in *Imperata cylindrica*, can reduce root growth and mycorrhizal colonization [137]. There are many non-protein amino acids involving allelopathic interferences with co-occurring species in grasslands (Figure 3). *meta*-Tyrosine of fine fescue grasses (*Festuca rubra*) can interfere with the root development of competing plants [138]. Mimosine of *Leucaena leucocephala* can retard plant growth by blocking the cell division of protoplasts and disturbing the associated enzyme activity [139]. L-Canavanine of *Vicia villosa* not only exerts the phytotoxic effect by disrupting the arginine metabolism in the plants but also significantly alters the microbial community composition and diversity in soil [140,141]. A novel cyclic dipeptide (6-Hydroxy-1,3-dimethyl-8-nonadecyl-[1,4]-diazocane-2,5-diketone) (Figure 3) has been found in Chinese fir; it is a highly active allelochemical to be responsible for serious replanting problems in plantations [142]. Benzoxazinoids are a class of well-known nitrogen-containing allelochemicals, among which 2,4-dihydroxy-7-methoxy-1,4-benzoxazin-3-one (DIMBOA) and 2,4-dihydroxy-(2H)-1,4-benzoxazin-3(4H)-one (DIBOA) (Figure 3) can be released by rye and exert strong suppression of plant growth [143,144]. Cyanogenic glycosides are specialized metabolites derived from amino acids. Sinigrin (Figure 3), as an allelochemical of cyanogenic glycosides from garlic mustard and broccoli (*Brassica oleracea*); it can lead to the poor establishment of North American forests by disrupting the AMF symbionts [145].

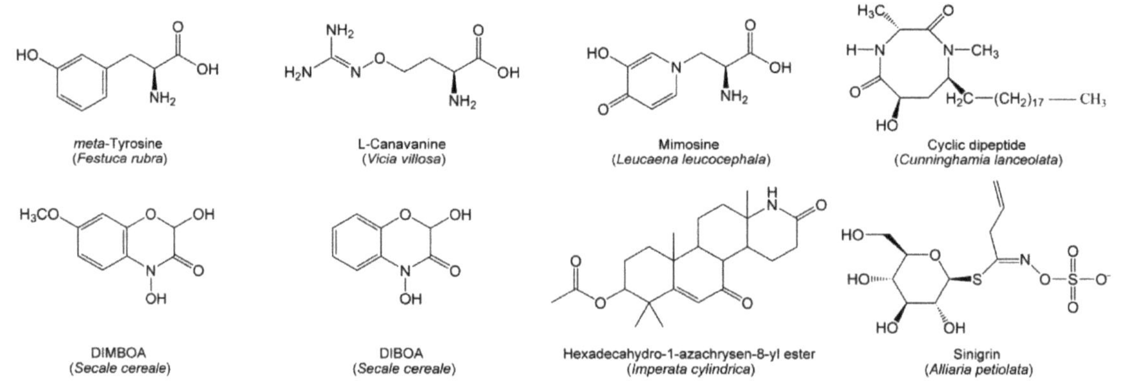

Figure 3. Nitrogen-containing allelochemicals from herbaceous and woody species in grasslands and forests.

4.2. Identification and Detection of Allelochemicals

Allelochemicals can be either unknown or known in plants and their environments. Unknown allelochemicals have to be identified by non-targeted analysis, while known allelochemicals can be detected by targeted analysis. The identification of unknown allelochemicals first isolates pure individuals from sample components, and then the individuals can be determined and analyzed by mass spectrum, infrared spectrum and nuclear magnetic resonance [11]. Such non-targeted analysis may investigate which allelochemicals are responsible for the allelopathic interactions in a given system. Therefore, applying non-targeted analysis for identification of unknown allelochemicals has been key to understanding the ecological role of allelopathy in grasslands and forests.

Compared with the identification of unknown allelochemicals, detection of known allelochemicals is straight forward. Targeted analysis of known allelochemicals is usually conducted by means of gas or liquid chromatography coupled with tandem mass spectrometry (GC-MS/MS, LC-MS/MS). GC-MS/MS is the most preferred technique for qualitative and quantitative assessment of volatile allelochemicals. In contrast, non-volatile allelochemicals with relatively high molecular weight, mainly produced and released from root exudation and plant decomposition, can be analyzed with LC-MS/MS.

Understanding the functional significance of allelopathic plant–plant interactions and processes occurring in grasslands and forests requires accurate information about the quantity, quality and spatiotemporal dynamics of allelochemicals. The best way to trap and detect allelochemicals in vivo, in situ and real time from living plants and their environments remains a problem. Accordingly, it is warranted to develop analytical methods that are more realistic or closer to the actual field situation [146]. Phillips et al. (2008) designed an experimental system employed to trap root exudates from intact tree roots in situ. This method can account for the spatial heterogeneity and temporal dynamics of forest soils and root systems [147]. A recent study has developed quick and in situ detection of allelochemicals in Taxus soil by microdialysis combined with UPLC-MS/MS [148], providing a more finely tuned picture of allelochemical dynamics in grasslands and forests.

4.3. Activity-Concentration Relationship of Allelochemicals

The action of allelochemicals is concentration-dependent. Thus, the activity–concentration relationship is crucial for allelochemical interference, particularly for their presence at the phytotoxic level in the soil. Although soil abiotic factors such as pH, enzyme activities, organic matter and nutrient availability contribute to the change of allelopathic activity, microbial effects are undoubtedly the most crucial factor that affect allelochemicals in the soil. Soil microbes determine the below-ground transportation and intensity of allelochemicals. Accordingly, the fate and dynamics of allelochemicals are mainly attributed to soil biological processes, and potential abiotic controls. For instance, flavonoid allelochemicals have high persistence in soil because they are decomposed very slowly and last a long time in soil, which favors suppression of the emergence and growth of plants and modification of the soil's properties, even at the low levels [149]. However, a recent study has found that soil organic carbon decreases the lifetime of flavonoids underlying plant–microbe interactions. In particular, the dissolved organic carbon in soils can repress flavonoid bioavailability and attenuates the efficacy of flavonoid-based plant–microbe communication [150].

Allelochemicals in grasslands and forests have differential concentrations, activities and categories. Most phenolics at a high concentration show allelopathic activities. Terpenoids and nitrogen-containing allelochemicals may impact plant species at a low concentration (Table 1). Accordingly, the action of phenolics involved in allelopathy requires a considerable amount of them, representing quantitative allelochemicals. In contrast, the act of terpenoids and nitrogen-containing allelochemicals greatly depends on their category rather than their amounts, representing qualitative allelochemicals. In addition, qualitative terpenoids with a diversity of allelochemicals frequently occur in annual grasslands, while quantitative phenolics with a few specialized allelochemicals occur in perennial forests. This is due to the production and release of allelochemicals in perennial forests by large-scale litter decomposition.

Table 1. The phytotoxic level of important allelochemicals in grasslands and forests.

Allelochemicals	Class	Targeted Species	IC50/μM *	Sources
Vanillic acid	Phenolic acids	Lactuca sativa	950.2	[151]
Hydroxybenzoic acid	Phenolic acids	Lactuca sativa	2470.0	[152]
Coumaric acid	Phenolic acids	Lepidum sativum	1120.0	[153]
Ferulic acid	Phenolic acids	Arabidopsis thaliana	1099.0	[154]
Juglone	Quinones	Lactuca sativa	50.0	[155]
Coumarin	Coumarins	Lactuca sativa	23.3	[55]
Umbelliferone	Coumarins	Lactuca sativa	430.0	[124]
Daphnoretin	Coumarins	Lactuca sativa	1558.3	[124]
Isoliquiritigenin	Flavonoids	Lactuca sativa	823.4	[125]
Camphor	Monoterpenes	Lactuca sativa	50.0	[156]
(−)-α-Thujone	Monoterpenes	Taraxacum mongolicum	140.2	[131]
Betulinic acid	Triterpenes	Lactuca sativa	78.8	[136]
Oleanolic acid	Triterpenes	Lactuca sativa	94.2	[136]
Ursolic acid	Triterpenes	Lactuca sativa	101.6	[136]
Cyclic dipeptide	Nonprotein amino acids	Cunninghamia lanceolata	12.5	[141]
meta-Tyrosine	Nonprotein amino acids	Lactuca sativa	17.0	[138]
Mimosine	Nonprotein amino acids	Lactuca sativa	300.0	[155]
DIMBOA	Benzoxazinoids	Lepidium sativum	542.3	[157]
DIBOA	Benzoxazinoids	Lepidium sativum	493.1	[157]

* Half-maximal inhibitory concentrations.

In many studies, the applied concentrations of allelochemicals were greater than those detected in the environment. This issue was because the concentrations of allelochemicals detected would be locally much higher in intact soils. Extractions would have diluted the allelochemicals over large soil volumes. Additionally, frequent allelochemicals provided over a long term at low concentrations can have powerful effects. Thus, even if the actual concentration of allelochemicals in the environment was still substantially lower than the necessary concentration to impact neighboring plant species, an effect would still be expected, because in the environment, there would be a constant release of allelochemicals.

5. Allelochemicals Mediate Below-Ground Interactions and Plant–Soil Feedback

5.1. Below-Ground Chemical Interactions

The action of allelochemicals requires their presence in the environment. Environmental factors such as temperature, light, soil nutrients and microorganisms may strengthen or alleviate the allelochemical activity. This adjustment of the action of allelochemicals in response to the environment reflects the adaptability of the allelopathic plants [9]. Most allelochemicals shift from plants into the soil following root exudation, decomposition, volatilization and leaching. These allelochemicals dispersing in the soil inevitably interact with a variety of below-ground components particularly for root placement pattern [158], soil nutrient availability, microbial community structure, mycorrhizal fungi colonization, and subsequent plant–soil feedback [8]. Therefore, the biodiversity, productivity and sustainability of grasslands and forests may be driven by allelochemical-mediated below-ground interactions and plant–soil feedback. Such a conceptual framework is outlined in Figure 4.

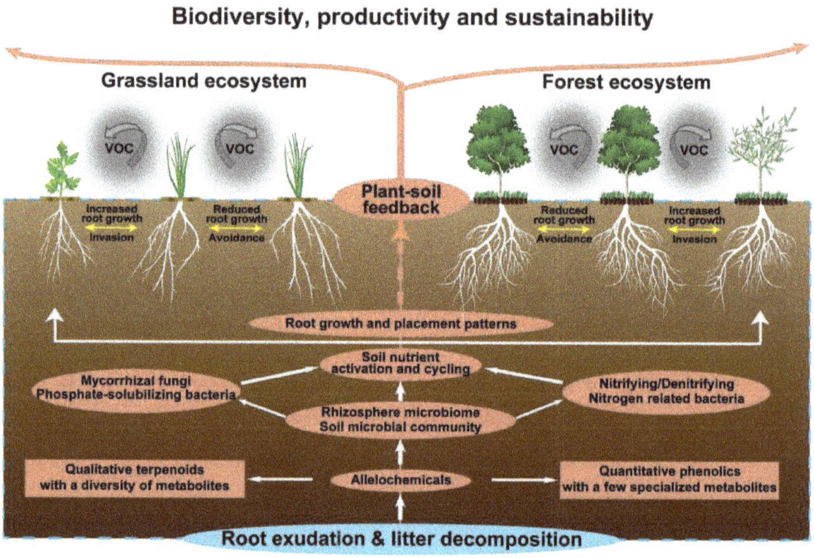

Figure 4. Allelochemical-mediated below-ground interactions and plant–soil feedback.

The root is a vital organ interacted with soil. In response to the soil environment, a plant may place its roots in intrusive (approaching), avoidant (repelling) or unresponsive patterns [159]. Such root placement patterns, particularly for intrusive and avoidant patterns, may be driven by allelochemicals [160,161], altering below-ground ecological interactions and ultimately affecting plant performance and productivity (Figure 4). A recent study has revealed that pairwise allelopathic plant–plant interactions generate all possible combinations of intrusive, avoidant and unresponsive root placement [158]. Allelopathic species showed a general tendency toward root intrusion, while most target species adjusted root placement to avoid root-secreted allelochemicals from allelopathic species [158]. Similar allelochemical-mediated root responses have been observed in forage grass and tree species, such as avoidant response of annual ryegrass (*Lolium rigidum*) roots to neighboring allelopathic canola (*Brassica napus*) [162], and root avoidance in mixed-species plantations of Chinese fir and *Michelia macclurei* [93].

Allelochemical-mediated root placement patterns may contribute to plant–microbe interactions that control vital below-ground processes [158,163]. Allelochemicals are important carbon sources of soil microorganisms that determine the changes in microbial composition and community, and then affect the activation and circulation of soil nutrients (Figure 4). Cinnamic acid can significantly alter soil microbial community functional diversity and genetic diversity [164]. The hyphal branching of AMF is induced and stimulated by flavonoids, and flavonoid-associated microorganisms can colonize the roots of a very wide range of plants in order to increase nutrient uptake, especially that of P, and enhance the plant health [165]. Allelochemicals also directly participate in the activation and cycling of soil nutrients. p-Hydroxybenzoic acid can alter the form of soil N, causing Chinese fir seedlings to shift their N uptake preference from NO_3^- to NH_4^+ [166].

In grassland ecosystems, allelochemicals exuded by toxic weeds may trigger a series of changes in soil enzyme activities, pH, nutrient availability and mycorrhizal fungal colonization [12,13,167]. In particular, the exudate-induced alteration of the soil microbial community heavily promotes the expansion of toxic weeds by supplying higher rhizosphere nutrients [14,45]. Compared with the soil free of toxic weed *Stellera chamaejasme*, the soil infested with *S. chamaejasme* exhibited lower nutrition, organic matter, fungal alpha diversity, and relative abundance of AMF, but a higher abundance of pathogenic fungi [13]. Moreover, *S. chamaejasme* root exudates were alkalescent (pH = 9.28) and had a negative effect on the rate of mycorrhiza infection and spore density of the AMF [167]. Together, allelochemicals exuded from *S. chamaejasme* might increase the soil pH, reduce the soil nutrient availability, damage the AMF of other plants and recruit more pathogenic fungi, thereby posing a great threat to grassland vegetation [167].

In forest ecosystems, the roots of most tree species are extensively infested with obligately soil-borne fungi and mycorrhizas that assist plants in nutrient acquisition, pathogen resistance and carbon transportation [168,169]. Several studies found the critical role of soil microorganisms in the maintenance of *Eucalyptus* plantations, which may mitigate the allelopathic effect of *E. grandisis* leachates [170]. Specifically, a lower content of total phenolics occurred in nonsterile soils than in sterile soils when both were exposed to the *E. grandisis* leachates [171]. In addition, root-associated fungi probably utilize *Eucalyptus* allelochemicals as a carbon source to decompose, ultimately alleviating the allelopathic effect of *Eucalyptus*. AMF colonized in *Eucalyptus* roots could better protect woody species from the allelopathic interference of *Eucalyptus* [172,173]. Allelochemicals of Chinese fir not only exert a direct phytotoxic effect on plant roots but also indirectly disturb the soil microbial community's composition and structure. Compared with the first rotation plantation, allelochemicals of Chinese fir probably suppress beneficial mycorrhizal species while promoting harmful fungi in the second rotation plantation, resulting in the deterioration of the soil microbial community [174]. Interestingly, the hyphal network enables allelochemicals of *Juglans nigra* to extend their bioactive zone and promote the effectiveness of allelopathy, indicating the importance of AMF in the movement of allelochemicals [175,176]. In another example, the leaf litter of nonmycorrhizal willows (*Salix glauca* and *Salix brachycarpa*) cannot reduce AMF colonization of understory herbaceous plants, but transplanted ectomycorrhizal willows can suppress AMF colonization of herbaceous hosts through the interaction of leaf litter and ectomycorrhizal fungi [177]. In addition, some soil fungi function as a 'shield' to protect plant roots from the attack of allelochemicals. *E. urophylla* root-associated fungi have the ability to partly offset the autotoxicity of phenolic acids [172].

On the other hand, allelochemicals are able to either promote or reduce the abundance and diversity of soil microbes. The leachate of *Acacia dealbata* can modify the soil microbial community's assembly, leading particularly to a prominent decline in bacterial richness and diversity in pine forest soil [109]. Likewise, extracts of *Eupatorium adenophorum*, especially from its leaves, can reduce bacterial richness and diversity in soils, [178]. Additionally, root exudates of *V. villosa* can shift the soil microbial community's composition, particularly increasing the abundance of Firmicutes and Actinobacteria while decreasing that of Proteobacteria and Acidobacteria [141]. In contrast, the litter of *P. juliflora* benefits the growth and reproduction of some soil microbes and can stimulate the soil microbial biomass carbon and soil metabolic quotient [179]. Similarly, the litter of *Mikania micrantha* can increase soil bacterial richness, yet decrease fungal richness, which enhances immediate nutrient availability and provides ecological advantages to *M. micrantha* [133]. Plus, allelochemicals may facilitate the reduction of soil pathogens. Aqueous root extracts of *Diplotaxis tenuifolia* can inhibit the activity of *Phytophthora cinnamomi* [180], illustrating that *D. tenuifolia* can be exploited for biological control in pathogen suppression. These studies indicate a shift in bacterial diversity, or a shift from fungal richness toward bacterial richness. However, there is a lack of data on the functional shifts' impact on the affected grasslands and forests, which calls for further studies.

5.2. Below-Ground Chemical Interactions Drive Plant–Soil Feedback

Plant–soil feedbacks (PSFs) are interactions among plants, soil organisms, and abiotic soil conditions that influence plant performance, plant species diversity and community structure, ultimately driving ecosystem processes [8]. Allelochemical-mediated below-ground interactions may alter PSFs and their potential consequences for ecosystem functioning. Allelochemicals influence PSFs through the performance of interacting species and altered community composition resulting from changes in species distributions. Allelochemicals affect plant inputs into the soil subsystem via litter and rhizodeposits. Further, root-exuded and litter-decomposed allelochemicals modulate microbial succession. These interactive effects may cause specific PSFs where the match between the species identity of living roots and litter can modify decomposition and feed back to plant nutrition [7,8].

Allelochemical-mediated below-ground interactions drive plant–soil feedback in grasslands and forests (Figure 4). In grasslands, *S. chamaejasme* exudes allelochemicals that incur the change of soil pH and nutrient availability, which partly contributes to the inhibition of adjacent *L. chinensis* [165]. Similarly, spotted knapweed can reduce the total soil carbon and nitrogen content and alter the soil elemental composition via allelochemicals, subsequently impacting soil ecosystem function and impeding the native plant growth [181]. Allelochemicals of *M. micrantha* can enhance the abundance of soil ammonia-oxidizing bacteria and promote the N cycling process. This plant–soil feedback by which *M. micrantha* improves soil N transformation facilitates its invasion in natural environments [182]. In forests, litter of *Robinia pseudoacacia* through allelopathy decrease understory soil nutrient availability, especially of P, and then hinder the growth of *Phytolacca americana*. This negative plant–soil feedback might underlie the limiting factors in the invasion of exotic plants [183]. Likewise, *Juniperus virginiana* exudes allelochemicals into the soil that allow the collapse or transformation of soil microbial communities, followed by inhibiting the growth of certain grass species through negative plant–soil feedback [184].

Importantly, many of these plant–soil feedback are species-specific and are greatly affected by the identity of co-occurring plant species. The presence of co-occurring plant species can alter the direction of plant–soil feedback as a result of long-lasting effects on below-ground interactions and plant responses to subsequent allelochemicals (Figure 4). In successful mixed-species tree plantations, an appropriate species can enhance autotoxic species growth through below-ground chemical interactions. For example, the presence of larch and *M. macclurei* can improve the establishment and productivity of autotoxic Manchurian walnut and Chinese fir in their mixed-species plantations. This is due to the fact that root exudates of larch and *M. macclurei* can facilitate the growth of autotoxic species and increase the degradation of allelochemicals from autotoxic species [92,93]. Accordingly, the allelochemical context alters the consequences of the below-ground ecological interactions, resulting in positive plant–soil feedback in mixed-species plantations.

6. Challenges and Opportunities

The importance of allelopathy and allelochemicals cannot be overemphasized in grasslands and forests. Recent efforts have made considerable progress toward understanding allelopathy and allelochemicals in grasslands and forests. Nevertheless, the functional consequences of allelopathy for plant communities in natural and managed grasslands and forests remain unsolved.

To confirm whether plant–plant allelopathic interactions occur in a grassland or a forest, there are four steps: (1) to find and select ecologically relevant plant species through field observation and investigation; (2) to determine that the selected plant species can produce and release allelochemicals into the environment through appropriate pathways (volatilization, leaching, root exudation or/and residues decomposition); (3) to qualify and quantify allelochemicals and their migration and transformation in soil; (4) to verify the effect of allelochemicals at effective states and concentrations on neighboring plants. These steps and processes involve multiple biotic and abiotic factors, but the focus and central driver must be allelochemicals. However, identifying when and how plant species produce and release allelochemicals is challenging.

An increasing number of studies have shown that the production of allelochemicals depends on the identity of neighboring plants. Allelopathic plants are capable of discriminating between their neighboring competitors and collaborators, adjusting their production of allelochemicals accordingly [185]. In particular, allelopathic plants may detect competing neighbors and respond by increasing allelochemicals to inhibit them, thereby maximizing their own growth. Accordingly, allelopathic interference involves two inseparable processes of plant neighbor detection and allelochemical response via signaling interactions [185], and even intraspecific kin recognition [186,187]. Particularly intriguing is intraspecific kin recognition's contribution to interspecific allelopathy and improving plant productivity [188]. (−)-Loliolide, jasmonic acid and several chemicals are responsible for these signaling interactions [161,189–191]. Importantly, these signaling chemicals are ubiquitous in plant species. Such plant neighbor detection and allelochemical response, as well as their underlying mechanisms, will provide a wealth of research opportunities in grasslands and forests.

In fact, plant species occurring in grasslands and forests can take advantage of both allelochemicals and signaling chemicals released by neighbors, regulating intraspecific and interspecific interactions. Allelochemicals and signaling chemicals synergistically interact to influence plant coexistence, diversity and community structure in grasslands and forests. Plant neighbor detection and allelochemical response have been found in several mixed-species tree plantations [90,91]. Interestingly, kin recognition could even help forests regenerate. A family of firs may grow faster than unrelated trees by tracing flows of nutrients and chemical signals between trees connected by underground fungi [186,192]. Therefore, revealing the intraspecific and interspecific interactions mediated by allelochemicals and signaling chemicals in grasslands and forests can not only broaden our insight into the key processes and mechanisms of the land surface, but also enhance our ability to predict terrestrial ecosystems' responses to global changes.

Author Contributions: Conceptualization, C.-H.K.; investigation, Y.X. and X.C.; resources, Y.X., X.C. and L.D.; writing—original draft preparation, Y.X. and C.-H.K.; writing—review and editing, C.-H.K.; visualization, Y.X., L.D. and X.C.; supervision, C.-H.K. All authors have read and agreed to the published version of the manuscript.

Funding: This research received no external funding.

Data Availability Statement: Not applicable.

Acknowledgments: The authors sincerely thank the anonymous referees for their constructive comments and suggestions.

Conflicts of Interest: The authors declare no conflict of interest.

References

1. Dangal, S.R.; Tian, H.; Lu, C.; Pan, S.; Pederson, N.; Hessl, A. Synergistic effects of climate change and grazing on net primary production of Mongolian grasslands. *Ecosphere* **2016**, *7*, e1274. [CrossRef]
2. De Frenne, P.; Lenoir, J.; Luoto, M.; Scheffers, B.R.; Zellweger, F.; Aalto, J.; Ashcroft, M.B.; Christiansen, D.M.; Decocq, G.; De Pauw, K. Forest microclimates and climate change: Importance, drivers and future research agenda. *Global Change Biol.* **2021**, *27*, 2279–2297. [CrossRef] [PubMed]
3. Sajjad, H.; Kumar, P.; Masroor, M.; Rahaman, M.H.; Rehman, S.; Ahmed, R.; Sahana, M. Forest vulnerability to climate change: A review for future research framework. *Forests* **2022**, *13*, 917.
4. Lamarque, P.; Tappeiner, U.; Turner, C.; Steinbacher, M.; Bardgett, R.D.; Szukics, U.; Schermer, M.; Lavorel, S. Stakeholder perceptions of grassland ecosystem services in relation to knowledge on soil fertility and biodiversity. *Reg. Environ. Chang.* **2011**, *11*, 791–804. [CrossRef]
5. Ninan, K.N.; Inoue, M. Valuing forest ecosystem services: What we know and what we don't. *Ecol. Econ.* **2013**, *93*, 137–149. [CrossRef]
6. Chesson, P. Mechanisms of maintenance of species diversity. *Annu. Rev. Ecol. Evol. Syst.* **2000**, *31*, 343–366. [CrossRef]
7. Bennett, J.A.; Maherali, H.; Reinhart, K.O.; Lekberg, Y.; Hart, M.M.; Klironomos, J. Plant-soil feedbacks and mycorrhizal type influence temperate forest population dynamics. *Science* **2017**, *355*, 181–184. [CrossRef] [PubMed]
8. Bennett, J.A.; Klironomos, J. Mechanisms of plant–soil feedback: Interactions among biotic and abiotic drivers. *New Phytol.* **2019**, *222*, 91–96. [CrossRef]
9. Hierro, J.L.; Callaway, R.M. The ecological importance of allelopathy. *Annu. Rev. Ecol. Evol. Syst.* **2021**, *52*, 25–45. [CrossRef]
10. Meiners, S.J.; Kong, C.; Ladwig, L.M.; Pisula, N.L.; Lang, K.A. Developing an ecological context for allelopathy. *Plant Ecol.* **2012**, *213*, 1221–1227. [CrossRef]
11. Kong, C.H.; Xuan, T.D.; Khanh, T.D.; Tran, H.; Trung, N.T. Allelochemicals and signaling chemicals in plants. *Molecules* **2019**, *24*, 2737. [CrossRef]
12. Sun, G.; Luo, P.; Wu, N.; Qiu, P.F.; Gao, Y.H.; Chen, H.; Shi, F.S. *Stellera chamaejasme* L. increases soil N availability, turnover rates and microbial biomass in an alpine meadow ecosystem on the eastern Tibetan Plateau of China. *Soil Biol. Biochem.* **2009**, *41*, 86–91. [CrossRef]
13. He, W.; Detheridge, A.; Liu, Y.; Wang, L.; Wei, H.; Griffith, G.W.; Scullion, J.; Wei, Y. Variation in soil fungal composition associated with the invasion of *Stellera chamaejasme* L. in Qinghai–Tibet plateau grassland. *Microorganisms* **2019**, *7*, 587. [CrossRef]
14. Jin, H.; Guo, H.; Yang, X.; Xin, A.; Liu, H.; Qin, B. Effect of allelochemicals, soil enzyme activity and environmental factors from *Stellera chamaejasme* L. on rhizosphere bacterial communities in the northern Tibetan Plateau. *Arch. Agron. Soil Sci.* **2022**, *68*, 547–560. [CrossRef]
15. Soderquist, C.J. Juglone and allelopathy. *J. Chem. Edu.* **1973**, *50*, 782. [CrossRef] [PubMed]
16. Muller, C.H.; Muller, W.H.; Haines, B.L. Volatile growth inhibitors produced by aromatic shrubs. *Science* **1964**, *143*, 471–473. [CrossRef] [PubMed]
17. Vilà, M.; Espinar, J.L.; Hejda, M.; Hulme, P.E.; Jarošík, V.; Maron, J.L.; Pergl, J.; Schaffner, U.; Sun, Y.; Pyšek, P. Ecological impacts of invasive alien plants: A meta-analysis of their effects on species, communities and ecosystems. *Ecol. Lett.* **2011**, *14*, 702–708. [CrossRef]
18. Bais, H.P.; Vepachedu, R.; Gilroy, S.; Callaway, R.M.; Vivanco, J.M. Allelopathy and exotic plant invasion: From molecules and genes to species interactions. *Science* **2003**, *301*, 1377–1380. [CrossRef] [PubMed]
19. Pollock, J.L.; Kogan, L.A.; Thorpe, A.S.; Holben, W.E. (±)-Catechin, a root exudate of the invasive *Centaurea stoebe* Lam. (spotted knapweed) exhibits bacteriostatic activity against multiple soil bacterial populations. *J. Chem. Ecol.* **2011**, *37*, 1044–1053. [CrossRef]
20. Reinhart, K.O.; Rinella, M. Comparing susceptibility of eastern and western US grasslands to competition and allelopathy from spotted knapweed [*Centaurea stoebe* L. subsp. micranthos (Gugler) Hayek]. *Plant Ecol.* **2011**, *212*, 821–828. [CrossRef]
21. Aschehoug, E.T.; Callaway, R.M.; Newcombe, G.; Tharayil, N.; Chen, S. Fungal endophyte increases the allelopathic effects of an invasive forb. *Oecologia* **2014**, *175*, 285–291. [CrossRef] [PubMed]
22. Callaway, R.M.; Ridenour, W.M. Novel weapons: Invasive success and the evolution of increased competitive ability. *Front. Ecol. Environ.* **2004**, *2*, 436–443. [CrossRef]
23. Thorpe, A.S.; Thelen, G.C.; Diaconu, A.; Callaway, R.M. Root exudate is allelopathic in invaded community but not in native community: Field evidence for the novel weapons hypothesis. *J. Ecol.* **2009**, *97*, 641–645. [CrossRef]
24. Barto, E.K.; Powell, J.R.; Cipollini, D. How novel are the chemical weapons of garlic mustard in North American forest understories? *Biol. Invasions* **2010**, *12*, 3465–3471. [CrossRef]
25. Kim, Y.O.; Lee, E.J. Comparison of phenolic compounds and the effects of invasive and native species in East Asia: Support for the novel weapons hypothesis. *Ecol. Res.* **2011**, *26*, 87–94. [CrossRef]
26. Pinzone, P.; Potts, D.; Pettibone, G.; Warren, R. Do novel weapons that degrade mycorrhizal mutualisms promote species invasion? *Plant Ecol.* **2018**, *219*, 539–548. [CrossRef]
27. Blair, A.C.; Nissen, S.J.; Brunk, G.R.; Hufbauer, R.A. A lack of evidence for an ecological role of the putative allelochemical (±)-catechin in spotted knapweed invasion success. *J. Chem. Ecol.* **2006**, *32*, 2327–2331. [CrossRef]
28. Duke, S.O.; Blair, A.C.; Dayan, F.E.; Johnson, R.D.; Meepagala, K.M.; Cook, D.; Bajsa, J. Is (−)-catechin a novel weapon of spotted knapweed (*Centaurea stoebe*)? *J. Chem. Ecol.* **2009**, *35*, 141–153. [CrossRef] [PubMed]

29. Yannelli, F.A.; Novoa, A.; Lorenzo, P.; Rodríguez, J.; Le Roux, J.J. No evidence for novel weapons: Biochemical recognition modulates early ontogenetic processes in native species and invasive acacias. *Biol. Invasions* **2020**, *22*, 549–562. [CrossRef]
30. Renne, I.J.; Sinn, B.T.; Shook, G.W.; Sedlacko, D.M.; Dull, J.R.; Villarreal, D.; Hierro, J.L. Eavesdropping in plants: Delayed germination via biochemical recognition. *J. Ecol.* **2014**, *102*, 86–94. [CrossRef]
31. Kalisz, S.; Kivlin, S.N.; Bialic-Murphy, L. Allelopathy is pervasive in invasive plants. *Biol. Invasions* **2021**, *23*, 367–371. [CrossRef]
32. Ning, L.; Yu, F.H.; van Kleunen, M. Allelopathy of a native grassland community as a potential mechanism of resistance against invasion by introduced plants. *Biol Invasions* **2016**, *18*, 3481–3493. [CrossRef]
33. Akiyama, T.; Kawamura, K. Grassland degradation in China: Methods of monitoring, management and restoration. *Grassl. Sci.* **2007**, *53*, 1–17. [CrossRef]
34. Gang, C.; Zhou, W.; Chen, Y.; Wang, Z.; Sun, Z.; Li, J.; Qi, J.; Odeh, I. Quantitative assessment of the contributions of climate change and human activities on global grassland degradation. *Environ. Earth. Sci.* **2014**, *72*, 4273–4282. [CrossRef]
35. Bardgett, R.D.; Bullock, J.M.; Lavorel, S.; Manning, P.; Schaffner, U.; Ostle, N.; Chomel, M.; Durigan, G.; Fry, E.L.; Johnson, D.; et al. Combatting global grassland degradation. *Nat. Rev. Earth Environ.* **2021**, *2*, 720–735. [CrossRef]
36. Li, X.F.; Wang, J.; Huang, D.; Wang, L.X.; Wang, K. Allelopathic potential of *Artemisia frigida* and successional changes of plant communities in the northern China steppe. *Plant Soil* **2011**, *341*, 383–398. [CrossRef]
37. Zuo, Z.J.; Zhang, R.M.; Gao, P.J.; Wen, G.S.; Hou, P.; Gao, Y. Allelopathic effects of *Artemisia frigida* Willd. on growth of pasture grasses in Inner Mongolia, China. *Biochem. Syst. Ecol.* **2011**, *39*, 377–383.
38. Guo, H.; Cui, H.; Jin, H.; Yan, Z.; Ding, L.; Qin, B. Potential allelochemicals in root zone soils of *Stellera chamaejasme* L. and variations at different geographical growing sites. *Plant Growth Regul.* **2015**, *77*, 335–342. [CrossRef]
39. Holechek, J.L. Do most livestock losses to poisonous plants result from "poor" range management? *J Range Manag. Arch.* **2002**, *55*, 270–276. [CrossRef]
40. Tokarnia, C.H.; Döbereiner, J.; Peixoto, P.V. Poisonous plants affecting livestock in Brazil. *Toxicon* **2002**, *40*, 1635–1660. [CrossRef]
41. Zhang, Y.; Tang, S.; Liu, K.; Li, X.; Huang, D.; Wang, K. The allelopathic effect of *Potentilla acaulis* on the changes of plant community in grassland, northern China. *Ecol. Res.* **2015**, *30*, 41–47. [CrossRef]
42. Goldschmidt, F.; Regoes, R.R.; Johnson, D.R. Metabolite toxicity slows local diversity loss during expansion of a microbial cross-feeding community. *ISME J.* **2018**, *12*, 136–144. [CrossRef]
43. Humphries, T.; Florentine, S.K.; Dowling, K.; Turville, C.; Sinclair, S. Weed management for landscape scale restoration of global temperate grasslands. *Land Degrad. Dev.* **2021**, *32*, 1090–1102. [CrossRef]
44. Ma, L.; Wu, H.; Bai, R.; Zhou, L.; Yuan, X.; Hou, D. Phytotoxic effects of *Stellera chamaejasme* L. root extract. *Afri. J. Agric. Res.* **2011**, *6*, 1170–1176.
45. Wang, W.; Jia, T.; Qi, T.; Li, S.; Degen, A.A.; Han, J.; Bai, Y.; Zhang, T.; Qi, S.; Huang, M. Root exudates enhanced rhizobacteria complexity and microbial carbon metabolism of toxic plants. *iScience* **2022**, *25*, 105243. [CrossRef]
46. Jiang, Z.; Tanaka, T.; Sakamoto, T.; Kouno, I.; Duan, J.; Zhou, R. Biflavanones, diterpenes, and coumarins from the roots of *Stellera chamaejasme* L. *Chem. Pharm. Bull.* **2002**, *50*, 137–139. [CrossRef] [PubMed]
47. Yan, Z.; Guo, H.; Yang, J.; Liu, Q.; Jin, H.; Xu, R.; Cui, H.; Qin, B. Phytotoxic flavonoids from roots of *Stellera chamaejasme* L.(Thymelaeaceae). *Phytochemistry* **2014**, *106*, 61–68. [CrossRef] [PubMed]
48. Zhang, R.; Zhang, W.; Zuo, Z.; Li, R.; Wu, J.; Gao, Y. Inhibition effects of volatile organic compounds from *Artemisia frigida* Willd. on the pasture grass intake by lambs. *Small Ruminant Res.* **2014**, *121*, 248–254. [CrossRef]
49. Lu, H.; Wang, S.S.; Zhou, Q.W.; Zhao, Y.N.; Zhao, B.Y. Damage and control of major poisonous plants in the western grasslands of China—A review. *Rangel. J.* **2012**, *34*, 329–339. [CrossRef]
50. Zhigzhitzhapova, S.V.; Randalova, T.E.; Radnaeva, L.D.; Dylenova, E.P.; Chen, S.; Zhang, F. Chemical composition of essentials oils of *Artemisia frigida* Willd.(Asteraceae) grown in the North and Central Asia. *J. Essent. Oil Bear Pl.* **2017**, *20*, 915–926. [CrossRef]
51. Zhang, R.M.; Zuo, Z.J.; Gao, P.J.; Hou, P.; Wen, G.S.; Gao, Y. Allelopathic effects of VOCs of *Artemisia frigida* Willd. on the regeneration of pasture grasses in Inner Mongolia. *J. Arid Environ.* **2012**, *87*, 212–218. [CrossRef]
52. Wang, Q.; Zhang, H.; Yang, Q.; Wang, T.; Zhang, J.; Liu, J.; Shi, M.; Ping, X. The impact of grazing intensity on the allelopathic effect of *Artemisia frigida* in a temperate grassland in northeasssrn China. *Flora* **2022**, *288*, 152005. [CrossRef]
53. Liu, Q.; Xu, R.; Yan, Z.; Jin, H.; Cui, H.; Lu, L.; Zhang, D.; Qin, B. Phytotoxic allelochemicals from roots and root exudates of *Trifolium pratense*. *J. Agric. Food Chem.* **2013**, *61*, 6321–6327. [CrossRef] [PubMed]
54. Braun, R.C.; Patton, A.J.; Watkins, E.; Koch, P.L.; Anderson, N.P.; Bonos, S.A.; Brilman, L.A. Fine fescues: A review of the species, their improvement, production, establishment, and management. *Crop Sci.* **2020**, *60*, 1142–1187. [CrossRef]
55. Tabaglio, V.; Marocco, A.; Schulz, M. Allelopathic cover crop of rye for integrated weed control in sustainable agroecosystems. *Ital. J. Agron.* **2013**, *8*, e5. [CrossRef]
56. Schulz, M.; Marocco, A.; Tabaglio, V.; Macias, F.A.; Molinillo, J.M. Benzoxazinoids in rye allelopathy-from discovery to application in sustainable weed control and organic farming. *J. Chem. Ecol.* **2013**, *39*, 154–174. [CrossRef]
57. Serajchi, M.; Schellenberg, M.P.; Lamb, E.G. The potential of seven native North American forage species to suppress weeds through allelopathy. *Can. J. Plant Sci.* **2017**, *97*, 881–890.
58. Zhao, H.H.; Kong, C.H.; Xu, X.H. Herbicidal efficacy and ecological safety of an allelochemical-based benzothiazine derivative. *Pest Manag. Sci.* **2019**, *75*, 2690–2697. [CrossRef]

59. Shang, Z.; Hou, Y.; Li, F.; Guo, C.; Jia, T.; Degen, A.A.; White, A.; Ding, L.; Long, R. Inhibitory action of allelochemicals from *Artemisia nanschanica* to control *Pedicularis kansuensis*, an annual weed of alpine grasslands. *Aust. J. Bot.* **2017**, *65*, 305–314. [CrossRef]
60. Adhikari, L.; Mohseni-Moghadam, M.; Missaoui, A. Allelopathic effects of cereal rye on weed suppression and forage yield in Alfalfa. *Amer. J. Plant Sci.* **2018**, *9*, 685. [CrossRef]
61. Chon, S.U.; Jennings, J.A.; Nelson, C.J. Alfalfa (*Medicago sativa* L.) autotoxicity: Current status. *Allelopathy J.* **2006**, *18*, 57–80.
62. Ghimire, B.K.; Ghimire, B.; Yu, C.Y.; Chung, I. Allelopathic and autotoxic effects of *Medicago sativa*—Derived allelochemicals. *Plants* **2019**, *8*, 233. [CrossRef] [PubMed]
63. Wang, C.; Liu, Z.; Wang, Z.; Pang, W.; Zhang, L.; Wen, Z.; Zhao, Y.; Sun, J.; Wang, Z.; Yang, C. Effects of autotoxicity and allelopathy on seed germination and seedling growth in *Medicago truncatula*. *Front. Plant Sci.* **2022**, *13*, 908426. [CrossRef]
64. Zhang, X.; Shi, S.; Li, X.; Li, C.; Zhang, C.; Kang, W.; Yin, G. Effects of autotoxicity on alfalfa (*Medicago sativa*): Seed germination, oxidative damage and lipid peroxidation of seedlings. *Agronomy* **2021**, *11*, 1027. [CrossRef]
65. Li, Q.; Song, Y.; Li, G.; Yu, P.; Wang, P.; Zhou, D. Grass-legume mixtures impact soil N, species recruitment, and productivity in temperate steppe grassland. *Plant Soil* **2015**, *394*, 271–285. [CrossRef]
66. Mischkolz, J.M.; Schellenberg, M.P.; Lamb, E.G. Assembling productive communities of native grass and legume species: Finding the right mix. *Appl. Veg. Sci.* **2016**, *19*, 111–121. [CrossRef]
67. Quan, X.; Qiao, Y.; Chen, M.; Duan, Z.; Shi, H. Comprehensive evaluation of the allelopathic potential of *Elymus nutans*. *Ecol. Evol.* **2021**, *11*, 12389–12400. [CrossRef]
68. Rakoczy Trojanowska, M.; Święcicka, M.; Bakera, B.; Kowalczyk, M.; Stochmal, A.; Bolibok, L. Cocultivating rye with berseem clover affects benzoxazinoid production and expression of related genes. *Crop Sci.* **2020**, *60*, 3228–3246. [CrossRef]
69. Tsubo, M.; Nishihara, E.; Nakamatsu, K.; Cheng, Y.; Shinoda, M. Plant volatiles inhibit restoration of plant species communities in dry grassland. *Basic Appl. Ecol.* **2012**, *13*, 76–84. [CrossRef]
70. Yu, R.P.; Li, X.X.; Xiao, Z.H.; Lambers, H.; Li, L. Phosphorus facilitation and covariation of root traits in steppe species. *New Phytol.* **2020**, *226*, 1285–1298. [CrossRef]
71. Wu, M.; Yao, L.; Ai, X.; Zhu, J.; Zhu, Q.; Wang, J.; Huang, X.; Hong, J. The reproductive characteristics of core germplasm in a native *Metasequoia glyptostroboides* population. *Biodiver. Sci.* **2020**, *28*, 303.
72. Xu, L.; Yao, L.; Ai, X.; Guo, Q.; Wang, S.; Zhou, D.; Deng, C.; Ai, X. Litter autotoxicity limits natural regeneration of *Metasequoia glyptostroboides*. *New Forest* **2022**. [CrossRef]
73. Huang, X.; Chen, J.; Liu, J.; Li, J.; Wu, M.; Tong, B. Autotoxicity hinders the natural regeneration of *Cinnamomum migao* HW Li in Southwest China. *Forests* **2019**, *10*, 919. [CrossRef]
74. Lang, T.; Wei, P.; Chen, X.; Fu, Y.; Tam, N.F.; Hu, Z.; Chen, Z.; Li, F.; Zhou, H. Microcosm study on allelopathic effects of leaf litter leachates and purified condensed tannins from *Kandelia obovata* on germination and growth of *Aegiceras corniculatum*. *Forests* **2021**, *12*, 1000. [CrossRef]
75. Hane, E.N. Indirect effects of beech bark disease on sugar maple seedling survival. *Can. J. Forest Res.* **2003**, *33*, 807–813. [CrossRef]
76. Collin, A.; Messier, C.; Kembel, S.W.; Bélanger, N. Low light availability associated with American beech is the main factor for reduced sugar maple seedling survival and growth rates in a hardwood forest of Southern Quebec. *Forests* **2017**, *8*, 413. [CrossRef]
77. Taylor, K.; Rowland, A.P.; Jones, H.E. *Molinia caerulea* (L.) Moench. *J. Ecol.* **2001**, *89*, 126–144. [CrossRef]
78. Fernandez, M.; Malagoli, P.; Gallet, C.; Fernandez, C.; Vernay, A.; Ameglio, T.; Balandier, P. Investigating the role of root exudates in the interaction between oak seedlings and purple moor grass in temperate forest. *Forest Ecol. Manag.* **2021**, *491*, 119175. [CrossRef]
79. Nilsen, E.T.; Huebner, C.D.; Carr, D.E.; Bao, Z. Interaction between *Ailanthus altissima* and native *Robinia pseudoacacia* in early succession: Implications for forest management. *Forests* **2018**, *9*, 221. [CrossRef]
80. Demeter, A.; Saláta, D.; Tormáné Kovács, E.; Szirmai, O.; Trenyik, P.; Meinhardt, S.; Rusvai, K.; Verbényiné Neumann, K.; Schermann, B.; Szegleti, Z. Effects of the Invasive tree species *Ailanthus altissima* on the floral diversity and soil properties in the Pannonian region. *Land* **2021**, *10*, 1155. [CrossRef]
81. Stone, R. Nursing China's ailing forests back to health. *Science* **2009**, *325*, 556–558. [CrossRef] [PubMed]
82. Williams, R.A. Mitigating biodiversity concerns in Eucalyptus plantations located in South China. *J. Biosci. Med.* **2015**, *3*, 1–8. [CrossRef]
83. Sasikumar, K.; Vijayalakshmi, C.; Parthiban, K.T. Allelopathic effects of four Eucalyptus species on redgram (*Cajanus cajan* L.). *J. Trop. Agric.* **2006**, *39*, 134–138.
84. Ahmed, R.; Hoque, A.; Hossain, M.K. Allelopathic effects of leaf litters of *Eucalyptus camaldulensis* on some forest and agricultural crops. *J. Forestry Res.* **2008**, *19*, 19–24. [CrossRef]
85. Chapuis-Lardy, L.; Contour-Ansel, D.; Bernhard-Reversat, F. High-performance liquid chromatography of water-soluble phenolics in leaf litter of three Eucalyptus hybrids (Congo). *Plant Sci.* **2002**, *163*, 217–222. [CrossRef]
86. Song, Q.; Qin, F.; He, H.; Wang, H.; Yu, S. Allelopathic potential of rain leachates from *Eucalyptus urophylla* on four tree species. *Agroforest Syst.* **2019**, *93*, 1307–1318. [CrossRef]
87. Zhang, C.; Li, X.; Chen, Y.; Zhao, J.; Wan, S.; Lin, Y.; Fu, S. Effects of Eucalyptus litter and roots on the establishment of native tree species in Eucalyptus plantations in South China. *Forest Ecol. Manag.* **2016**, *375*, 76–83. [CrossRef]

88. Qin, F.; Liu, S.; Yu, S. Effects of allelopathy and competition for water and nutrients on survival and growth of tree species in *Eucalyptus urophylla* plantations. *Forest Ecol. Manag.* **2018**, *424*, 387–395. [CrossRef]
89. Chen, L.; Wang, S. Allelopathic behaviour of Chinese fir from plantations of different ages. *Forestry* **2013**, *86*, 225–230. [CrossRef]
90. Chen, L.; Wang, S.; Wang, P.; Kong, C. Autoinhibition and soil allelochemical (cyclic dipeptide) levels in replanted Chinese fir (*Cunninghamia lanceolata*) plantations. *Plant Soil* **2014**, *374*, 793–801. [CrossRef]
91. Forrester, D.I.; Bauhus, J.; Cowie, A.L.; Vanclay, J.K. Mixed-species plantations of Eucalyptus with nitrogen-fixing trees: A review. *Forest Ecol. Manag.* **2006**, *233*, 211–230. [CrossRef]
92. Yang, L.X.; Wang, P.; Kong, C.H. Effect of larch (*Larix gmelini* Rupr.) root exudates on Manchurian walnut (*Juglans mandshurica* Maxim.) growth and soil juglone in a mixed-species plantation. *Plant Soil* **2010**, *329*, 249–258. [CrossRef]
93. Xia, Z.C.; Kong, C.H.; Chen, L.C.; Wang, P.; Wang, S.L. A broadleaf species enhances an autotoxic conifers growth through belowground chemical interactions. *Ecology* **2016**, *97*, 2283–2292. [CrossRef] [PubMed]
94. Hashoum, H.; Santonja, M.; Gauquelin, T.; Saatkamp, A.; Gavinet, J.; Greff, S.; Lecareux, C.; Fernandez, C.; Bousquet-Mélou, A. Biotic interactions in a Mediterranean oak forest: Role of allelopathy along phenological development of woody species. *Eur. J. Forest Res.* **2017**, *136*, 699–710. [CrossRef]
95. Khaled, A.; Sleiman, M.; Goupil, P.; Richard, C. Phytotoxic effect of Macerates and Mulches from *Cupressus leylandii* leaves on clover and cress: Role of chemical composition. *Forests* **2020**, *11*, 1177. [CrossRef]
96. Lorenzo, P.; Palomera-Pérez, A.; Reigosa, M.J.; González, L. Allelopathic interference of invasive *Acacia dealbata* link on the physiological parameters of native understory species. *Plant Ecol.* **2011**, *212*, 403–412. [CrossRef]
97. Constán-Nava, S.; Soliveres, S.; Torices, R.; Serra, L.; Bonet, A. Direct and indirect effects of invasion by the alien tree *Ailanthus altissima* on riparian plant communities and ecosystem multifunctionality. *Biol. Invasions* **2015**, *17*, 1095–1108. [CrossRef]
98. Warren, R.J.; Labatore, A.; Candeias, M. Allelopathic invasive tree (*Rhamnus cathartica*) alters native plant communities. *Plant Ecol.* **2017**, *218*, 1233–1241. [CrossRef]
99. de Las Heras, P.; Medina-Villar, S.; Pérez-Corona, M.E.; Vázquez-de-Aldana, B.R. Leaf litter age regulates the effect of native and exotic tree species on understory herbaceous vegetation of riparian forests. *Basic Appl. Ecol.* **2020**, *48*, 11–25. [CrossRef]
100. Huang, W.; Hu, H.; Hu, T.; Chen, H.; Wang, Q.; Chen, G.; Tu, L. Impact of aqueous extracts of *Cinnamomum septentrionale* leaf litter on the growth and photosynthetic characteristics of *Eucalyptus grandis* seedlings. *New Forest* **2015**, *46*, 561–576. [CrossRef]
101. Muturi, G.M.; Poorter, L.; Bala, P.; Mohren, G.M.J. Unleached Prosopis litter inhibits germination but leached stimulates seedling growth of dry woodland species. *J. Arid Environ.* **2017**, *138*, 44–50. [CrossRef]
102. Orr, S.P.; Rudgers, J.A.; Clay, K. Invasive plants can inhibit native tree seedlings: Testing potential allelopathic mechanisms. *Plant Ecol.* **2005**, *181*, 153–165. [CrossRef]
103. Sayer, E.J. Using experimental manipulation to assess the roles of leaf litter in the functioning of forest ecosystems. *Biol. Rev.* **2006**, *81*, 1–31. [CrossRef]
104. Meiners, S.J. Functional correlates of allelopathic potential in a successional plant community. *Plant Ecol.* **2014**, *215*, 661–672. [CrossRef]
105. Da Silva, E.R.; Da Silveira, L.H.R.; Overbeck, G.E.; Soares, G.L.G. Inhibitory effects of *Eucalyptus saligna* leaf litter on grassland species: Physical versus chemical factors. *Plant Ecol. Divers.* **2018**, *11*, 55–67. [CrossRef]
106. Bonanomi, G.; Incerti, G.; Barile, E.; Capodilupo, M.; Antignani, V.; Mingo, A.; Lanzotti, V.; Scala, F.; Mazzoleni, S. Phytotoxicity, not nitrogen immobilization, explains plant litter inhibitory effects: Evidence from solid-state ^{13}C NMR spectroscopy. *New Phytol.* **2011**, *191*, 1018–1030. [CrossRef] [PubMed]
107. Cummings, J.A.; Parker, I.M.; Gilbert, G.S. Allelopathy: A tool for weed management in forest restoration. *Plant Ecol.* **2012**, *213*, 1975–1989. [CrossRef]
108. Matuda, Y.; Iwasaki, A.; Suenaga, K.; Kato-Noguchi, H. Allelopathy and allelopathic substances of fossil tree species *Metasequoia glyptostroboides*. *Agronomy* **2022**, *12*, 83. [CrossRef]
109. Lorenzo, P.; Pereira, C.S.; Rodríguez-Echeverría, S. Differential impact on soil microbes of allelopathic compounds released by the invasive *Acacia dealbata* Link. *Soil Biol. Biochem.* **2013**, *57*, 156–163. [CrossRef]
110. Bao, Z.; Nilsen, E.T. Interactions between seedlings of the invasive tree *Ailanthus altissima* and the native tree *Robinia pseudoacacia* under low nutrient conditions. *J. Plant Interact.* **2015**, *10*, 173–184. [CrossRef]
111. Metlen, K.L.; Aschehoug, E.T.; Callaway, R.M. Competitive outcomes between two exotic invaders are modified by direct and indirect effects of a native conifer. *Oikos* **2013**, *122*, 632–640. [CrossRef]
112. Kaur, R.; Callaway, R.M.; Inderjit. Soils and the conditional allelopathic effects of a tropical invader. *Soil Biol. Biochem.* **2014**, *78*, 316–325. [CrossRef]
113. Kaur, R.; Gonzáles, W.L.; Llambi, L.D.; Soriano, P.J.; Callaway, R.M.; Rout, M.E.; Gallaher, T.J. Community impacts of *Prosopis Juliflora* invasion: Biogeographic and congeneric comparisons. *PLoS ONE* **2012**, *7*, e44966. [CrossRef] [PubMed]
114. Gunarathne, R.; Perera, G. Does the invasion of *Prosopis juliflora* cause the die-back of the native *Manilkara hexandra* in seasonally dry tropical forests of Sri Lanka. *Trop. Ecol.* **2016**, *57*, 475–488.
115. Nuzzo, V. Invasion pattern of herb garlic mustard (*Alliaria petiolata*) in high quality forests. *Biol. Invasions* **1999**, *1*, 169–179. [CrossRef]
116. Callaway, R.M.; Cipollini, D.; Barto, K.; Thelen, G.C.; Hallett, S.G.; Prati, D.; Stinson, K.; Klironomos, J. Novel weapons: Invasive plant suppresses fungal mutualists in America but not in its native Europe. *Ecology* **2008**, *89*, 1043–1055. [CrossRef]

117. Hale, A.N.; Lapointe, L.; Kalisz, S. Invader disruption of belowground plant mutualisms reduces carbon acquisition and alters allocation patterns in a native forest herb. *New Phytol.* **2016**, *209*, 542–549. [CrossRef]
118. Lankau, R.A.; Nuzzo, V.; Spyreas, G.; Davis, A.S. Evolutionary limits ameliorate the negative impact of an invasive plant. *Proc. Natl. Acad. Sci. USA* **2009**, *106*, 15362–15367. [CrossRef]
119. Barto, E.K.; Antunes, P.M.; Stinson, K.; Koch, A.M.; Klironomos, J.N.; Cipollini, D. Differences in arbuscular mycorrhizal fungal communities associated with sugar maple seedlings in and outside of invaded garlic mustard forest patches. *Biol. Invasions* **2011**, *13*, 2755–2762. [CrossRef]
120. Lankau, R.A. Coevolution between invasive and native plants driven by chemical competition and soil biota. *Proc. Natl. Acad. Sci. USA* **2012**, *109*, 11240–11245. [CrossRef]
121. Huang, F.; Lankau, R.; Peng, S. Coexistence via coevolution driven by reduced allelochemical effects and increased tolerance to competition between invasive and native plants. *New Phytol.* **2018**, *218*, 357–369. [CrossRef]
122. Liu, J.G.; Liao, H.X.; Chen, B.M.; Peng, S.L. Do the phenolic acids in forest soil resist the exotic plant invasion. *Allelopathy J.* **2017**, *41*, 167–175. [CrossRef]
123. Takemura, T.; Kamo, T.; Sakuno, E.; Hiradate, S.; Fujii, Y. Discovery of coumarin as the predominant allelochemical in *Gliricidia sepium*. *J. Trop. For. Sci.* **2013**, *25*, 268–272.
124. Yan, Z.; Wang, D.; Cui, H.; Zhang, D.; Sun, Y.; Jin, H.; Li, X.; Yang, X.; Guo, H.; He, X. Phytotoxicity mechanisms of two coumarin allelochemicals from *Stellera chamaejasme* in lettuce seedlings. *Acta Physiol. Plant* **2016**, *38*, 1–10. [CrossRef]
125. Zhang, S.; Sun, S.; Shi, H.; Zhao, K.; Wang, J.; Liu, Y.; Liu, X.; Wang, W. Physiological and biochemical mechanisms mediated by allelochemical isoliquiritigenin on the growth of lettuce seedlings. *Plants* **2020**, *9*, 245. [CrossRef] [PubMed]
126. Li, J.; Ye, Y.; Huang, H.; Dong, L. Kaempferol-3-O-β-D-glucoside, a potential allelochemical isolated from *Solidago Canadensis*. *Allelopathy J.* **2011**, *28*, 259–266.
127. Perry, L.G.; Johnson, C.; Alford, É.R.; Vivanco, J.M.; Paschke, M.W. Screening of grassland plants for restoration after spotted knapweed invasion. *Restor. Ecol.* **2005**, *13*, 725–735. [CrossRef]
128. Weir, T.L.; Bais, H.P.; Stull, V.J.; Callaway, R.M.; Thelen, G.C.; Ridenour, W.M.; Bhamidi, S.; Stermitz, F.R.; Vivanco, J.M. Oxalate contributes to the resistance of *Gaillardia grandiflora* and *Lupinus sericeus* to a phytotoxin produced by *Centaurea maculosa*. *Planta* **2006**, *223*, 785–795. [CrossRef]
129. Willis, R.J. *Juglans* spp., juglone and allelopathy. *Allelopathy J.* **2000**, *7*, 1–5.
130. Strugstad, M.; Despotovski, S. A summary of extraction, synthesis, properties, and potential uses of juglone: A literature review. *J. Ecosyst. Manag.* **2012**, *13*, 72–86. [CrossRef]
131. Bai, L.; Wang, W.; Hua, J.; Guo, Z.; Luo, S. Defensive functions of volatile organic compounds and essential oils from northern white-cedar in China. *BMC Plant Biol.* **2020**, *20*, 1–9. [CrossRef] [PubMed]
132. Santonja, M.; Bousquet Mélou, A.; Greff, S.; Ormeño, E.; Fernandez, C. Allelopathic effects of volatile organic compounds released from *Pinus halepensis* needles and roots. *Ecol. Evol.* **2019**, *9*, 8201–8213. [CrossRef] [PubMed]
133. Yu, H.; Le Roux, J.J.; Zhao, M.; Li, W. Mikania sesquiterpene lactones enhance soil bacterial diversity and fungal and bacterial activities. *Biol. Invasions* **2022**, *25*, 1–14. [CrossRef]
134. Braine, J.W.; Curcio, G.R.; Wachowicz, C.M.; Hansel, F.A. Allelopathic effects of *Araucaria angustifolia* needle extracts in the growth of *Lactuca sativa* seeds. *J. Forest Res. Jpn.* **2012**, *17*, 440–445. [CrossRef]
135. Kato-Noguchi, H.; Kimura, F.; Ohno, O.; Suenaga, K. Involvement of allelopathy in inhibition of understory growth in red pine forests. *J. Plant Physiol.* **2017**, *218*, 66–73. [CrossRef]
136. Wang, C.; Chen, H.; Li, T.; Weng, J.; Jhan, Y.; Lin, S.; Chou, C. The role of pentacyclic triterpenoids in the allelopathic effects of *Alstonia scholaris*. *J. Chem. Ecol.* **2014**, *40*, 90–98. [CrossRef]
137. Hagan, D.L.; Jose, S.; Lin, C. Allelopathic exudates of cogongrass (*Imperata cylindrica*): Implications for the performance of native pine savanna plant species in the southeastern US. *J. Chem. Ecol.* **2013**, *39*, 312–322. [CrossRef]
138. Bertin, C.; Weston, L.A.; Huang, T.; Jander, G.; Owens, T.; Meinwald, J.; Schroeder, F.C. Grass roots chemistry: *meta*-Tyrosine, an herbicidal nonprotein amino acid. *Proc. Natl. Acad. Sci. USA* **2007**, *104*, 16964–16969. [CrossRef]
139. Kato-Noguchi, H.; Kurniadi, D. Allelopathy and allelochemicals of *Leucaena leucocephala* as an invasive plant species. *Plants* **2022**, *11*, 1672. [CrossRef]
140. Nakajima, N.; Hiradate, S.; Fujii, Y. Plant growth inhibitory activity of L-canavanine and its mode of action. *J. Chem. Ecol.* **2001**, *27*, 19–31. [CrossRef]
141. Mardani-Korrani, H.; Nakayasu, M.; Yamazaki, S.; Aoki, Y.; Kaida, R.; Motobayashi, T.; Kobayashi, M.; Ohkama-Ohtsu, N.; Oikawa, Y.; Sugiyama, A. L-Canavanine, a root exudate from hairy vetch (*Vicia villosa*) drastically affecting the soil microbial community and metabolite pathways. *Front. Microbiol.* **2021**, *12*, 701796. [CrossRef] [PubMed]
142. Kong, C.H.; Chen, L.C.; Xu, X.H.; Wang, P.; Wang, S.L. Allelochemicals and activities in a replanted Chinese fir (*Cunninghamia lanceolata* (Lamb.) Hook) tree ecosystem. *J. Agric. Food Chem.* **2008**, *56*, 11734–11739. [CrossRef] [PubMed]
143. Hashimoto, Y.; Shudo, K. Chemistry of biologically active benzoxazinoids. *Phytochemistry* **1996**, *43*, 551–559. [CrossRef] [PubMed]
144. Niemeyer, H.M. Hydroxamic acids derived from 2-hydroxy-2H-1,4- benzoxazin-3(4H)-one: Key defense chemicals of cereals. *J. Agric. Food Chem.* **2009**, *57*, 1677–1696. [CrossRef]
145. Lankau, R.A.; Strauss, S.Y. Mutual feedbacks maintain both genetic and species diversity in a plant community. *Science* **2007**, *317*, 1561–1563. [CrossRef]

46. Oburger, E.; Schmidt, H. New methods to unravel rhizosphere processes. *Trends Plant Sci.* **2016**, *21*, 243–255. [CrossRef]
47. Phillips, R.P.; Erlitz, Y.; Bier, R.; Bernhardt, E.S. New approach for capturing soluble root exudates in forest soils. *Funct. Ecol.* **2008**, *22*, 990–999. [CrossRef]
48. Qiao, B.; Nie, S.; Li, Q.; Majeed, Z.; Cheng, J.; Yuan, Z.; Li, C.; Zhao, C. Quick and *in situ* detection of different polar allelochemicals in Taxus soil by microdialysis combined with UPLC-MS/MS. *J. Agric. Food Chem.* **2022**, *70*, 16435–16445. [CrossRef]
49. Sosa, T.; Valares, C.; Alías, J.C.; Chaves Lobón, N. Persistence of flavonoids in *Cistus ladanifer* soils. *Plant Soil* **2010**, *337*, 51–63. [CrossRef]
50. Del Valle, I.; Webster, T.M.; Cheng, H.; Thies, J.E.; Kessler, A.; Miller, M.K.; Ball, Z.T.; MacKenzie, K.R.; Masiello, C.A.; Silberg, J.J. Soil organic matter attenuates the efficacy of flavonoid-based plant-microbe communication. *Sci. Adv.* **2020**, *6*, x8254. [CrossRef]
51. Hoang, L.; Joo, G.; Kim, W.; Jeon, S.; Choi, S.; Kim, J.; Rhee, I.; Hur, J.; Song, K. Growth inhibitors of lettuce seedlings from *Bacillus cereus* EJ-121. *Plant Growth Regul.* **2005**, *47*, 149–154. [CrossRef]
52. Tuyen, P.T.; Xuan, T.D.; Tu Anh, T.T.; Mai Van, T.; Ahmad, A.; Elzaawely, A.A.; Khanh, T.D. Weed suppressing potential and isolation of potent plant growth inhibitors from *Castanea crenata* Sieb. et Zucc. *Molecules* **2018**, *23*, 345. [CrossRef]
53. Ida, N.; Iwasaki, A.; Teruya, T.; Suenaga, K.; Kato-Noguchi, H. Tree fern *Cyathea lepifera* may survive by its phytotoxic property. *Plants* **2019**, *9*, 46. [CrossRef] [PubMed]
54. Reigosa, M.J.; Pazos-Malvido, E. Phytotoxic effects of 21 plant secondary metabolites on *Arabidopsis thaliana* germination and root growth. *J. Chem. Ecol.* **2007**, *33*, 1456–1466. [CrossRef] [PubMed]
55. Hiradate, S.; Ohse, K.; Furubayashi, A.; Fujii, Y. Quantitative evaluation of allelopathic potentials in soils: Total activity approach. *Weed Sci.* **2010**, *58*, 258–264. [CrossRef]
56. Kang, G.; Mishyna, M.; Appiah, K.S.; Yamada, M.; Takano, A.; Prokhorov, V.; Fujii, Y. Screening for plant volatile emissions with allelopathic activity and the identification of L-Fenchone and 1, 8-Cineole from star anise (*Illicium verum*) leaves. *Plants* **2019**, *8*, 457. [CrossRef] [PubMed]
57. Macías, F.A.; Marín, D.; Oliveros-Bastidas, A.; Castellano, D.; Simonet, A.M.; Molinillo, J.M. Structure—activity relationships (SAR) studies of benzoxazinones, their degradation products and analogues. Phytotoxicity on standard target species (STS). *J. Agric. Food Chem.* **2005**, *53*, 538–548. [CrossRef]
58. Wang, C.Y.; Li, L.L.; Meiners, S.J.; Kong, C.H. Root placement patterns in allelopathic plant-plant interactions. *New Phytol.* **2023**, *237*, 563–575. [CrossRef]
59. Semchenko, M.; John, E.A.; Hutchings, M.J. Effects of physical connection and genetic identity of neighbouring ramets on root-placement patterns in two clonal species. *New Phytol.* **2007**, *176*, 644–654. [CrossRef]
60. Semchenko, M.; Saar, S.; Lepik, A. Plant root exudates mediate neighbour recognition and trigger complex behavioural changes. *New Phytol.* **2014**, *204*, 631–637. [CrossRef]
61. Wang, N.Q.; Kong, C.H.; Wang, P.; Meiners, S.J. Root exudate signals in plant–plant interactions. *Plant Cell Environ.* **2021**, *44*, 1044–1058. [CrossRef] [PubMed]
62. Asaduzzaman, M.; An, M.; Pratley, J.E.; Luckett, D.J.; Lemerle, D.; Coombes, N. The seedling root response of annual ryegrass (*Lolium rigidum*) to neighbouring seedlings of a highly-allelopathic canola (*Brassica napus*). *Flora* **2016**, *219*, 18–24. [CrossRef]
63. van Dam, N.M.; Bouwmeester, H.J. Metabolomics in the rhizosphere: Tapping into belowground chemical communication. *Trends Plant Sci.* **2016**, *21*, 256–265. [CrossRef]
64. Wu, F.; Wang, X.; Xue, C. Effect of cinnamic acid on soil microbial characteristics in the cucumber rhizosphere. *Eur. J. Soil Biol.* **2009**, *45*, 356–362. [CrossRef]
65. Buee, M.; Rossignol, M.; Jauneau, A.; Ranjeva, R.; Bécard, G. The pre-symbiotic growth of arbuscular mycorrhizal fungi is induced by a branching factor partially purified from plant root exudates. *Mol. Plant Microbe. Interact.* **2000**, *13*, 693–698. [CrossRef]
66. Yu, L.; Zhao, H.; Chen, G.; Yuan, S.; Lan, T.; Zeng, J. Allelochemical-driven N preference switch from NO_3^- to NH_4^+ affecting plant growth of *Cunninghamia lanceolata* (lamb.) hook. *Plant Soil* **2020**, *451*, 419–434. [CrossRef]
67. Zhu, X.; Li, X.; Xing, F.; Chen, C.; Huang, G.; Gao, Y. Interaction between root exudates of the poisonous plant *Stellera chamaejasme* L. and arbuscular mycorrhizal fungi on the growth of *Leymus Chinensis* (Trin.) Tzvel. *Microorganisms* **2020**, *8*, 364. [CrossRef]
68. Brundrett, M.C. Mycorrhizal associations and other means of nutrition of vascular plants: Understanding the global diversity of host plants by resolving conflicting information and developing reliable means of diagnosis. *Plant Soil* **2009**, *320*, 37–77. [CrossRef]
69. Garcia, K.; Zimmermann, S.D. The role of mycorrhizal associations in plant potassium nutrition. *Front. Plant Sci.* **2014**, *5*, 337. [CrossRef] [PubMed]
70. Merilä, P.; Malmivaara-Lämsä, M.; Spetz, P.; Stark, S.; Vierikko, K.; Derome, J.; Fritze, H. Soil organic matter quality as a link between microbial community structure and vegetation composition along a successional gradient in a boreal forest. *Appl. Soil Ecol.* **2010**, *46*, 259–267. [CrossRef]
71. Lu, F.; Zheng, L.; Chen, Y.; Li, D.; Zeng, R.; Li, H. Soil microorganisms alleviate the allelopathic effect of *Eucalyptus grandis*× *E. urophylla* leachates on *Brassica chinensis*. *J. Forestry Res.* **2017**, *28*, 1203–1207. [CrossRef]
72. Liu, S.; Qin, F.; Yu, S. *Eucalyptus urophylla* root-associated fungi can counteract the negative influence of phenolic acid allelochemicals. *Appl. Soil Ecol.* **2018**, *127*, 1–7. [CrossRef]
73. Qin, F.; Yu, S. Arbuscular mycorrhizal fungi protect native woody species from novel weapons. *Plant Soil* **2019**, *440*, 39–52. [CrossRef]

174. Xia, Z.C.; Kong, C.H.; Chen, L.C.; Wang, S.L. Allelochemical-mediated soil microbial community in long-term monospecific Chinese fir forest plantations. *Appl. Soil Ecol.* **2015**, *96*, 52–59. [CrossRef]
175. Achatz, M.; Rillig, M.C. Arbuscular mycorrhizal fungal hyphae enhance transport of the allelochemical juglone in the field. *Soil Boil. Biochem.* **2014**, *78*, 76–82. [CrossRef]
176. Achatz, M.; Morris, E.K.; Müller, F.; Hilker, M.; Rillig, M.C. Soil hypha-mediated movement of allelochemicals: Arbuscular mycorrhizae extend the bioactive zone of juglone. *Funct. Ecol.* **2014**, *28*, 1020–1029. [CrossRef]
177. Becklin, K.M.; Pallo, M.L.; Galen, C. Willows indirectly reduce arbuscular mycorrhizal fungal colonization in understorey communities. *J. Ecol.* **2012**, *100*, 343–351. [CrossRef]
178. Zhu, X.; Li, Y.; Feng, Y.; Ma, K. Response of soil bacterial communities to secondary compounds released from *Eupatorium adenophorum*. *Biol. Invasions* **2017**, *19*, 1471–1481. [CrossRef]
179. Mahdhi, M.; Tounekti, T.; Khemira, H. Effects of *Prosopis juliflora* on germination, plant growth of *Sorghum bicolor*, mycorrhiza and soil microbial properties. *Allelopathy J.* **2019**, *46*, 121–132. [CrossRef]
180. Rodríguez-Romero, M.; Godoy-Cancho, B.; Calha, I.M.; Passarinho, J.A.; Moreira, A.C. Allelopathic effects of three herb species on *phytophthora cinnamomi*, a pathogen causing severe oak decline in mediterranean wood pastures. *Forests* **2021**, *12*, 285. [CrossRef]
181. Singh, J.P.; Kuang, Y.; Ploughe, L.; Coghill, M.; Fraser, L.H. Spotted knapweed (*Centaurea stoebe*) creates a soil legacy effect by modulating soil elemental composition in a semi-arid grassland ecosystem. *J. Environ. Manag.* **2022**, *317*, 115391. [CrossRef] [PubMed]
182. Chen, W.B.; Chen, B.M.; Liao, H.X.; Su, J.Q.; Peng, S.L. Leaf leachates have the potential to influence soil nitrification via changes in ammonia—oxidizing archaea and bacteria populations. *Eur. J. Soil Sci.* **2020**, *71*, 119–131. [CrossRef]
183. Chen, P.; Hou, Y.; Zhuge, Y.; Wei, W.; Huang, Q. The effects of soils from different forest types on the growth of the invasive plant *Phytolacca americana*. *Forests* **2019**, *10*, 492. [CrossRef]
184. Bennion, L.D.; Ward, D. Plant–soil feedback from eastern redcedar (*Juniperus virginiana*) inhibits the growth of grasses in encroaching range. *Ecol. Evol.* **2022**, *12*, e9400. [CrossRef] [PubMed]
185. Kong, C.H.; Zhang, S.Z.; Li, Y.H.; Xia, Z.C.; Yang, X.F.; Meiners, S.J.; Wang, P. Plant neighbor detection and allelochemical response are driven by root-secreted signaling chemicals. *Nat. Commun.* **2018**, *9*, 3867. [CrossRef] [PubMed]
186. Pennisi, E. Do plants favor their kin? *Science* **2019**, *363*, 15–16. [CrossRef]
187. Yang, X.F.; Li, L.L.; Xu, Y.; Kong, C.H. Kin recognition in rice (*Oryza sativa* L.) lines. *New Phytol.* **2018**, *220*, 567–578. [CrossRef]
188. Xu, Y.; Cheng, H.F.; Kong, C.H.; Meiners, S.J. Intra-specific kin recognition contributes to inter-specific allelopathy: A case study of allelopathic rice interference with paddy weeds. *Plant Cell Environ.* **2021**, *44*, 3709–3721. [CrossRef]
189. Li, L.L.; Zhao, H.H.; Kong, C.H. (−)-Loliolide, the most ubiquitous lactone, is involved in barnyardgrass-induced rice allelopathy. *J. Exp. Bot.* **2020**, *71*, 1540–1550. [CrossRef]
190. Li, F.L.; Chen, X.; Luo, H.M.; Meiners, S.J.; Kong, C.H. Root-secreted (−)-loliolide modulates both belowground defense and aboveground flowering in Arabidopsis and tobacco. *J. Exp. Bot.* **2023**, *74*, 964–975. [CrossRef]
191. Li, L.L.; Li, Z.; Lou, Y.G.; Meiners, S.J.; Kong, C.H. (−)-Loliolide is a general signal of plant stress that activates jasmonate-related responses. *New Phytol.* **2022**. [CrossRef] [PubMed]
192. Asay, A.K.; Simard, S.W.; Dudley, S.A. Altering neighborhood relatedness and species composition affects interior Douglas-fir size and morphological traits with context-dependent responses. *Front. Ecol. Evol.* **2020**, *8*, 578524. [CrossRef]

Disclaimer/Publisher's Note: The statements, opinions and data contained in all publications are solely those of the individual author(s) and contributor(s) and not of MDPI and/or the editor(s). MDPI and/or the editor(s) disclaim responsibility for any injury to people or property resulting from any ideas, methods, instructions or products referred to in the content.

Article

Effects of UVA on Flavonol Accumulation in *Ginkgo biloba*

Qun Zhao [1], Zheng Wang [1], Gaiping Wang [1,*], Fuliang Cao [1], Xiaoming Yang [1], Huiqin Zhao [1] and Jinting Zhai [2]

[1] Co-Innovation Center for Sustainable Forestry in Southern China, College of Forestry and Grassland, College of Soil and Water Conservation, Nanjing Forestry University, Nanjing 210037, China; zhaoqun@njfu.edu.cn (Q.Z.); wangzheng921520@163.com (Z.W.); fuliangcaonjfu@163.com (F.C.); xmyang@njfu.edu.cn (X.Y.); huiqinzhao1968@gmail.com (H.Z.)

[2] Yancheng Forest Farm, Yancheng 224057, China; zjt13770152123@126.com

* Correspondence: wanggaiping@njfu.edu.cn

Abstract: Ginkgo is an economic tree species with high medicinal value, and flavonols are its main medicinal components. This research was conducted to investigate the molecular mechanism underlying the influence of Ultraviolet A (UVA) treatment on the synthesis of ginkgo flavonols with the aim of increasing their content. Ginkgo full-sib hybrid offspring were used as test materials. The phenylalanine ammonialyase (PAL), cinnamate 4-hydroxylase (C4H), and 4-coumarate: CoA ligase (4CL) enzyme activities, as well as flavonol contents, were measured under the same intensity of white light (300 µmol·m^{-2}·s^{-1}) with the addition of 20, 40, and 60 µmol·m^{-2}·s^{-1} UVA separately after 20 days of treatment. The control check (CK) and treatment with the highest flavonol content were chosen for transcriptome sequencing analysis. The results showed that the PAL, C4H, and 4CL enzyme activities, as well as the flavonol and totalflavonol glycoside contents, of ginkgo hybrid progeny differed significantly under different UVA treatments. They showed a tendency to increase and then decrease, reaching a maximum value under UVA-4 (40 µmol·m^{-2}·s^{-1} ultraviolet UVA light intensity) treatment. Ribonucleic acid (RNA) sequencing revealed the presence of 4165 genes with differential expression, and Kyoto Encyclopedia of Genes and Genomes (KEGG) enrichment analysis revealed that the metabolic pathways commonly enriched across all four comparison groups included 'phenylpropanoid biosynthesis', while the pathways commonly enriched in green-leaf ginkgo UVA-4 treatment (TL), yellow-leaf ginkgo mutant CK treatment (CKY), and green-leaf ginkgo CK treatment (CKL) were related to 'flavonoid biosynthesis'. Treatment with UVA light led to the increased expression of PAL and 4CL enzymes in the phenylpropanoid biosynthesis pathway, as well as increased expression of chalcone synthase (CHS), Flavanone 3-hydroxylase (F3H), and flavonol synthase (FLS) enzymes in the flavonoid biosynthesis pathway, thereby promoting the synthesis of ginkgo flavonols. In summary, the use of 40 µmol·m^{-2}·s^{-1} UVA treatment for 20 days significantly increased the flavonol content and the expression of related enzyme genes in ginkgo hybrid offspring, enhancing ginkgo flavonoids and increasing the medicinal value of ginkgo.

Keywords: UVA; *Ginkgo biloba*; flavonol

Citation: Zhao, Q.; Wang, Z.; Wang, G.; Cao, F.; Yang, X.; Zhao, H.; Zhai, J. Effects of UVA on Flavonol Accumulation in *Ginkgo biloba*. *Forests* **2024**, *15*, 909. https://doi.org/10.3390/f15060909

Academic Editor: Claudia Mattioni

Received: 11 April 2024
Revised: 16 May 2024
Accepted: 21 May 2024
Published: 23 May 2024

Copyright: © 2024 by the authors. Licensee MDPI, Basel, Switzerland. This article is an open access article distributed under the terms and conditions of the Creative Commons Attribution (CC BY) license (https://creativecommons.org/licenses/by/4.0/).

1. Introduction

Ginkgo biloba L., revered as a "living fossil" in China [1], boasts remarkable medicinal value, particularly in treating cardiovascular and cerebrovascular diseases, owing to its abundant flavonoids [2]. *Ginkgo biloba* 'Wannianjin' is a budding variety of ordinary *G. biloba*. Its leaves are golden in spring, turn green in summer, and turn yellow in autumn, possessing good ornamental value and medicinal value [3]. Compared with common ginkgo, the yellow coloration of the yellow-leaf ginkgo mutant is due to chlorophyll degradation and an abnormal chloroplast ultrastructure [4]. It has been shown that a high ratio of carotenoids to chlorophyll b is the main reason for the yellowing of the leaves [5]. Except for leaf color, the phenotype of yellow-leaf ginkgo mutant is similar to that of common ginkgo. China has unique advantages in terms of ginkgo drug sourcing, but the quality standard of *G. biloba* extracts in China needs to be improved [6]. Therefore, taking certain measures to improve the total flavonoid content of *G. biloba* is of great significance.

These flavonoids, which are pivotal secondary metabolites in plants, are essential for growth, development, and stress resilience [7]. The synthesis of flavonoids first occurs through the phenylalanine metabolism pathway; PAL, C4H, and 4CL are the key enzymes of the phenylalanine metabolism pathway, and they play a critical role in the synthesis of flavonoid precursor substances [8]. Enzyme genes such as CHS, chalcone isomerase (CHI), and F3H are also essential for the biosynthesis of flavonols. Under stress conditions, plants promote the synthesis of secondary metabolites by increasing the gene expression of the phenylalanine metabolism pathway and enzymes related to the flavonoid synthesis pathway in order to protect themselves from damage [9,10]. It has been shown that the flavonoid content of *G. biloba* significantly increases to resist adversity stresses such as ultraviolet B (UVB) [11] and NaCl [12]. MYB constitutes one of the most extensive families of plant transcription factors, exerting significant influence on the pathways involved in the flavonoid synthesis pathway [13]. Premathilake et al. found that MYB transcription factors positively regulate flavonoid biosynthesis in pears (*Pyrus* spp.) by stimulating key enzyme genes in flavonoid biosynthesis pathways, such as CHS, CHI, F3H, and FLS [14].

Light serves as the primary energy source required for growth through photosynthesis, the process by which plants produce nutrients that are essential for the production of flavonoid secondary metabolites [15]. Ultraviolet (UV) radiation is a crucial element of solar radiation, consisting of UVA (320–400 nm), UVB (ultraviolet B, 280–320 nm), and UVC (ultraviolet C, 200–280 nm), of which UVA accounts for 95% [16,17]. UV radiation is often considered an abiotic stress, and plants can rapidly synthesize antioxidant enzymes and antioxidants such as flavonoids to prevent cellular damage caused by reactive oxygen species [18,19]. He et al. [20] showed that UVA significantly increased the total flavonoid content of kale (*Brassica oleracea* L.). Lim et al. [21] found that UVA treatment favored the accumulation of flavonoids in soybean sprouts (*Glycine max*). Zhao et al. [11] demonstrated that prolonged irradiation with UVB was beneficial to the ginkgo flavonoid content. At present, there is a solid foundation of research on increasing the flavonoid content of *G. biloba*. This mainly focuses on fertilization [22], salt stress [12], drought [23], temperature [24], and the intensity of illumination [25]. However, studies on the effect of UVA radiation on ginkgo are limited; most of the studies on yellow-leaf ginkgo mutant are based on grafted seedlings, and few studies focus on real seedlings. Taking this issue as the starting point, this study used different intensities of UVA under the same intensity of white light to investigate its impact on flavonol accumulation in *G. biloba*. We mined the relevant functional genes using RNA sequencing (RNA-seq) and quantitative real-time polymerase chain reaction (qRT-PCR) analysis and initially probed the molecular mechanism of flavonol biosynthesis in *G. biloba* under UVA to provide certain bases for the cultivation and management of *G. biloba*.

2. Materials and Methods

2.1. Experimental Materials

On 20 September 2021, seeds of the full-sib hybrid offspring of *G. biloba* were collected from the Xiashu Forest Farm of Nanjing Forestry University (118°58′–119°58′ E, 31°37′–32°19′ N). The mother was a *Ginkgo biloba* 'Wannianjin' tree from Anlu, Hubei province, and the father was a healthy green-leaf ginkgo tree from Nanjing Forestry University. The seeds were stripped of their hulls and placed in a cold freezer at 4 °C for sand storage. On March 20 of the subsequent year, the seeds were immersed in clean water for 48 h and then transferred to a light incubator, set at a temperature of 25 °C, to induce germination. After germination, they were differentiated into yellow-leaf and green-leaf phenotypes and planted in non-woven bags. The substrate consisted of a blend of peat, vermiculite, and perlite in equal volumes (1:1:1 ratio).

2.2. Experimental Design

On 1 June 2023, second-year ginkgo full-sib hybrid progeny seedlings with uniform growth and good performance were selected and placed in an artificial climate chamber for one month for acclimatization and then treated with UVA. White light (8 white light tubes, 300 µmol·m^{-2}·s^{-1}) was used as the control, and 2, 4, and 6 UVA tubes (395~405 nm) were uniformly added to the 8 white light tubes, respectively. The intensities of UVA were 20, 40, and 60 µmol·m^{-2}·s^{-1}, and the three treatments were denoted by UVA-2, UVA-4, and UVA-6, respectively, with the control noted as CK. The experimental setup included a 12 h light and 12 h dark cycle within an artificial climate chamber, and the incubation temperature was 25 °C. Each treatment group consisted of 18 yellow-leaf ginkgo mutants and 18 green-leaf ginkgo plants. A completely randomized trial was conducted, with three biological replicates used for each treatment, and samples were taken for the determination of the relevant indexes after 20 days of treatment with different intensities of UVA. From the treatments, CK and UVA-4 (with the highest total flavonol glycoside content) were identified from both yellow-leaf ginkgo mutant and green-leaf *G. biloba* samples. These selected samples were carefully preserved in RNase-free cryopreservation tubes. They were initially flash-frozen in liquid nitrogen for 30 min, and subsequently stored at −80 °C until further analysis. The leaves were labeled as yellow-leaf ginkgo mutant CK treatment (CKY), green-leaf ginkgo CK treatment (CKL), yellow-leaf ginkgo mutant UVA-4 treatment (TY), and green-leaf ginkgo UVA-4 treatment (TL), respectively. LED UVA lamps were provided by Xiamen Sannonghui Photoelectric Technology Limited, Xiamen, China. The spectra of UVA treatments at different intensities were determined using an OHSP-350 UV spectrometer (Hangzhou Hongpu Photochromic Technology Co. Ltd., Hangzhou, China), as shown in Figure 1.

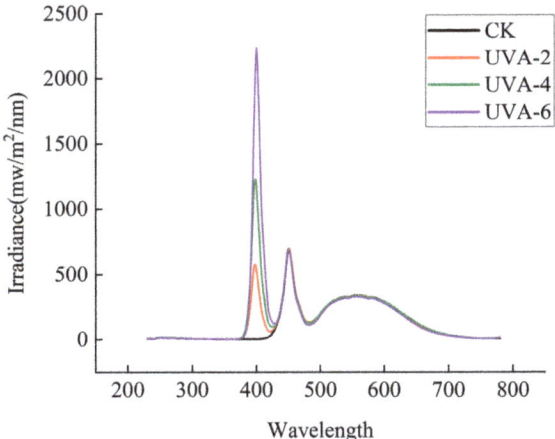

Figure 1. Spectra of UVA treatments at different intensities.

2.3. Test Method

2.3.1. Determination of the Metabolic Enzyme Activity of Phenylpropane

The activities of PAL, C4H, and 4CL were determined using the method outlined by Wang et al. [26]. PAL is expressed as 1 unit (U) of enzyme activity for every 0.1 change in OD per minute. C4H and 4CL are expressed as 1 unit (U) of enzyme activity for every 0.01 change in OD per minute.

2.3.2. Determination of the Flavonol Content

The quercetin, kaempferol, and isorhamnetin contents were determined using high-performance liquid chromatography (HPLC) (Alliance e2695, Waters, Milford, MA, USA), following the method outlined by Zhang et al. [27]. The contents of quercetin, kaempferol, and isorhamnetin were calculated based on the peak areas in the liquid chromatograms.

2.3.3. RNA Extraction, Transcriptome Sequencing, and Library Construction

RNA extraction, transcriptome sequencing, and cDNA library construction were conducted by Shanghai Meiji Biomedical Technology Co., Ltd. (Shanghai, China). The raw transcriptome sequencing data can be viewed in the Sequence Read Archive (SRA) at the National Center for Biotechnology Information (NCBI) under the following Bioproject number: PRJNA1097839.

2.3.4. Real-Time Quantitative PCR (qRT-PCR)

In order to validate the transcriptome sequencing results, 8 genes of the flavonoid synthesis pathway and 5 related transcription factors were randomly selected for validation. RNA reverse transcription was performed with an EvoM-MLV RT Mix Kit. Real-time quantitative fluorescence PCR (qRT-PCR) was measured by an SYBR Green Premix Pro Taq HS qPCR Kit (ROX Plus) kit. Glyceraldehyde-3-phosphate dehydrogenase (GAPDH) was used as the internal reference gene, and the design and synthesis of primers were performed by Sangon Bioengineering (Shanghai, China) Co. The specific primer sequences are shown in Table 1.

Table 1. Primer sequences for the qRT-PCR test.

Gene ID	Forward Primer (5′-3′)	Reverse Primer (3′-5′)
GAPDH	ATCCACGGGAGTATTCAC	CTCATTCACGCCAACAAC
PAL (Gb_09812)	TCCTGACCTCGGCGTAGATTATGG	GGTGACTGGGTTTGCGAGATACTG
4CL (Gb_40571)	AACAGAAGCGGATGAGAGCGAATG	TGTGAGTTAGCATGACGCCCTTTG
CHS (Gb_20355)	GCATGTGCCACCACTGGAGAAG	CGCTTCGCAAGACAACAGTTTCG
F3H (Gb_05058)	GGCGGCGTGCGAGGAATG	CTGGCGGGAGGGCAAAGAAATC
FLS (Gb_14030)	TGCCATCTCTCCCCTCGCTCTTC	CATGCCAGTTTAGTGCCGTAGCC
DFR (Gb_26458)	GGCTGGTTATGCGTTTGCTTCAAC	TTCATCGTCCAAGTCGGCTTTCC
LAR (Gb_08481)	ATTGGTAATCGCAGCAGCAGAGTC	TGAGCGTACAAGAGCGTAAGTTGG
ANS (Gb_21859)	GTGCCTGGTCTCCAACTCTTCAAG	GCCCACTCTTGTATTTGCCATTGC
MYB (Gb_39081)	ATGGAGAATGGAAACACGGACTTG	ACCACGCCACTGCCTTGAG
MYB (Gb_06451)	AGCACAAGAAGCACGCACAAG	GATGGTAAGGCAGTTGGAGTGAAG
b ZIP (Gb_28107)	GCCAGCTTGTGCAGACTTTGAC	TTCAGCATTCGAGACCTCCCATC
bHLH (Gb_35908)	TCAGCAACAGATACAGTCACATTCC	AGCAGATTTGATGATCCACACTCAG
ERF (Gb_12588)	ATCGGCGGCGTCTGTAGC	TTGGGTCGTGCTTGATTCTTGAG

3. Result Analysis

3.1. Effects of Different UVA Intensities on Phenylpropane Metabolizing Enzyme Activities in Ginkgo Hybrid Offspring

PAL, C4H, and 4CL are key enzymes in the flavonoid synthesis pathway. The enzyme activities of PAL, C4H, and 4CL in the yellow-leaf ginkgo mutant differed significantly ($p < 0.05$) under different intensities of UVA treatments. As shown in Figure 2, the enzyme activities significantly increased ($p < 0.05$) under UVA treatment compared with CK and reached a maximum level under UVA-4 treatment. The maximum PAL, C4H, and 4CL enzyme activities in the yellow-leaf ginkgo mutant were 35.57 μmol/min·g, 162.95 μmol/min·g, and 295.05 μmol/min·g, which were 44.58%, 51.94%, and 307.24% higher than the levels seen in CK treatment, respectively. There were highly significant differences ($p < 0.01$) in PAL, C4H, and 4CL enzyme activities under different intensities of UVA treatment in green-leaf ginkgo. The enzyme activities of PAL, C4H, and 4CL were significantly increased ($p < 0.05$) under UVA treatment compared with CK and reached a maximum level under UVA-4 treatment. The maximum PAL, C4H, and 4CL enzyme activities in the green-leaf ginkgo were 45.34 μmol/min·g, 193.77 μmol/min·g, and 353.5 μmol/min·g, which were 80.43%, 71.28%, and 144.55% higher than those in the CK treatment, respectively. This suggests that moderate amounts of UVA are beneficial in increasing PAL, C4H, and 4CL enzyme activity, whereas excessive amounts of UVA lead to enzyme inactivation. The overall PAL, C4H, and 4CL enzyme activities were lower in the yellow-leaf ginkgo mutant than in green-leaf ginkgo.

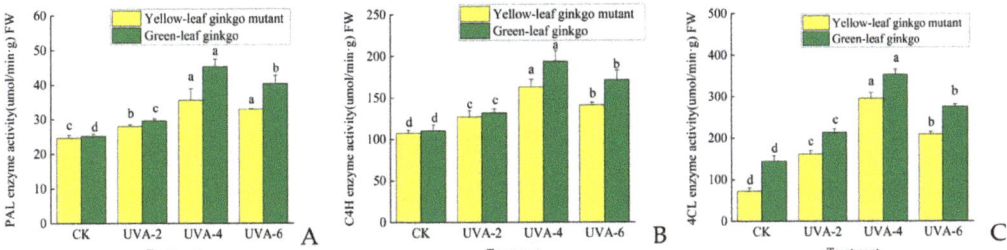

Figure 2. Phenylpropane-metabolizing enzyme activities of *G. biloba* under different intensities of UVA. (**A**) PAL enzyme activity. (**B**) C4H enzyme activity. (**C**) 4CL enzyme activity. Different lowercase letters in the graphs indicate significant differences ($p < 0.05$) at different UVA intensities for the gingko of the same leaf color.

3.2. Effects of Different Intensities of UVA on Flavonol Content of G. biloba Hybrid Offspring

The effect of UVA of different intensities on flavonols of ginkgo sibling hybrid progeny was significant ($p < 0.05$). As shown in Figure 3, the contents of quercetin, kaempferol, and isorhamnetin of yellow-leaf ginkgo mutant under UVA treatment were significantly higher than those of CK treatment ($p < 0.05$). Their contents reached their maximum values under UVA-4 treatment, which were 61.92%, 115.50%, and 98.28% higher than those of CK, respectively. The contents of quercetin, kaempferol, and isorhamnetin in green-leaf ginkgo under UVA treatment were significantly ($p < 0.01$) higher than those of CK treatment and reached their maximum values under UVA-4 treatment, which were 56.46%, 53.97%, and 40.71% higher than those of CK, respectively. This indicates that UVA-4 treatment significantly increased the ginkgo flavonol contents, and that excessive UVA inhibited the synthesis of ginkgo flavonol. The overall contents of quercetin, kaempferol, and isorhamnetin were lower in gold-leaved ginkgo than in green-leaf ginkgo.

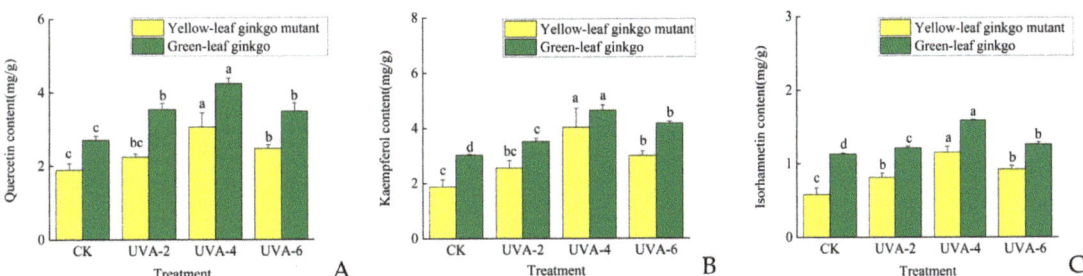

Figure 3. Flavonol content of *G. biloba* under different UVA intensities. (**A**) Quercetin content. (**B**) Kaempferol content. (**C**) Isorhamnetin content. Different lowercase letters in the graphs indicate significant differences ($p < 0.05$) at different UVA intensities for the gingko of the same leaf color.

3.3. Transcriptome Sequencing Statistics and Quality Evaluation

Using Illumina second-generation sequencing technology, RNA sequencing of samples from CK-treated and UVA-4-treated yellow-leaf ginkgo mutants and green-leaf *G. biloba* was performed, and a total of 12 cDNA libraries were created. After quality control, the average Cleanreads of CKY, CKL, TY, and TL were 47,779,668, 46,278,104.67, 48,858,654, and 55,623,168, respectively. The Cleanreads of the samples were aligned with the reference genome (http://gigadb.org/dataset/100613, accessed on 4 June 2019) of *G. biloba*. The total alignment rate was above 92.88%, and the unique position alignment rate ranged from 87.26% to 88.69%, which was high. The percentage of Q30 bases was above 94.94%, and the percentage of GC bases ranged from 45.76% to 46.31%, which indicated that this sequencing had a low error rate and a high overall quality and could be used for the analysis of the subsequent data. Detailed sequencing information is shown in Table 2.

Table 2. Statistics table of sequencing data.

Sample	Raw Reads	Clean Reads	Total Mapped	Multiple Mapped	Uniquely Mapped	Q30 (%)	GC Content (%)
CKY_1	45,408,162	44,808,992	42,120,429 (94.0%)	2,378,996 (5.31%)	39,741,433 (88.69%)	95.14	46.19
CKY_2	54,468,294	53,771,538	50,289,981 (93.53%)	2,810,629 (5.23%)	47,479,352 (88.3%)	94.94	46.25
CKY_3	45,184,742	44,758,474	41,882,169 (93.57%)	2,322,728 (5.19%)	39,559,441 (88.38%)	95.13	46.09
CKL_1	49,435,520	48,592,064	45,742,011 (94.13%)	2,952,929 (6.08%)	42,789,082 (88.06%)	95.09	46.31
CKL_2	44,788,866	44,100,438	41,473,752 (94.04%)	2,571,246 (5.83%)	38,902,506 (88.21%)	94.97	46.28
CKL_3	46,647,470	46,141,812	43,506,464 (94.29%)	2,767,065 (6.0%)	40,739,399 (88.29%)	95.72	46.12
TY_1	41,700,364	41,246,156	38,580,392 (93.54%)	2,113,162 (5.12%)	36,467,230 (88.41%)	95.83	45.81
TY_2	52,013,850	51,369,770	48,034,026 (93.51%)	2,681,649 (5.22%)	45,352,377 (88.29%)	95.76	45.88
TY_3	54,527,510	53,960,036	50,488,323 (93.57%)	2,847,049 (5.28%)	47,641,274 (88.29%)	95.76	45.9
TL_1	46,585,016	46,059,222	42,780,240 (92.88%)	2,588,474 (5.62%)	40,191,766 (87.26%)	95.67	45.87
TL_2	60,566,890	59,865,396	55,681,034 (93.01%)	3,426,978 (5.72%)	52,254,056 (87.29%)	95.96	45.85
TL_3	61,495,202	60,944,886	56,738,281 (93.1%)	3,509,698 (5.76%)	53,228,583 (87.34%)	95.96	45.76

3.4. Functional Annotation of the Six Major Databases

All the expressed genes (31,695 genes in total) were annotated using six functional databases, namely, GO, KEGG, EggCOG, NR, Swiss-Prot, and Pfam, and a total of 28,360 genes were annotated, which accounted for 88.18% of the total number of genes. As shown in Figure 4, 24,577 (0.7641), 11,848 (0.3684), 25,792 (0.8019), 28,171 (0.8759), 22,582 (0.7021), and 18,903 (0.5877) genes were annotated in the six databases, respectively. Among them, the number of functional genes annotated in the six major functional databases was 8017, accounting for 28.27%.

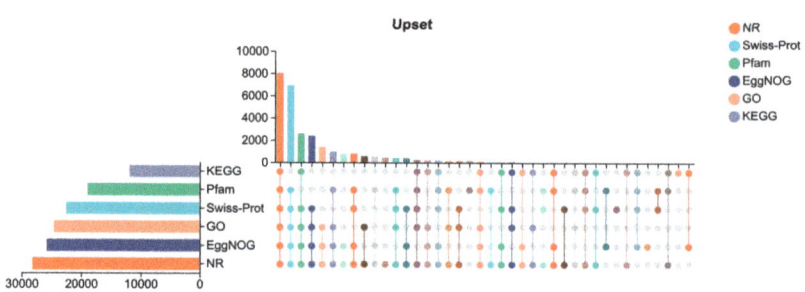

Figure 4. Functional annotation diagram of six databases.

3.5. Differential Expression Analysis of Ginkgo Genes under Different Intensities of UVA Treatment

Using the RPKM values, the variations in gene expression under various levels of UVA treatment were analyzed. The screening conditions were as follows: p-adjust < 0.05 and $|\log2FC| \geq 1$. A total of 4165 differentially expressed genes (DEGs) were screened in the four comparison groups, and 102 genes were found to be significantly different among the four groups, accounting for 2.45% of the total number of differentially expressed genes (Figure 5A). Among them, there were 1328 DEGs between CKY and CKL, including 670 significantly up-regulated genes and 658 down-regulated genes (Figure 5B); there were 1088 DEGs between TY and TL, including 495 significantly up-regulated and 593 down-regulated genes (Figure 5C); and there were 1983 DEGs between TY and CKY, including 1263 significantly up-regulated and 675 down-regulated genes (Figure 5D). There were a total of 2406 DEGs between TL and CKL, of which 1674 were significantly up-regulated and 832 were significantly down-regulated (Figure 5E).

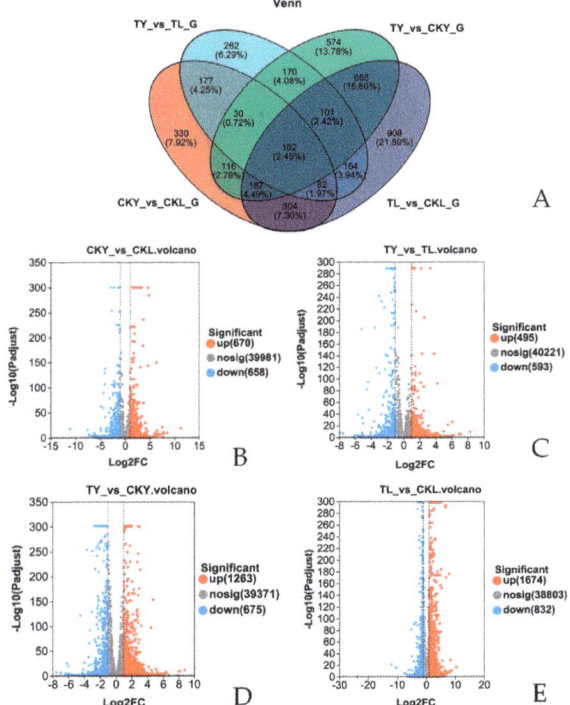

Figure 5. Differential gene expression map. (**A**) Differential gene Venn diagram. (**B**) Differential gene volcano map between CKY and CKL. (**C**) Differential gene volcano map between TY and TL. (**D**) Differential gene volcano map between TY and CKY. (**E**) Differential gene volcano map between TL and CKL.

3.6. GO Annotation Analysis of DEGs

All the above DEGs were analyzed using GO annotation, and a total of 49 metabolic processes were annotated. Among them, 14 metabolic processes belonged to 'molecular function', 13 belonged to 'cellular component', and 22 belonged to 'biological process'. Regarding the top 20 entries in terms of abundance, it can be seen from Figure 6 that there were more entries annotated by DEGs related to 'biological processes', such as 'cellular process' and 'metabolic process'. The entries with more DEGs annotated in terms of 'cellular process' included the 'membrane part' and 'cell part'. The most frequently annotated DEGs in terms of molecular functions were catalytic activity and binding.

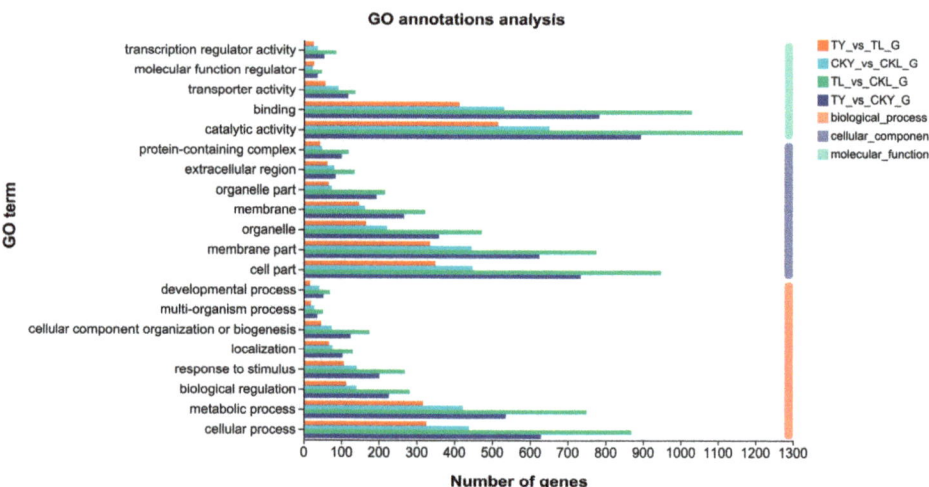

Figure 6. GO function annotation of DEGs.

3.7. KEGG Enrichment Analysis of Differentially Expressed Genes

With padjust ≤ 0.05, the top 10 metabolic pathways were selected for KEGG enrichment analysis, and 17 metabolic pathways were enriched in the DEGs of the four comparison groups (Figure 7). Among the comparisons made, 10 metabolic pathways exhibited enrichment in DEGs between TL and CKL. Similarly, when comparing TY and CKY, seven metabolic pathways showed enrichment in DEGs, with photosynthesis-antenna proteins being the most highly enriched pathways. For the comparison between CKY and CKL, 10 metabolic pathways displayed enrichment in DEGs, with those relating to monoterpenoid biosynthesis being the most highly enriched pathways. When comparing TY and TL, four metabolic pathways exhibited enrichment in DEGs, with cutin, suberine, and wax biosynthesis being the most highly enriched pathways. Notably, the phenylpropanoid biosynthesis pathway emerged as a commonly enriched metabolic pathway among all four comparison groups. The metabolic pathway co-enriched by TL, CKL, and CKY was the flavonoid biosynthesis pathway.

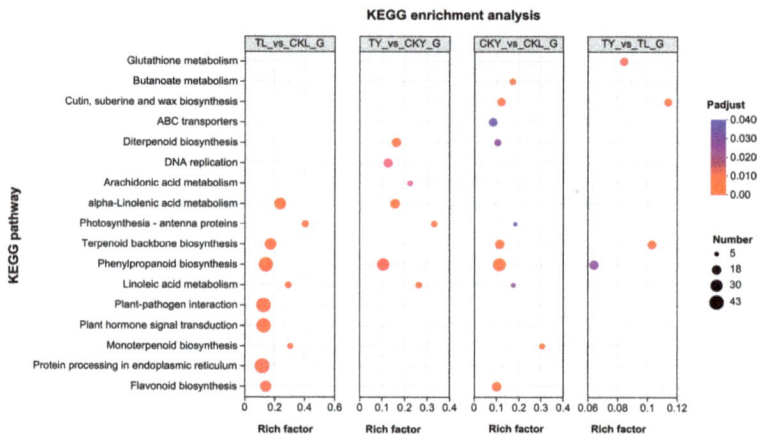

Figure 7. KEGG enrichment analysis of DEGs.

3.8. Differential Expression of Genes Related to Flavonoid Biosynthesis

Based on the KEGG enrichment outcomes, we conducted the integration and analysis of the phenylalanine biosynthesis pathway and flavonoid biosynthesis, and a metabolic pathway map of flavonol biosynthesis was obtained (Figure 8). The PAL enzyme was enriched to six DEGs, and, except for *Gb_10949* and *Gb_21115*, gene expression was significantly up-regulated under UVA treatment. The 4CL enzyme gene *Gb_34525* was significantly down-regulated under UVA treatment, while *Gb_40571* was up-regulated under UVA treatment. Chalcone synthase (CHS) genes (except for *Gb_20355*) were significantly up-regulated by UVA, and Flavanone 3-hydroxylase (F3H) genes (except *Gb_29563*) were up-regulated under UVA treatment. Flavonol synthase (FLS) *Gb_11130* and *Gb_06006* enzyme genes were significantly up-regulated under UVA treatment. Dihydroflavonol reductase (DFR) and anthocyanin synthase (ANS) enzyme genes were significantly up-regulated under UVA treatment. Leucoanthocyanidin reductase (LAR) enzyme genes (except *Gb_40651*) were significantly up-regulated under UVA treatment.

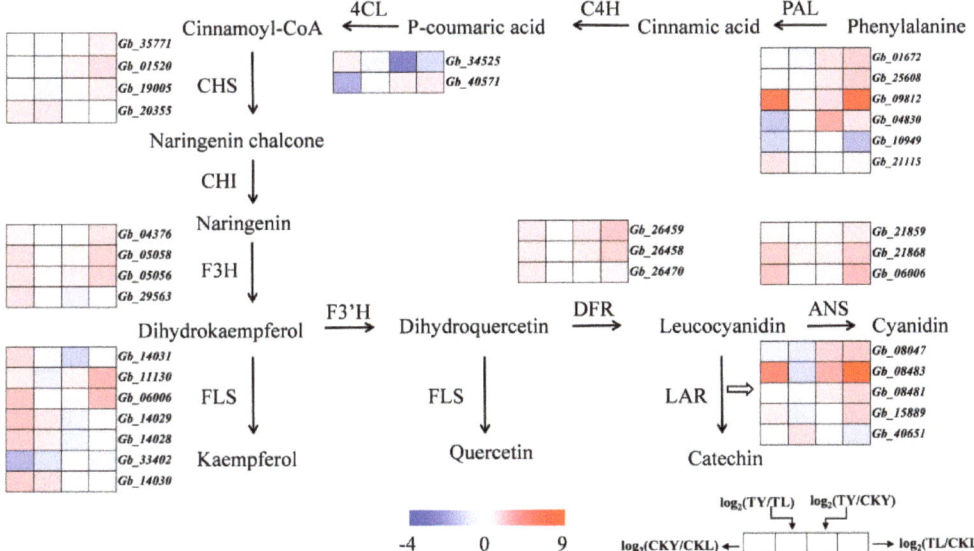

Figure 8. Differential expression of the enzyme genes of the flavonol biosynthetic pathway.

3.9. Validation of qRT-PCR for Key Differential Enzyme Genes of the Ginkgo Flavonoid Synthesis Pathway

To ensure the precision of the RNA-seq data, eight differentially expressed genes in the phenylpropane and flavonoid biosynthesis pathways were randomly selected for qRT-PCR experiments. The qRT-PCR relative expression levels of 4CL (Figure 9B) and LAR (Figure 9G) enzymes were consistent with the trends seen in their RNA-seq FPKM values. The relative expression levels of the PAL (Figure 9A), CHS (Figure 9C), F3H (Figure 9D), FLS (Figure 9E), DFR (Figure 9F), and ANS (Figure 9H) enzyme genes were consistent with the trends in their RNA-seq FPKM values, which, to a large extent, verified the credibility of the sequencing results.

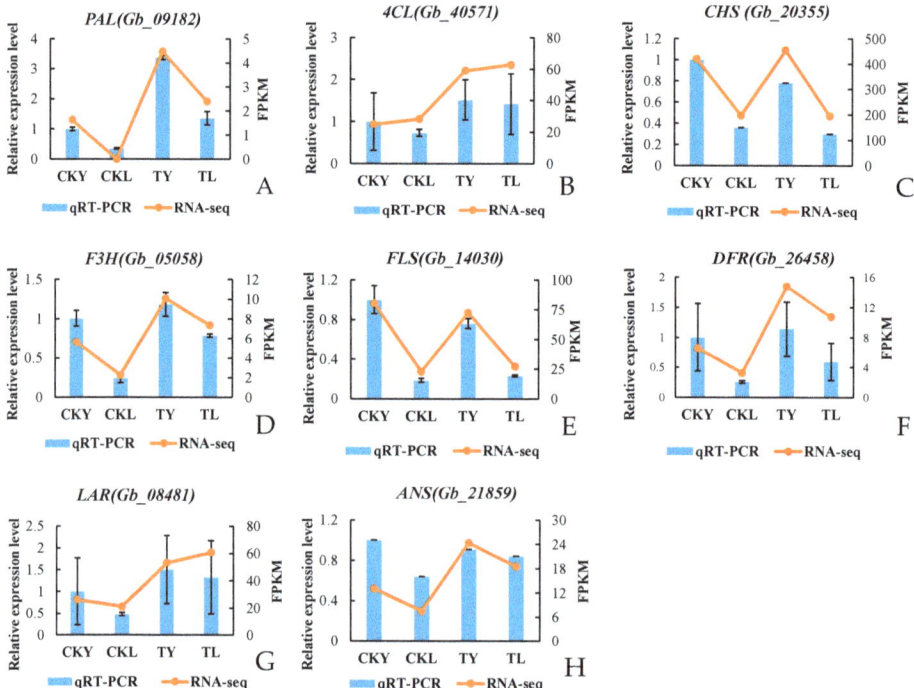

Figure 9. Validation of qRT-PCR for key differential enzyme genes of the ginkgo flavonoid synthesis pathway. The left ordinate represents the relative expression level in qRT-PCR, and the right ordinate represents the FPKM value in RNA-seq. (**A**) Validation of the PAL enzyme gene. (**B**) Validation of the 4CL enzyme gene. (**C**) Validation of the CHS enzyme gene. (**D**) Validation of the F3H enzyme gene. (**E**) Validation of the FLS enzyme gene. (**F**) Validation of the DFR enzyme gene. (**G**) Validation of the LAR enzyme gene. (**H**) Validation of the ANS enzyme gene.

3.10. Differential Expression of Transcription Factors

Transcription factor prediction was performed for all DEGs (1840) using the Transcription Factor Database (PlantTFDB) with a set E-value $\leq 1.0 \times 10^{-5}$ (Figure 10). A total of 121 differentially expressed transcription factors were identified, involving 26 transcription factor families. Among these, the proportions of ERF, MYB, and bHLH family transcription factors were highest, accounting for 15%, 14%, and 13%, respectively. As shown in Figure 11, most of the MYB family transcription factors (except for *Gb_09986*, *Gb_34386*, *Gb_06045*, and *Gb_08692*) were up-regulated under UVA treatment. Two transcription factors of the MYB-related family (*Gb_38499*, *Gb_23424*) were up-regulated under UVA treatment. All transcription factors of the bHLH family were up-regulated, except for *Gb_32351* and *Gb_39758*, which were significantly down-regulated under UVA treatment. The bZIP family transcription factors *Gb_00122* and *Gb_29784* were significantly up-regulated under UVA treatment, and *Gb_28107* was significantly down-regulated under UVA treatment. Six transcription factors (*Gb_36136*, *Gb_12588*, *Gb_08437*, *Gb_36842*, *Gb_26438*, and *Gb_26738*) of the stress-associated ERF family were significantly up-regulated under UVA treatment. The NAC family was up-regulated, except for three transcription factors (*Gb_33280*, *Gb_02849*, and *Gb_22607*), which were significantly down-regulated under UVA treatment. The WRKY family was significantly up-regulated under UVA treatment, except for the *Gb_05026* transcription factor.

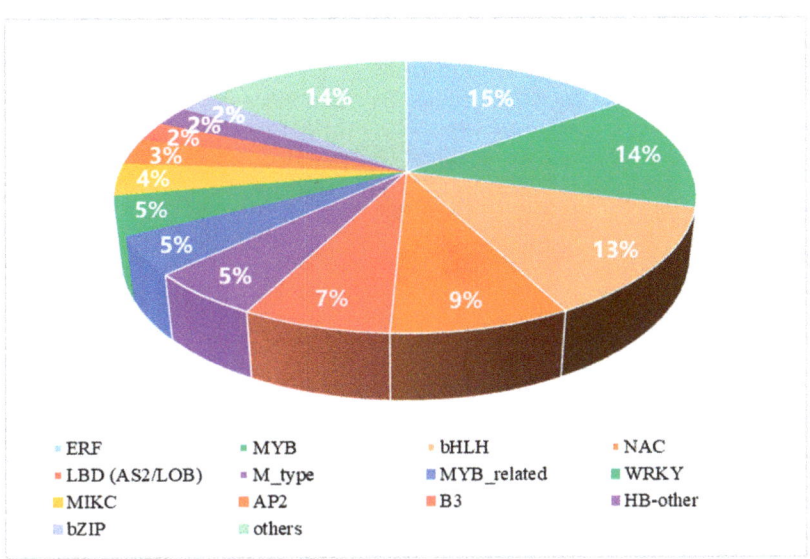

Figure 10. Transcription factor prediction of DEGs.

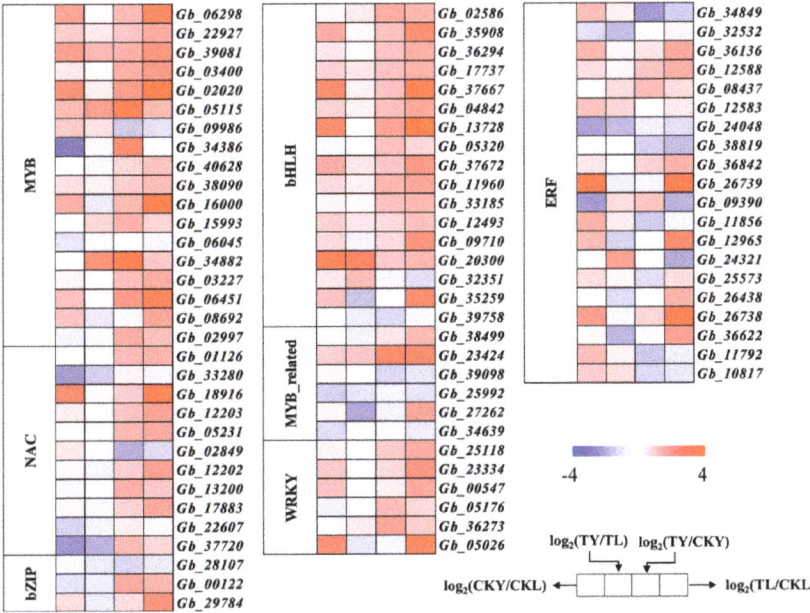

Figure 11. Expression levels of transcription factors. Up-regulated (red) and down-regulated (blue) genes are shown.

3.11. Validation of qRT-PCR for Key Transcription Factors of Ginkgo Flavonoid Synthesis

The outcomes from qRT-PCR validation of the pivotal transcription factors in the ginkgo flavonoid synthesis pathway are presented in Figure 12. The expression of two transcription factors belonging to the MYB family, *Gb_39081* and *Gb_06451*, was significantly up-regulated under UVA treatment relative to white light treatment, as shown by TY > CKY > TL > CKL. Regarding the expression of bZIP family transcription factors, *Gb_28107* was significantly down-regulated under UVA treatment relative to white light, while the expression of *Gb_00122* was significantly up-regulated under UVA treatment relative to white light. The expression of bHLH family *Gb_35908* was significantly up-regulated under UVA treatment relative to white light, as shown by TY > CKY > TL > CKL. The expression of the ERF family *Gb_12588* was significantly up-regulated under UVA treatment relative to white light, as shown by TY > CKY > TL > CKL. Except for the bZIP family's transcription factor *Gb_28107*, whose qRT-PCR relative expression level was more consistent with the trend in its RNA-seq FPKM value, the relative qRT-PCR expression levels of the other transcription factors were completely consistent with the trend in the RNA-seq FPKM values, which, to a large extent, verified the credibility of the sequencing results presented.

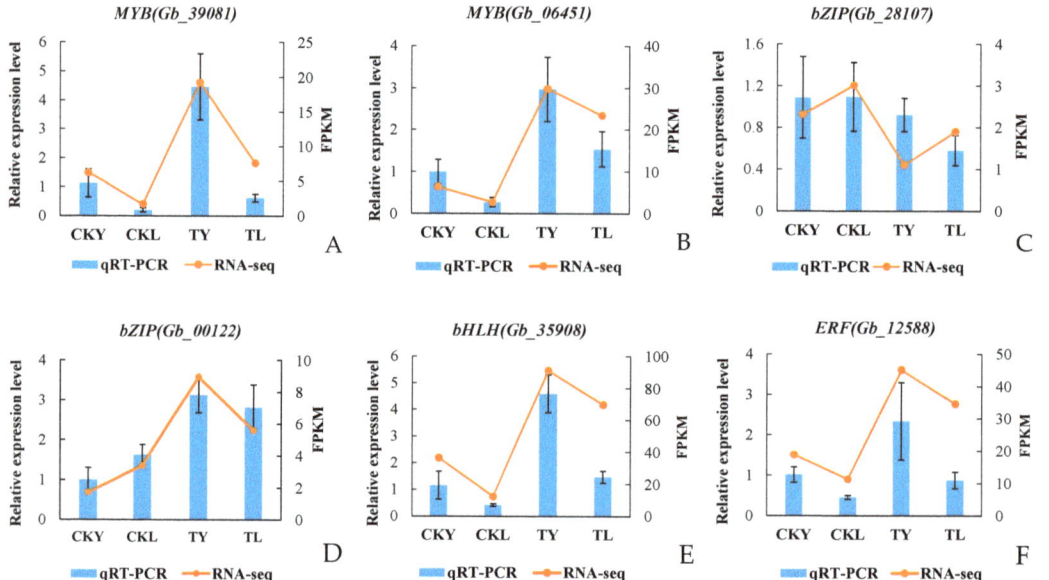

Figure 12. Validation of qRT-PCR for key transcription factors of ginkgo flavonoid synthesis. The left ordinate represents the relative expression level in qRT-PCR, and the right ordinate represents the FPKM value in RNA-seq. (**A,B**) Validation of MYB transcription factors. (**C,D**) Validation of bZIP transcription factors. (**E**) Validation of bHLH transcription factors. (**F**) Validation of EFR transcription factors.

4. Discussion

4.1. Effects of Different Intensities of UVA on the Enzyme Activity of the Phenylpropane Pathway in Hybrid Ginkgo Offspring

PAL, C4H, and 4CL play crucial roles in the phenylpropane pathway [26,28]. Their activities can influence the synthesis of secondary metabolites in plants, thereby enhancing plant stress tolerance [29]. Lee et al. [30] demonstrated that continuous UVA radiation induced PAL enzyme activity and gene expression in sowthistle (*Ixeris dentata* Nakai). Miao et al. [31] observed an increase in PAL, C4H, and 4CL enzyme activities in the above-ground parts of scutellaria (*Scutellaria baicalensis* Georgi) under UVA radiation. Kang et al. [32] showed that sweet basil (*Ocimum basilicum* L.) exhibited the highest PAL enzyme activity when treated with UVA (20 W·m^{-2}) for 14 days. In the present study, it was found that different intensities of UVA had significant effects on the PAL, C4H, and 4CL enzyme activities of *G. biloba*, and that the enzyme activities were greatest under UVA-4 treatment.

This suggests that moderate amounts of UVA are beneficial for increasing PAL, C4H, and 4CL enzyme activity, whereas excessive amounts of UVA lead to enzyme inactivation.

4.2. Effects of Different Intensities of UVA on Flavonols in Ginkgo Hybrid Offspring

UVA can be absorbed by substances such as flavonoids in the epidermis of plant leaves, and flavonoids can mitigate cellular damage induced by UVA [33]. Fu et al. [34] found that a high proportion of UVA could increase the activity of flavonoid glycosyltransferase and the content of quercetin and kaempferol derivatives. Chen et al. [35] found that supplemental UVA (365 nm) could augment the overall flavonoid content in lettuce (*Lactuca sativa* L.) leaves, with a corresponding increase in total flavonoid content with the increase in UVA intensity. Li et al. [36] found that UVA treatment increased the flavonol content of quercetin and kaempferol in brassicaceae (*Brassicaceae* Burnett) baby leaves. Miao et al. [37] showed that the total flavonoid content of postharvest scutellaria (*Scutellaria baicalensis* Georgi) roots increased with the duration of UVA radiation. In the present study, UVA-4 treatment was found to significantly increase the content of ginkgo flavonols, which suggests that moderate UVA irradiation can promote the synthesis of ginkgo flavonols, with excessive UVA irradiation having an inhibitory effect.

4.3. Effects of Different Intensities of UVA on Enzyme Genes Related to Flavonoid Metabolism of Hybrid G. biloba

Flavonoids are biosynthesized by the phenylpropane metabolic pathway, and PAL, C4H, 4CL, and CHS are the key enzyme genes of the flavonoid synthesis pathway, regulating and controlling the biosynthesis of secondary metabolites [38]. Lee et al. [39] found that short-term irradiation of kale (*Brassica oleracea* L.) leaves with UVA promoted the expression of PAL, CHS, and F3H genes, and Qian et al. [40] found that UVA growth light led to a significant rise in PAL and CHS gene expression in basil (*Ocimum basilicum*) leaves. Li et al. [41] observed the up-regulation of structural genes such as F3H, DFR, and ANS in the dark-purple tea cultivar 'Ziyan' (*Camellia sinensis*) under UVA treatment. Liu et al. [42] demonstrated that many photoreceptors and potential genes implicated in UVB signaling (UVR8_L, HY5, COP1, and RUP1/2) exhibited decreased expression patterns consistent with those of structural genes (flavonoid 3'-hydroxylase (F3'H), FLS, anthocyanin synthase (ANS), anthocyanidin reductase (ANR), leucoanthocyanidin reductase (LAR), Dihydroflavonol 4-reductase (DFR), and CHSs), as well as potential TFs (MYB4, MYB12, MYB14, and MYB111) involved in flavonoid biosynthesis. The transcriptome sequencing results of this experiment identified six differentially expressed PAL genes and two 4CL genes, of which the PAL enzyme genes (except *Gb_10949* and *Gb_21115*) were up-regulated and expressed under UVA treatment, while the 4CL enzyme gene *Gb_34525* was significantly down-regulated under UVA treatment, and the expression of *Gb_40571* was up-regulated under UVA treatment. UVA treatment enhanced the expression of CHS, F3H, FLS, DFR, ANS, LAR, and other structural genes in the flavonoid biosynthetic pathway.

4.4. Analysis of the Transcriptional Regulation Mechanism of Ginkgo biloba Flavonoid Synthesis by Different Intensities of UVA

MYB transcription factors play a pivotal role in modulating the expression of key genes involved in plant flavonoid metabolism, thus effectively regulating flavonoid biosynthesis [43]. It has been shown that the heterologous expression of AtMYB11, a flavonol-specific transcription factor from *Arabidopsis thaliana*, in tobacco (*Nicotiana tabacum* L.) is favorable for flavonoid biosynthesis [44]. The q-PCR validation results of this experiment showed that MYB (*Gb_39081* and *Gb_06451*), bZIP (*Gb_00122* and *Gb_28107*), bHLH (*Gb_35908*), and ERF (*Gb_12588*) transcription factors were up-regulated and expressed in response to UVA treatment. Guo et al. [45] investigated the flavonoid accumulation of highly related modules in which GbERFs (*Gb_12588*) were positive in the regulation of upstream structural genes, consistent with the results of this experiment. In recent years, MYB binding sites have been successively identified in the promoter regions of Ginkgo flavonoid metabolism functional genes such as GbFLS, GbCHS, GbANS, and GbCHI [46–48]. MYB and bHLH transcription factors form a complex and regulate the expression of structural genes through the binding of MYB to the promoter regions of the structural genes, regulating the biosynthesis of flavonols in the legume *Entada phaseoloides* [49]. In addition to MYB and bHLH, three transcription factors, ERF, WRKY, and bZIP, are also significant contributors to the light-mediated regulation of flavonoid biosynthesis in mango (*Mangifera indica* L.) [50]. The BHLH gene can bind directly to the DFR promoter, thereby triggering anthocyanin accumulation [51]. Hartmann et al. [52] showed that the bZIP transcription factor induces expression of FLS gene by binding to other transcription factors, which enhances flavonol production. Morishita et al. [53] found that the NAC transcription factor (*ANAC078*) is crucial for regulating flavonoid biosynthesis in *Arabidopsis thaliana* under high light intensity. In this experiment, we found that the expression levels of transcription factors such as MYB (*Gb_02997*), bHLH (*Gb_05320*), bZIP (*Gb_00122*), and NAC (*Gb_13200, Gb_37720*) were consistent with the trend of the total flavonol glycoside contents of *G. biloba*. These contents were all arranged in the order of TY > TL > CKY > CKL, and they could positively regulate the synthesis of ginkgo flavonols, while MYB (*Gb_09986*), bHLH (*Gb_32351*), and ERF (*Gb_10817, Gb_34849*) transcription factors showed the opposite trend to total ginkgo flavonol glycoside and could negatively regulate the synthesis of ginkgo flavonols.

5. Conclusions

The enzyme activities of the phenylpropane pathway and the flavonol contents of the whole sibling hybrid offspring of *G. biloba* showed significant differences under different intensities of UVA treatment and reached the maximum level in the UVA-4 treatment. This indicates that moderate UVA intensity can promote the enzyme activities related to the flavonoid synthesis pathway and the accumulation of flavonol in ginkgo, while excessive UVA expression play an inhibitory role. UVA treatment promoted the expression of PAL (*Gb_01672, Gb_25608*) and 4CL (*Gb_40571*) enzymes in the phenylpropane biosynthesis pathway, alongside the expression of CHS (*Gb_35771, Gb_01520*), F3H (*Gb_05058, Gb_05056*), and FLS (*Gb_11130*) enzymes in the flavonoid synthesis pathway, which in turn promoted the synthesis of ginkgo flavonols. Transcription factors such as MYB (*Gb_02997*), bHLH (*Gb_05320*), bZIP (*Gb_00122*), and NAC (*Gb_13200, Gb_37720*) likely exert positive regulatory effects on ginkgo flavonol synthesis, while the MYB (*Gb_09986*), bHLH (*Gb_32351*), and ERF (*Gb_10817, Gb_34849*) transcription factors may negatively regulate ginkgo flavonol synthesis. In conclusion, 20 days of UVA-4 treatment significantly increased the flavonol content and the expression of related enzyme genes in the ginkgo hybrid offspring, which was favorable for the accumulation of ginkgo flavonol and the improvement of medicinal value.

Author Contributions: Writing—original draft preparation, Q.Z.; data curation, Q.Z.; formal analysis, Z.W. and H.Z.; writing—review and editing, G.W. and X.Y.; funding acquisition, F.C.; resources, J.Z. All authors have read and agreed to the published version of the manuscript.

Funding: This research was funded by the Natural Science Foundation of Jiangsu Province (BK20210611), and the Jiangsu Science and Technology Plan Project (BE2021367).

Data Availability Statement: The raw data have been submitted to NCBI database with the accession number PRJNA1097839.

Acknowledgments: We express our gratitude to Shanghai Meiji Biomedical Technology Co., Ltd. for their invaluable assistance in conducting the RNA-sequencing analysis.

Conflicts of Interest: We declare that there is no conflict of interest.

References

1. Smarda, P.; Vesely, P.; Smerda, J.; Bures, P.; Knápek, O.; Chytrá, M. Polyploidy in a "living fossil" *Ginkgo biloba*. *New Phytol.* **2016**, *212*, 11–14. [CrossRef] [PubMed]
2. Liu, P.; Pan, S. Advance in study of ginkgolic acid contained in *Ginkgo biloba* preparations. *China J. Chin. Mater. Medica* **2012**, *37*, 274–277.
3. Liu, X.; Yu, W.; Wang, G.; Cao, F.; Cai, J.; Wang, H. Comparative Proteomic and Physiological Analysis Reveals the Variation Mechanisms of Leaf Coloration and Carbon Fixation in a Xantha Mutant of *Ginkgo biloba* L. *Int. J. Mol. Sci.* **2016**, *17*, 1794. [CrossRef] [PubMed]
4. Li, W.-X.; Yang, S.-B.; Lu, Z.; He, Z.-C.; Ye, Y.-L.; Zhao, B.-B.; Wang, L.; Jin, B. Cytological, physiological, and transcriptomic analyses of golden leaf coloration in *Ginkgo biloba* L. *Hortic. Res.* **2018**, *5*, 32. [CrossRef]
5. Sun, Y.; Bai, P.-P.; Gu, K.-J.; Yang, S.-Z.; Lin, H.-Y.; Shi, C.-G.; Zhao, Y.-P. Dynamic transcriptome and network-based analysis of yellow leaf mutant Ginkgo biloba. *BMC Plant Biol.* **2022**, *22*, 465. [CrossRef] [PubMed]
6. Kressmann, S.; Biber, A.; Wonnemann, M.; Schug, B.; Blume, H.H.; Müller, W.E. Influence of pharmaceutical quality on the bioavailability of active components from *Ginkgo biloba* preparations. *J. Pharm. Pharmacol.* **2002**, *54*, 1507–1514. [CrossRef] [PubMed]
7. Wu, J.T.; Lv, S.D.; Zhao, L.; Gao, T.; Yu, C.; Hu, J.N.; Ma, F. Advances in the study of the function and mechanism of the action of flavonoids in plants under environmental stresses. *Planta* **2023**, *257*, 108. [CrossRef] [PubMed]
8. Fan, C.X.; Hu, H.Q.; Wang, L.H.; Zhou, Q.; Huang, X.H. Enzymological mechanism for the regulation of lanthanum chloride on flavonoid synthesis of soybean seedlings under enhanced ultraviolet-B radiation. *Environ. Sci. Pollut. Res.* **2014**, *21*, 8792–8800. [CrossRef]
9. Kanazawa, K.; Hashimoto, T.; Yoshida, S.; Sungwon, P.; Fukuda, S. Short Photoirradiation Induces Flavonoid Synthesis and Increases Its Production in Postharvest Vegetables. *J. Agric. Food Chem.* **2012**, *60*, 4359–4368. [CrossRef]
10. Noda, N.; Kanno, Y.; Kato, N.; Kazuma, K.; Suzuki, M. Regulation of gene expression involved in flavonol and anthocyanin biosynthesis during petal development in lisianthus (*Eustoma grandiflorum*). *Physiol. Plant.* **2004**, *122*, 305–313. [CrossRef]
11. Zhao, B.B.; Wang, L.; Pang, S.Y.; Jia, Z.C.; Wang, L.; Li, W.X.; Jin, B. UV-B promotes flavonoid synthesis in *Ginkgo biloba* leaves. *Ind. Crop. Prod.* **2020**, *151*, 112483. [CrossRef]
12. Ni, J.; Hao, J.; Jiang, Z.F.; Zhan, X.R.; Dong, L.X.; Yang, X.L.; Sun, Z.H.; Xu, W.Y.; Wang, Z.K.; Xu, M.J. NaCl Induces Flavonoid Biosynthesis through a Putative Novel Pathway in Post-harvest Ginkgo Leaves. *Front. Plant Sci.* **2017**, *8*, 920. [CrossRef]
13. Xu, W.J.; Dubos, C.; Lepiniec, L. Transcriptional control of flavonoid biosynthesis by MYB-bHLH-WDR complexes. *Trends Plant Sci.* **2015**, *20*, 176–185. [CrossRef] [PubMed]
14. Premathilake, A.T.; Ni, J.B.; Bai, S.L.; Tao, R.Y.; Ahmad, M.; Teng, Y.W. R2R3-MYB transcription factor PpMYB17 positively regulates flavonoid biosynthesis in pear fruit. *Planta* **2020**, *252*, 59. [CrossRef] [PubMed]
15. Hou, J.L.; Li, W.D.; Zheng, Q.Y.; Wang, W.Q.; Xiao, B.; Xing, D. Effect of low light intensity on growth and accumulation of secondary metabolites in roots of *Glycyrrhiza uralensis* Fisch. *Biochem. Syst. Ecol.* **2010**, *38*, 160–168. [CrossRef]
16. Sanchez-Campos, Y.; Cárcamo-Fincheira, P.; González-Villagra, J.; Jorquera-Fontena, E.; Acevedo, P.; Soto-Cerda, B.; Nunes-Nesi, A.; Inostroza-Blancheteau, C.; Tighe-Neira, R. Physiological and molecular effects of TiO$_2$ nanoparticle application on UV-A radiation stress responses in *Solanum lycopersicum* L. *Protoplasma* **2023**, *260*, 1527–1537. [CrossRef] [PubMed]
17. Surjadinata, B.B.; Jacobo-Velázquez, D.A.; Cisneros-Zevallos, L. UVA, UVB and UVC Light Enhances the Biosynthesis of Phenolic Antioxidants in Fresh-Cut Carrot through a Synergistic Effect with Wounding. *Molecules* **2017**, *22*, 668. [CrossRef] [PubMed]
18. Kotilainen, T.; Venäläinen, T.; Tegelberg, R.; Lindfors, A.; Julkunen-Tiitto, R.; Sutinen, S.; O'Hara, R.B.; Aphalo, P.J. Assessment of UV Biological Spectral Weighting Functions for Phenolic Metabolites and Growth Responses in Silver Birch Seedlings. *Photochem. Photobiol.* **2009**, *85*, 1346–1355. [CrossRef] [PubMed]
19. Barnes, P.W.; Tobler, M.A.; Keefover-Ring, K.; Flint, S.D.; Barkley, A.E.; Ryel, R.J.; Lindroth, R.L. Rapid modulation of ultraviolet shielding in plants is influenced by solar ultraviolet radiation and linked to alterations in flavonoids. *Plant Cell Environ.* **2016**, *39*, 222–230. [CrossRef]
20. He, R.; Gao, M.F.; Li, Y.M.; Zhang, Y.T.; Song, S.W.; Su, W.; Liu, H.C. Supplemental UV-A Affects Growth and Antioxidants of Chinese Kale Baby-Leaves in Artificial Light Plant Factory. *Horticulturae* **2021**, *7*, 294. [CrossRef]
21. Lim, Y.J.; Lyu, J.I.; Kwon, S.J.; Eom, S.H. Effects of UV-A radiation on organ-specific accumulation and gene expression of isoflavones and flavonols in soybean sprout. *Food Chem.* **2021**, *339*, 128080. [CrossRef] [PubMed]
22. Son, Y. Effect of nitrogen fertilization on foliar nutrient dynamics of ginkgo seedlings. *J. Plant Nutr.* **2002**, *25*, 93–102. [CrossRef]
23. Yu, W.W.; Liu, H.M.; Luo, J.Q.; Zhang, S.Q.; Xiang, P.; Wang, W.; Cai, J.F.; Lu, Z.G.; Zhou, Z.D.; Hu, J.J.; et al. Partial root-zone simulated drought induces greater flavonoid accumulation than full root-zone simulated water deficiency in the leaves of Ginkgo biloba. *Environ. Exp. Bot.* **2022**, *201*, 104998. [CrossRef]
24. Wu, Z.F.; Wang, S.X.; Fu, Y.S.H.; Gong, Y.F. Spatial variation in the interaction between temperature and sunlight on ginkgo germination. *Front. Plant Sci.* **2022**, *13*. [CrossRef] [PubMed]
25. Xu, Y.; Wang, G.B.; Cao, F.L.; Zhu, C.C.; Wang, G.Y.; El-Kassaby, Y.A. Light intensity affects the growth and flavonol biosynthesis of Ginkgo (*Ginkgo biloba* L.). *New For.* **2014**, *45*, 765–776. [CrossRef]
26. Wang, G.B.; Cao, F.L.; Chang, L.; Guo, X.Q.; Wang, J. Temperature has more effects than soil moisture on biosynthesis of flavonoids in Ginkgo (*Ginkgo biloba* L.) leaves. *New For.* **2014**, *45*, 797–812. [CrossRef]

27. Wang, G.P.; Zhang, L.; Wang, G.B.; Cao, F.L. Growth and flavonol accumulation of *Ginkgo biloba* leaves affected by red and blue light. *Ind. Crop. Prod.* **2022**, *187*, 115488. [CrossRef]
28. Cheng, L.; Han, M.; Yang, L.M.; Yang, L.; Sun, Z.; Zhang, T. Changes in the physiological characteristics and baicalin biosynthesis metabolism of *Scutellaria baicalensis* Georgi under drought stress. *Ind. Crop. Prod.* **2018**, *122*, 473–482. [CrossRef]
29. Zhao, Q.Y.; Ma, Y.; Huang, X.Q.; Song, L.J.; Li, N.; Qiao, M.W.; Li, T.E.; Hai, D.; Cheng, Y.X. GABA Application Enhances Drought Stress Tolerance in Wheat Seedlings (*Triticum aestivum* L.). *Plants* **2023**, *12*, 2495. [CrossRef]
30. Lee, M.J.; Son, J.E.; Oh, M.M. Growth and phenolic content of sowthistle grown in a closed-type plant production system with a UV-A or UV-B lamp. *Hortic. Environ. Biotechnol.* **2013**, *54*, 492–500. [CrossRef]
31. Miao, N.; Yun, C.; Shi, Y.T.; Gao, Y.; Wu, S.; Zhang, Z.H.; Han, S.L.; Wang, H.M.; Wang, W.J. Enhancement of flavonoid synthesis and antioxidant activity in *Scutellaria baicalensis* aerial parts by UV-A radiation. *Ind. Crop. Prod.* **2022**, *187*, 115532. [CrossRef]
32. Kang, S.H.; Kim, J.E.; Zhen, S.Y.; Kim, J. Mild-Intensity UV-A Radiation Applied Over a Long Duration Can Improve the Growth and Phenolic Contents of Sweet Basil. *Front. Plant Sci.* **2022**, *13*, 858433. [CrossRef] [PubMed]
33. Julkunen-Tiitto, R.; Häggman, H.; Aphalo, P.J.; Lavola, A.; Tegelberg, R.; Veteli, T. Growth and defense in deciduous trees and shrubs under UV-B. *Environ. Pollut.* **2005**, *137*, 404–414. [CrossRef]
34. Fu, B.; Ji, X.M.; Zhao, M.Q.; He, F.; Wang, X.L.; Wang, Y.D.; Liu, P.F.; Niu, L. The influence of light quality on the accumulation of flavonoids in tobacco (*Nicotiana tabacum* L.) leaves. *J. Photochem. Photobiol. B-Biol.* **2016**, *162*, 544–549. [CrossRef]
35. Chen, Y.C.; Li, T.; Yang, Q.C.; Zhang, Y.T.; Zou, J.; Bian, Z.H.; Wen, X.Z. UVA Radiation Is Beneficial for Yield and Quality of Indoor Cultivated Lettuce. *Front. Plant Sci.* **2019**, *10*, 1563. [CrossRef]
36. Li, Y.M.; Zheng, Y.J.; Zheng, D.Q.; Zhang, Y.T.; Song, S.W.; Su, W.; Liu, H.C. Effects of Supplementary Blue and UV-A LED Lights on Morphology and Phytochemicals of *Brassicaceae* Baby-Leaves. *Molecules* **2020**, *25*, 5678. [CrossRef]
37. Miao, N.; Yun, C.; Han, S.L.; Shi, Y.T.; Gao, Y.; Wu, S.; Zhao, Z.W.; Wang, H.M.; Wang, W.J. Postharvest UV-A radiation affects flavonoid content, composition, and bioactivity of *Scutellaria baicalensis* root. *Postharvest Biol. Technol.* **2022**, *189*, 111933. [CrossRef]
38. Hu, T.; Gao, Z.Q.; Hou, J.M.; Tian, S.K.; Zhang, Z.X.; Yang, L.; Liu, Y. Identification of biosynthetic pathways involved in flavonoid production in licorice by RNA-seq based transcriptome analysis. *Plant Growth Regul.* **2020**, *92*, 15–28. [CrossRef]
39. Lee, J.H.; Oh, M.M.; Son, K.H. Short-Term Ultraviolet (UV)-A Light-Emitting Diode (LED) Radiation Improves Biomass and Bioactive Compounds of Kale. *Front. Plant Sci.* **2019**, *10*, 1042. [CrossRef]
40. Qian, M.J.; Kalbina, I.; Rosenqvist, E.; Jansen, M.A.K.; Strid, Å. Supplementary UV-A and UV-B radiation differentially regulate morphology in *Ocimum basilicum*. *Photochem. Photobiol. Sci.* **2023**, *22*, 2219–2230. [CrossRef]
41. Li, W.; Tan, L.; Zou, Y.; Tan, X.; Huang, J.; Chen, W.; Tang, Q. The Effects of Ultraviolet A/B Treatments on Anthocyanin Accumulation and Gene Expression in Dark-Purple Tea Cultivar 'Ziyan' (*Camellia sinensis*). *Molecules* **2020**, *25*, 354. [CrossRef] [PubMed]
42. Liu, L.L.; Li, Y.Y.; She, G.B.; Zhang, X.C.; Jordan, B.; Chen, Q.; Zhao, J.; Wan, X.C. Metabolite profiling and transcriptomic analyses reveal an essential role of UVR8-mediated signal transduction pathway in regulating flavonoid biosynthesis in tea plants (*Camellia sinensis*) in response to shading. *Bmc Plant Biol.* **2018**, *18*, 233. [CrossRef] [PubMed]
43. Muhammad, N.; Luo, Z.; Zhao, X.; Yang, M.; Liu, Z.G.; Liu, M.J. Transcriptome-wide expression analysis of *MYB* gene family leads to functional characterization of flavonoid biosynthesis in fruit coloration of *Ziziphus* Mill. *Front. Plant Sci.* **2023**, *14*, 1171288. [CrossRef] [PubMed]
44. Pandey, A.; Misra, P.; Trivedi, P.K. Constitutive expression of *Arabidopsis* MYB transcription factor, *AtMYB11*, in tobacco modulates flavonoid biosynthesis in favor of flavonol accumulation. *Plant Cell Rep.* **2015**, *34*, 1515–1528. [CrossRef]
45. Guo, Y.; Gao, C.; Wang, M.; Fu, F.-f.; El-Kassaby, Y.A.; Wang, T.; Wang, G. Metabolome and transcriptome analyses reveal flavonoids biosynthesis differences in *Ginkgo biloba* associated with environmental conditions. *Ind. Crop. Prod.* **2020**, *158*, 112963. [CrossRef]
46. Xu, F.; Cheng, H.; Cai, R.; Li, L.L.; Chang, J.; Zhu, J.; Zhang, F.X.; Chen, L.J.; Wang, Y.; Cheng, S.H.; et al. Molecular Cloning and Function Analysis of an Anthocyanidin Synthase Gene from *Ginkgo biloba*, and Its Expression in Abiotic Stress Responses. *Mol. Cells* **2008**, *26*, 536–547. [CrossRef] [PubMed]
47. Zhang, W.W.; Xu, F.; Cheng, S.Y.; Liao, Y.L. Characterization and functional analysis of a *MYB* gene (*GbMYBFL*) related to flavonoid accumulation in *Ginkgo biloba*. *Genes Genom.* **2018**, *40*, 49–61. [CrossRef] [PubMed]
48. Xu, F.; Li, L.L.; Zhang, W.W.; Cheng, H.; Sun, N.N.; Cheng, S.Y.; Wang, Y. Isolation, characterization, and function analysis of a flavonol synthase gene from *Ginkgo biloba*. *Mol. Biol. Rep.* **2012**, *39*, 2285–2296. [CrossRef]
49. Lin, M.; Zhou, Z.Q.; Mei, Z.A. Integrative Analysis of Metabolome and Transcriptome Identifies Potential Genes Involved in the Flavonoid Biosynthesis in *Entada phaseoloides* Stem. *Front. Plant Sci.* **2022**, *13*, 792674. [CrossRef]
50. Qian, M.J.; Wu, H.X.; Yang, C.K.; Zhu, W.C.; Shi, B.; Zheng, B.; Wang, S.B.; Zhou, K.B.; Gao, A.P. RNA-Seq reveals the key pathways and genes involved in the light-regulated flavonoids biosynthesis in mango (*Mangifera indica* L.) peel. *Front. Plant Sci.* **2023**, *13*, 1119384. [CrossRef]
51. Xiang, L.L.; Liu, X.F.; Li, X.; Yin, X.R.; Grierson, D.; Li, F.; Chen, K.S. A Novel bHLH Transcription Factor Involved in Regulating Anthocyanin Biosynthesis in Chrysanthemums (*Chrysanthemum morifolium* Ramat.). *PLoS ONE* **2015**, *10*, e0143892. [CrossRef] [PubMed]

52. Hartmann, U.; Sagasser, M.; Mehrtens, F.; Stracke, R.; Weisshaar, B. Differential combinatorial interactions of *cis*-acting elements recognized by R2R3-MYB, BZIP, and BHLH factors control light-responsive and tissue-specific activation of phenylpropanoid biosynthesis genes. *Plant Mol. Biol.* **2005**, *57*, 155–171. [CrossRef] [PubMed]
53. Morishita, T.; Kojima, Y.; Maruta, T.; Nishizawa-Yokoi, A.; Yabuta, Y.; Shigeoka, S. Arabidopsis NAC Transcription Factor, ANAC078, Regulates Flavonoid Biosynthesis under High-light. *Plant Cell Physiol.* **2009**, *50*, 2210–2222. [CrossRef] [PubMed]

Disclaimer/Publisher's Note: The statements, opinions and data contained in all publications are solely those of the individual author(s) and contributor(s) and not of MDPI and/or the editor(s). MDPI and/or the editor(s) disclaim responsibility for any injury to people or property resulting from any ideas, methods, instructions or products referred to in the content.

Article

Physiological Mechanisms of *Bretschneidera sinensis* Hemsl. Seed Dormancy Release and Germination

Lijun Zhong [1,2,†], Hongxing Dong [1,2,†], Zhijun Deng [1,2], Jitao Li [1,2], Li Xu [1,2], Jiaolin Mou [1,2] and Shiming Deng [1,2,*]

1. Hubei Key Laboratory of Biologic Resources Protection and Utilization, Hubei Minzu University, Enshi 445000, China; 202330431@hbmzu.edu.cn (L.Z.); 202011877@hbmzu.edu.cn (H.D.); dengzhijun@bhmzu.edu.cn (Z.D.); lijitao@hbmzu.edu.cn (J.L.); xuli@hbmzu.edu.cn (L.X.); moujiaolin@hbmzu.edu.cn (J.M.)
2. Research Center for Germplasm Engineering of Characteristic Plant Resources in Enshi Prefecture, Hubei Minzu University, Enshi 445000, China
* Correspondence: 2010003@hbmzu.edu.cn
† These authors contributed equally to this work.

Abstract: *Bretschneidera sinensis*, the sole species of *Bretschneidera*, belonging to the family *Akaniaceae*, is a tertiary paleotropical flora. It is considered an endangered species by the International Union for Conservation of Nature (IUCN). It has an important protective and scientific value. The study of its seed dormancy and germination mechanisms contributes to better protection. In this study, the dormancy of fresh mature *B. sinensis* seeds released via low-temperature wet stratification was studied. In addition, the endogenous phytohormone levels, antioxidant enzyme activity, soluble sugar content, and the key metabolic enzyme activities of seeds at different stratification time nodes were determined. The goal was to analyze the mechanisms of seed dormancy release and germination comprehensively. Results show that low-temperature wet stratification under 5 °C can release seed dormancy effectively. During the seed dormancy release, the seed germination rate was positively correlated with soluble sugar, GA_3, and IAA levels, as well as G-6-PDH, SOD, POD, CAT, and APX activity, but it was negatively correlated with MDH activity and ABA content. These imply that dormancy release might be attributed to the degradation of endogenous ABA and the oxidation of reactive oxygen species induced by low-temperature wet stratification. GA_3, IAA, and the metabolism of energy substrates may be correlated with the induction and promotion of germination.

Keywords: *Bretschneidera sinensis*; seed dormancy and germination; cold stratification; endogenous hormones; enzyme activity

Citation: Zhong, L.; Dong, H.; Deng, Z.; Li, J.; Xu, L.; Mou, J.; Deng, S. Physiological Mechanisms of *Bretschneidera sinensis* Hemsl. Seed Dormancy Release and Germination. *Forests* **2023**, *14*, 2430. https://doi.org/10.3390/f14122430

Academic Editor: Luz Valbuena

Received: 13 October 2023
Revised: 12 December 2023
Accepted: 12 December 2023
Published: 13 December 2023

Copyright: © 2023 by the authors. Licensee MDPI, Basel, Switzerland. This article is an open access article distributed under the terms and conditions of the Creative Commons Attribution (CC BY) license (https://creativecommons.org/licenses/by/4.0/).

1. Introduction

Seed dormancy refers to the inability of viable seeds (or other germination units) to germinate under favorable environmental conditions (e.g., water, temperature, light) during a specific period [1,2]. Dormancy enables plant seeds to time germination until environmental conditions become favorable for seedling survival and growth. The dormancy characteristics of seeds are of important ecological adaptive significance and notable agricultural value [3,4].

Dormancy is a characteristic of many seed plants. Seeds exhibit a diverse range and degree of dormancy for environmental adaptation [5]. Further, seed dormancy release and the physiological mechanism of germination vary among different species. Until now, seed dormancy release and the physiological mechanisms of germination have been mainly attributed to endogenous phytohormone regulation, reactive oxygen substance production, and antioxidant generation, as well as energy metabolism.

The phytohormone regulation hypothesis states that GA_3, cytokinin, ABA, and IAA collectively regulate seed dormancy and germination [6–9]. However, the regulation mechanisms are highly complex. Many studies have shown that abscisic acid (ABA) and gibberellin (GA) are the key factors for seed dormancy and germination. Mature seeds in a dormancy state contain high levels of ABA and low levels of GA. ABA induces and maintains seed dormancy, while GA antagonizes ABA and promotes seed germination [10–13]. Numerous studies have confirmed that the maintenance of seed dormancy depends on the ratio of ABA and GA_3 in seeds, and the release of dormancy is closely related to the decrease in ABA and the increase in GA_3 [14]. Studies also reported the role of other phytohormones in seed dormancy and germination directly or indirectly mediated via signaling of ABA and GA [15,16].

However, the role of IAA in seed dormancy remains controversial because IAA has shown opposite effects on seed germination. For example, IAA at low concentrations promoted the germination rate of unstratified and stratified Arabidopsis thaliana seeds, whereas IAA at high concentrations inhibited seed germination [9,17]. SA has been extensively reported to participate in the regulation of seed dormancy and germination. At low concentrations, SA can effectively inhibit oxidative damage to the cell membrane, protect membrane integrity, and promote seed dormancy release and germination under low-temperature stress [18,19]. Moreover, the interactions between phytohormones have a much stronger influence on seed dormancy and germination as compared with individual phytohormones [20].

Reactive oxygen species (ROS) production and antioxidant theory state that under normal conditions, plants maintain a balance between ROS production and elimination. However, plant cells under stress accumulate high levels of ROS, thereby disrupting the balance. The cellular antioxidant defense system rapidly eliminates surplus ROS at low levels. However, oxidative stress occurs when ROS production exceeds the short-period scavenging capacity, causing damage to the cytoplasmic membrane. The antioxidant defense system, which includes non-enzymatic antioxidants and antioxidant enzymes, scavenges excessive ROS through the water–water cycle, the ascorbate–glutathione cycle, the glutathione peroxidase cycle, and CAT [21,22]. Among these, SOD plays a pivotal role in the defense against oxidative damage caused by oxygen free radicals, effectively scavenging oxygen free radicals and protecting cells from oxidative damage [23]. Due to its high affinity for H_2O_2, APX is sensitive to minor changes in H_2O_2 content, thereby enabling the precise modulation of H_2O_2, which is of considerable importance in signal transduction [24]. CAT functions in scavenging high levels of ROS under oxidative stress [25,26].

According to the theory of energy metabolism, the release of plant seed dormancy is related to the internal respiratory metabolism of seeds and plant respiratory metabolism, including the Embden–Meyerhof Pathway (EMP), tricarboxylic acid cycle (TCA), and pentose phosphate pathway (PPP) [27]. A preliminary study reported that the process from seed dormancy release to germination depends on the transition from EMP to PPP and changes from NADH to NADPH [28]. PPP is an important pathway that is inhibited by TCA. Hence, seed dormancy release or germination is closely related to changes in the internal energy of seeds and the transformation of such internal energy into energy needed for seed germination via various metabolic pathways [29].

Bretschneidera sinensis Hemsl., the sole species in the genus Bretschneidera and the family Bretschneideraceae, is a perennial deciduous tree species endemic to China. As a paleotropical relict tree species that originated in the Tertiary Period, *B. sinensis* has considerable significance in the study of angiosperm phylogeny, flora, paleogeography, and paleoclimate [30]. However, *B. sinensis* is a critically endangered species due to habitat destruction and its difficulty in natural reproduction caused by a low growth rate, limited female tree flowering, and a low seed setting rate. Moreover, seed dormancy and a low natural germination rate seriously affect the regeneration of this species in vegetation communities and seedling production [31,32]. Therefore, it is important to clarify the physiological mechanisms underlying the release of their seed dormancy.

2. Materials and Methods

2.1. Materials

In October 2020 and 2021, seeds were collected from female trees aged 40 years in a natural *B. sinensis* population located in the Wufeng Houhe National Nature Reserve in Hubei Province, China (110°30′9.97″ E, 30°08′55.01″ N). Fruits with cracked skin were harvested and brought back to the laboratory. The red exocarp and endocarp were quickly removed. The seeds were rinsed with water and dried in the shade for 2 d at 25 °C.

2.2. Determination of Moisture Content

Test samples were crushed mechanically and mixed evenly. Four replicates (50 g each) were collected and recorded as G1. The samples were stored in a sample box with a diameter larger than 8 cm. The contents of the box were then dried in an oven, which was preheated to 110 °C for about (17 ± 1) h under (103 ± 2) °C. The box was covered before the samples were removed from the oven. The samples were transferred to a dryer, cooled, and weighed. This processed sample was recorded as G2. The mean of four replicate samples was used to calculate the water content (%) using the formula $(G1 - G2)/G1 \times 100\%$. Water content was calculated using fresh weight as the cardinal number.

2.3. Stratification Treatment

Fresh seeds were subjected to flotation, and the selected seeds were naturally dried. The seeds were mixed with moist perlite ($W_{perlite}/W_{seed}$ = 3:1). Stratification was performed at 5 °C and 15 °C in the dark. Two portions of seeds were collected at 10, 20, 30, 40, 50, and 60 d. One portion was used for the germination experiment according to the method described in Section 2.4. Four replicates were conducted, and each replicate contained 30 seeds. The seeds that germinated during stratification were excluded from the calculation of the germination rate. The other portion was sealed in tin foil and stored in liquid nitrogen for 24 h. Then, the seeds were quickly ground in liquid nitrogen and stored in an ultra-low-temperature refrigerator at −80 °C. Measurements were performed according to the methods described in Sections 2.5–2.7.

2.4. Germination Experiment

Seed viability at different stages of stratification was determined using a seed germination experiment. Seeds were evenly placed in a glass box containing moist perlite. The seeds were covered with a 3 cm layer of moist perlite (approximately 85% moisture content) and cultured at 25 °C under light conditions. Seed germination was observed, and the seed germination rate was calculated at 30 d of cultivation. Germination was defined as the emergence of the radicle through the seed coat for a distance of 2 mm.

2.5. Determination of Soluble Sugar Content

Seeds obtained after different durations of stratification were stored in liquid nitrogen and pulverized. Later, 0.15 g samples were collected and treated with 1 mL distilled water to obtain a homogenate. The homogenate was transferred to a centrifuge tube with a cover, followed by heating in a water bath for 10 min at 95 °C. After cooling, the homogenate was centrifuged for 10 min at 25 °C and $8000 \times g$. The supernatant (10 mL) was collected in a test tube and dissolved in distilled water to 10 mL, and vortexed for later use. The soluble sugar content was determined using visible spectrophotometry with an assay kit (Comin, Suzhou, China). Four biological replicates were performed for each sample.

2.6. Determination of Antioxidant Enzyme Activity

Seeds obtained after different periods of stratification were transferred to liquid nitrogen and pulverized. Later, 0.15 g samples were collected, and each sample was treated with 1 mL distilled water to prepare an ice-bath homogenate. The samples were centrifuged at the rate of $8000 \times g$ for 10 min at 4 °C. The supernatant was collected and placed on ice for the determination of antioxidant enzyme activity. The catalase (CAT), peroxidase (POD), superoxide dismutase (SOD), ascorbate peroxidase (APX), malate dehydrogenase (MDH), and glucose-6-phosphate dehydrogenase (G-6-PDH) activities were measured using visible spectrophotometry according to the methods of Li et al. [33]. Four biological replicates were performed for each sample.

2.7. Determination of Endogenous Hormone Contents

Samples at different stratification levels treated with liquid nitrogen were collected from the ultra-low-temperature freezer and then pulverized. A total of 50 mg powder was collected and treated with the appropriate internal standard. The samples were extracted using 500 μL acetonitrile aqueous solution, and then centrifuged. The supernatant was used for secondary extraction. The supernatant was mixed. Later, 10 μL TEA and 10 μL BPTAB were added to the sample supernatant for 1 h at 90 °C and then dried under nitrogen. Samples were dissolved again with 100 μL acetonitrile aqueous solution, filtered through a 0.22 μm film, and then transferred to the sample bottle. According to the methods of Su et al. [34], the endogenous phytohormone contents were determined using high-performance liquid chromatography–mass spectrometry. Four biological replicates were performed for each sample.

2.8. Statistical Analysis

The calculated data were expressed as the mean ± standard error. To ensure the homogeneity of variance, seed germination data were subjected to arcsine transformation, followed by one-way analysis of variance and the Student–Newman–Keuls test for multiple comparisons ($p = 0.05$). Data processing was performed using WPS 13.0.503.101. Statistical analysis and plotting were conducted in R i386 3.5.2.

3. Results and Analysis

3.1. Effects of Cold Stratification on Seed Dormancy Release and Germination

As shown in Figure 1, cold stratification at 5 °C significantly promoted the seed germination rate ($F = 700.46$, $p < 0.05$). Under cold stratification (5 °C), the germination rate increased significantly with the extension of the treatment time. The seed coat gradually cracked and the maximum germination rate was $92.7 \pm 0.3\%$ at 50 d. Under warm stratification (15 °C), the maximum seed germination rate was $44 \pm 1.1\%$, which was significantly lower than that under cold stratification ($p < 0.05$). These results suggest that stratification at 5 °C can effectively release the dormancy of *B. sinensis* seeds.

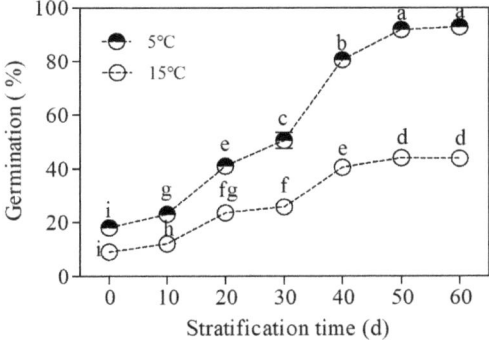

Figure 1. Changes in the seed germination rate of *Bretschneidera sinensis* during cold stratification. The 5 °C and 15 °C represent cold and warm stratification, respectively. On the *x* axis, 0 is the control (CK, CK represents control unstratified seeds) and the numbers 10, 20, 30, 40, 50, and 60 represent the duration (days) of stratification. Each treatment had four replicates, and each replicate contained 30 seeds. Germination was performed at 25 °C under light conditions. The germination rate is expressed as the mean ± standard error. The same lowercase letters indicate insignificant differences ($p = 0.05$).

3.2. Changes in Endogenous Hormone Content during Cold Stratification

The GA$_3$ content showed a trend of decreasing, increasing, and then decreasing during seed stratification ($F = 1083.03$, $p < 0.05$, Figure 2A). The GA$_3$ content was 0.03 ng/g DW in unstratified seeds, which significantly decreased to a minimum value of 0.01 ng/g DW at 10 d of stratification. The GA$_3$ content then increased sharply and reached a maximum value of 6.1 ng/g DW at 40 d of stratification. After that, the value decreased, but was always higher than that of the CK group. The GA$_3$ content remained at 3.81 ng/g DW at the end of stratification (60 d).

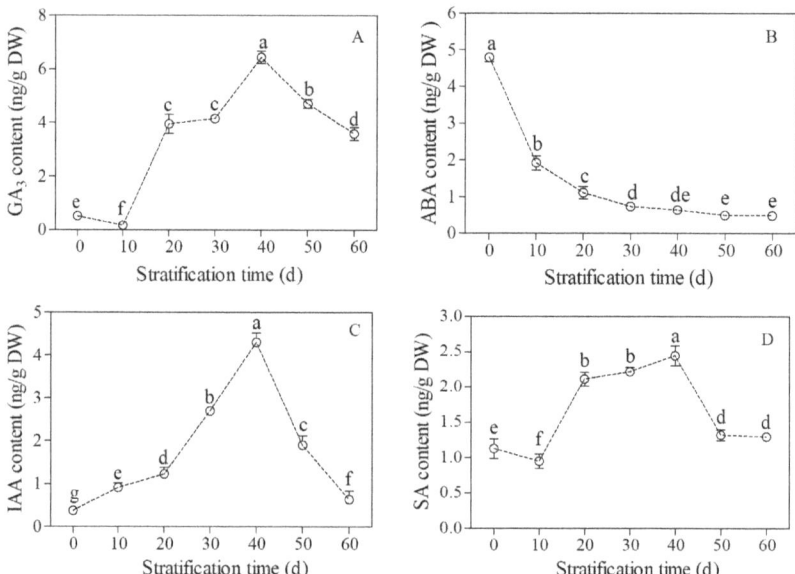

Figure 2. Changes in endogenous hormone content in *Bretschneidera sinensis* seeds during cold stratification. (**A–D**) represent gibberellic acid (GA$_3$), abscisic acid (ABA), indole-3-acetic acid (IAA), and salicylic acid (SA), respectively. Each treatment had four replicates. On the x axis, 0 is the control (CK) and the numbers 10, 20, 30, 40, 50, and 60 represent the duration (days) of stratification. The same lowercase letters indicate insignificant differences ($p = 0.05$).

The ABA content decreased continuously during seed stratification ($F = 1374.40$, $p < 0.05$, Figure 2B). The ABA content was the highest in unstratified seeds (4.83 ng/g DW). It decreased sharply during seed stratification, and reached the minimum value of 0.51 ng/g DW at 50 d.

The IAA content increased and then decreased during seed stratification ($F = 9.92$, $p < 0.05$, Figure 2C). The IAA content was the lowest in unstratified seeds (0.03 ng/g DW), increased sharply during stratification, and reached a maximum value of 4.21 ng/g DW at 40 d. Then, the value decreased dramatically, but was always higher than that of the CK group. The IAA content remained at 0.05 ng/g DW at the end of stratification (60 d).

The SA content showed a trend of decreasing, increasing, and then decreasing ($F = 10.087$, $p < 0.05$, Figure 2D). The SA content was 1.2 ng/g DW in unstratified seeds, which decreased to the lowest value of 0.08 ng/g DW at 10 d of stratification. The value then gradually increased, and reached the maximum of 2.43 ng/g DW at 40 d. After that, the value decreased, but always remained higher than that of the CK group. The SA content remained at 1.43 ng/g DW at the end of stratification (60 d).

These results suggested that the ABA content in the embryo decreased during cold stratification, while GA$_3$, IAA, and SA increased before 40 d of stratification and then gradually decreased. These findings imply the synergism or antagonism of these hormones during the dormancy release and germination of *B. sinensis* seeds.

3.3. Changes in Antioxidant Enzyme Activities during Cold Stratification

The SOD activity showed a trend of decreasing, increasing, and then decreasing (F = 209.81, p < 0.05, Figure 3A). The SOD activity was 622 μg/g DW in unstratified seeds, which significantly decreased to a minimum value of 40.25 μg/g DW at 10 d of stratification. The value then increased sharply and reached a maximum of 123.3 μg/g DW at 40 d of stratification. After that, a notable decrease in SOD activity was observed.

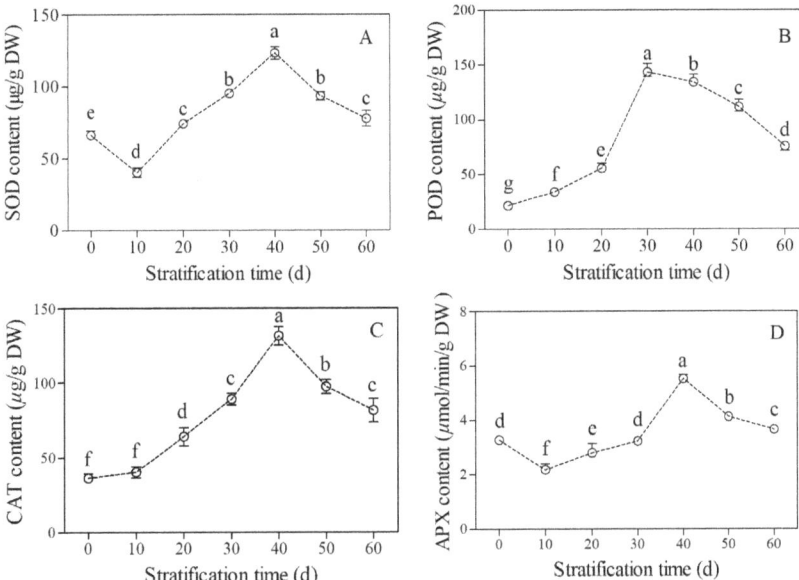

Figure 3. Changes in antioxidant enzyme activity in *Bretschneidera sinensis* seeds during cold stratification. (**A–D**) represent superoxide dismutase (SOD), peroxidase (POD), catalase (CAT), and ascorbate peroxidase (APX), respectively. Each treatment had four replicates. On the x axis, 0 is the control (CK) and the numbers 10, 20, 30, 40, 50, and 60 represent the duration (days) of stratification. The same lowercase letters indicate insignificant differences (p = 0.05).

The POD activity increased and then decreased (F = 724.52, p < 0.05, Figure 3B). The POD activity was the lowest in unstratified seeds (21.65 μg/g DW), then increased sharply with cold stratification and reached the maximum value of 144.2 μg/g DW at 30 d. The value then decreased, but was always higher than that of the CK group. The POD activity remained at 75.5 μg/g DW at the end of stratification (60 d).

The CAT activity increased and then decreased (F = 314.22, p < 0.05, Figure 3C). The CAT activity was the lowest in unstratified seeds (32 μg/g DW), then increased sharply with stratification and reached the maximum value of 131.47 μg/g DW at 40 d. The value then decreased, but always remained higher than that of the CK group. The CAT activity remained at 81.6 μg/g DW at the end of stratification (60 d).

The APX activity showed a trend of decreasing, increasing, and then decreasing (F = 115.58, p < 0.05, Figure 3D). The APX activity was 3.28 μmol/min/g DW in unstratified seeds, which decreased to the minimum value at 10 d of stratification, and then increased sharply and reached the maximum value of 5.51 μmol/min/g DW at 40 d of treatment. After that, the value declined, but was always higher than that of the CK group. The APX activity remained at 3.65 μg/g DW at the end of stratification (60 d).

These results suggest that the scavenging of excessive ROS mainly depends on SOD, POD, and CAT, while APX may play an important role in fine-tuning signal transduction during the dormancy release and germination of the fresh mature seeds of *B. sinensis*.

3.4. Changes in Soluble Sugar Content and Two Key Enzymes in Sugar Metabolism during Cold Stratification

With the extension of stratification time, the soluble sugar content increased and then decreased ($F = 402.65$, $p < 0.05$, Figure 4). The soluble sugar content was the lowest (20.17 mg/g dry weight (DW)) in unstratified seeds (CK), then increased rapidly with stratification, and peaked at 65.32 mg/g DW at 40 d of treatment. However, with the further extension of stratification, the soluble sugar content decreased but remained higher than that in the CK group. Notably, at the end of stratification (60 d), the soluble sugar content remained at a level of 42.23 ng/g DW. These observations indicate that *B. sinensis* seeds undergo a substantial consumption of soluble sugars during dormancy release and germination, and that soluble sugars facilitate seed germination. Nevertheless, the soluble sugar content gradually decreased with dormancy release at the late stage of stratification.

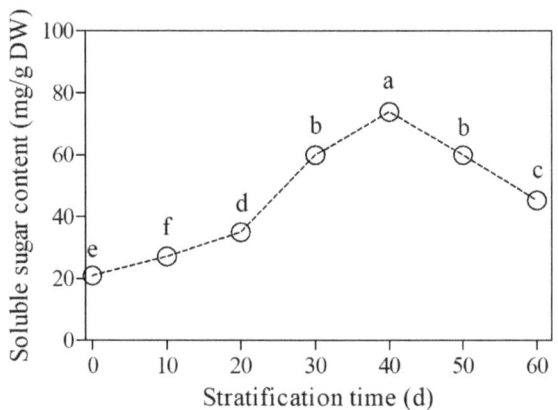

Figure 4. Changes in soluble sugar content in *Bretschneidera sinensis* seeds during cold stratification. Each treatment had four replicates. On the *x* axis, 0 is the control (CK) and the numbers 10, 20, 30, 40, 50, and 60 represent the duration (days) of stratification. The same lowercase letters indicate insignificant differences ($p = 0.05$).

With the extension of seed stratification, the MDH activity followed the trend of decreasing, increasing, and then decreasing ($F = 415.69$, $p < 0.05$, Figure 5A). The activity of MDH in unstratified seeds was 6.72 μmol/min/g DW, which decreased after 10 d of stratification, then increased and reached the highest value of 8.86 μmol/min/g DW at 30 d of stratification. With the further extension of stratification, the MDH activity decreased sharply to the lowest value of 4.36 μmol/min/g DW at 60 d of stratification.

The G-6-PDH activity decreased and then increased during seed stratification ($F = 440.62$, $p < 0.05$, Figure 5B). The G-6-PDH activity was 0.37 μmol/min/g DW in unstratified seeds, which decreased at 10 d of stratification. The G-6-PDH activity then increased and reached the maximum value of 0.48 μmol/min/g DW at 40 d of stratification. The G-6-PDH activity remained stable with the further extension of cold stratification.

Therefore, the MDH and G-6-PDH activities fluctuated considerably during 30–40 d of stratification, suggesting significant changes in the respiratory pathways of sugar metabolism during dormancy release and germination.

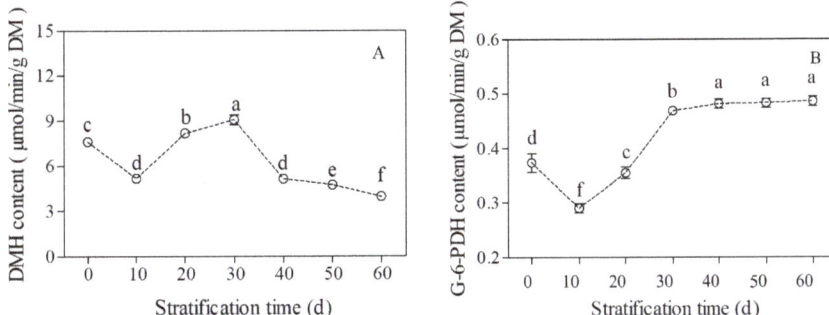

Figure 5. Changes in key enzyme activity in sugar metabolism in *Bretschneidera sinensis* seeds during cold stratification. (**A**,**B**) represent malate dehydrogenase (MDH) and glucose-6-phosphate dehydrogenase (G-6-PDH), respectively. Each treatment had four replicates. On the x axis, 0 is the control (CK) and the numbers 10, 20, 30, 40, 50, and 60 represent the duration (days) of stratification. The same lowercase letters indicate insignificant differences ($p = 0.05$).

3.5. Correlations of the Germination Rate with Physiological Indicators during Cold Stratification

As shown in Figure 6, the germination rate showed positive correlations with the soluble sugar content and G-6-PDH activity ($p < 0.05$), while demonstrating a negative correlation with MDH activity ($p < 0.05$). The germination rate showed positive correlations with GA_3 and IAA contents ($p < 0.05$), but a negative correlation with ABA ($p < 0.05$). Moreover, SOD, POD, CAT, and APX activities were positively correlated with the seed germination rate ($p < 0.05$). These results highlight the interactions of these physiological indicators with seed dormancy release and germination.

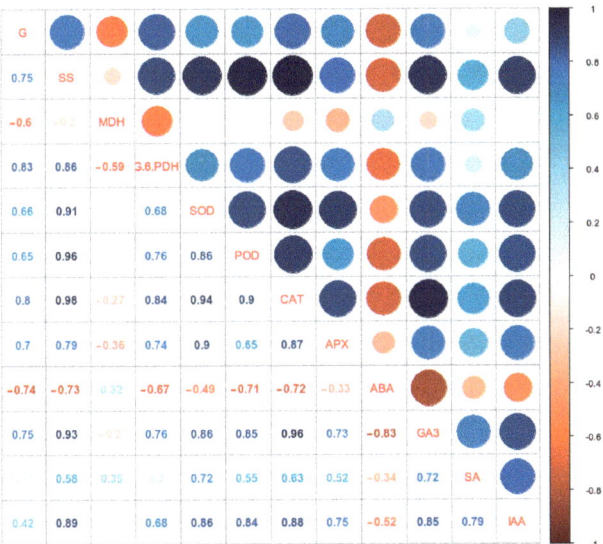

Figure 6. Correlations between the average germination rate and soluble sugars, phytohormones, and antioxidant enzymes during cold stratification. Note: The right vertical axis represents the correlation between two datasets. The color of the circle indicates the direction of the correlation (red, positive correlation; blue, negative correlation). The size of the circle indicates the magnitude of the correlation coefficient, which is displayed by the corresponding value. G is the seed germination rate, SS is soluble sugars, and other abbreviations are physiological indicators as described in the text.

4. Discussion

4.1. Regulation of Seed Dormancy Release and Germination by Endogenous Hormones

Many studies have demonstrated that GA, ABA, and IAA are the most important endogenous factors regulating seed dormancy and germination [10]. *B. sinensis* seeds exhibit dormancy. The changes in hormone levels at different stages from the dry seed stage to imbibition and then to germination reflect the internal mechanisms of dormancy regulation. It is well known that ABA plays a pivotal role in inducing and maintaining seed dormancy. ABA reduces the osmotic potential and water absorption of the seed embryo, impeding radicle elongation and seed germination. Consequently, the release of seed dormancy is usually accompanied by a decline in ABA. GA_3 is an important factor in releasing dormancy and promoting seed germination that mainly acts through increasing the vigor of the embryo, weakening the tissue around the embryo, and relieving the constraint imposed by the seed coat. The antagonism between ABA and GA, which are the major phytohormones, plays a key role in the regulation of seed dormancy. Hence, seed germination can be controlled by regulating the metabolism and signal transduction of ABA and GA [7,8]. The levels of ABA in *B. sinensis* seeds reached the maximum in the unstratified stage but decreased significantly with the increase in stratification time. As shown in Figure 2B, it decreased 9.47-fold at 50 d of seed stratification. The ABA content is negatively correlated with the germination rate of seeds (Figure 6). Conversely, the GA_3 content increased significantly by 33.3-fold by the end of seed germination (Figure 2A). These findings indicate that the increase in the GA_3/ABA ratio serves as a key determinant in seed dormancy release and germination, which is consistent with the classical theory highlighting the role of ABA and GA_3 in regulating seed dormancy and germination.

IAA is another important endogenous hormone, which promotes seed germination by inducing elongation and differentiation of blastocytes [35]. For example, IAA content during dormancy release and germination of Rhizoma Paridis Yunnanensis, Paris polyphylla, and Medicago sativa seeds increased rapidly. In this study, the IAA content increased significantly in different stages of germination of *B. sinensis* seeds. The IAA content at the end of seed germination was 14-fold higher than the initial value (Figure 2C). This indicates that IAA plays a positive role in inducing seed germination. Nevertheless, seed germination does not depend on the increase or decrease in a single phytohormone, such as ABA, GA, or IAA. The spatiotemporal balance of ABA, GA, and IAA is the major factor determining seed germination [36,37]. Based on the analysis of hormone ratios at different stages, the ABA/GA_3 and ABA/IAA ratios showed a significant decline. This is consistent with the regulation of germinating hormones in *Taxus yunnanensis* and *Garcinia paucinervis* seeds. This indicates that ABA might be the key inhibitor of germination of *B. sinensis* seeds. The antagonism with ABA was intensified by regulating the biosynthesis of the growth promoters GA_3 and IAA to promote seed germination [38,39]. Additionally, it is widely accepted that SA regulates seed growth and development, including cell expansion and axial elongation of seed radicles. In this experiment, the SA level reached the maximum after 40 d of stratification (Figure 2D). This result can be explained by the role of *B. sinensis* seeds in regulating the growth of plumular axis and radicals from imbibition to germination stages by increasing SA synthesis, thus promoting the seed germination [40–42]. According to the endogenous hormone regulation theory of seed dormancy, seed germination requires interaction between hormones. The changes in hormone levels in different stages of germination suggest that the increase in GA_3 and the decrease in ABA concentration promote seed germination in *B. sinensis*.

4.2. Antioxidative Stress during Seed Dormancy Release and Germination

Seed dormancy release and germination are related to changes in phytohormones. Further, different types of enzymes, such as antioxidant enzymes, may alter the response during seed dormancy release and germination [43]. The internal respiratory metabolism of seeds is enhanced from seed dormancy release to germination stages, which may generate substantial amounts of reactive oxygen species (ROS), leading to the accumulation of osmotic substances.

ROS transform mutually via spontaneous or catalytic reactions. They are strong oxidants and induce lipid peroxidation, resulting in oxidative damage to cellular structures [44,45]. This has been established in physiological studies of seed germination [46,47]. SOD can cleave O^{2-} into H_2O_2 and O_2 rapidly with minimal toxicity [48]. CAT is an enzyme that uses H_2O_2 as the substrate and represents an important index of metabolic changes in plants. CAT can be used to characterize cell growth and development. The interaction between CAT and POD can further eliminate H_2O_2, thereby decreasing and eliminating the free radical-induced damage to cell membranes [49,50]. The study findings suggest an increase in the levels of SOD, POD, CAT, and APX during the release from seed dormancy stage in *B. sinensis* (Figure 3A–D). They are positively correlated with the seed germination rate (Figure 6). Thus, the seeds experience oxidative stress during stratification. The high levels of SOD, POD, CAT, and APX perform rapid free radical-scavenging activity. This indicates that the combination of SOD, POD, CAT, and APX activities not only contributes to seed dormancy release in *B. sinensis*, but also facilitates seed cell growth and structural integrity, ensuring smooth germination and prompt scavenging of internal free radicals.

4.3. The Role of Energy Metabolism in Seed Dormancy Release and Germination

The complex physiological transition from seed dormancy to germination requires a significant energy supply. Energy reserves are key to the activation of seed germination and metabolism and the changes in their content also indirectly reflect the mechanism of seed dormancy release [51]. Specifically, carbohydrate, the major reserve substrate in most types of seeds [52], represents an important source of energy for seed germination. Seed dormancy release originates in the glycolysis pathway, which generates the primary energy in seeds [53]; therefore, glucose metabolism plays an important role in seed germination [54]. Soluble sugars in seeds mainly include glucose, maltose, and saccharose. As the substrate of respiratory metabolism in the embryo, soluble sugar is the major source of energy during germination [55]. Soluble sugar exists in the embryo or endosperm. However, soluble sugar can be transformed from fat, starch, and soluble proteins [56]. This study found a dramatic increase in the levels of soluble sugar in *B. sinensis* seeds after stratification at an early stage, and the sugar was consumed significantly during seed dormancy release and reinforced the respiratory effect of the embryo (Figure 4). Further, the seed germination rate increased accordingly, suggesting that the soluble sugar level regulates the release of seed dormancy in *B. sinensis*.

Roberts et al. [57] hypothesized that the shift of sugar metabolism from glycolysis to the pentose phosphate pathway (PPP) is critical to seed dormancy release. MDH and 6-G-PDH are the key enzymes in the tricarboxylic acid cycle (TCA) and PPP, respectively, and their activities reflect the strength of seed respiration. In the present investigation, with the extension of cold stratification, there was a rapid decrease in the activity of MDH in TCA (Figure 5A), alongside a considerable increase in the activity of 6-G-PDH in PPP, showing a significant negative correlation (Figure 6). These results indicate a shift in the predominant respiratory metabolic pathway in *B. sinensis* seeds from TCA to PPP during dormancy release, which is similar to the findings of Li et al. [58]. The results of the present study provide new evidence for the hypothesis proposed by Roberts et al. [57]. In addition, *B. sinensis* seeds need to be released from physiological dormancy before germination. The natural cracking of the seed coat at 30 d of stratification indicates the development of the seed embryo and the increase in metabolic activity. This explains the significant changes in MDH and G-6-PDH activities during 30–40 d of stratification. During the process of seed stratification, therefore, 30–40 d is an important period for the shift from dormancy to germination preparation.

This study demonstrates that soluble sugars play an important role in protecting seeds against oxidative stress. It can be used to regulate osmotic balance and stabilize protein structure and biological membrane. Soluble sugars also protect phospholipids in the biological membrane by inducing vitrification in cytoplasm. The accumulation of soluble sugar increases significantly during seed dormancy release in *B. sinensis* (Figure 4), possibly due to the increased glycosylation of cell proteins [59,60]. According to Coue et al., the soluble sugar content is related to metabolic pathways associated with ROS generation. Conversely, the soluble sugar can induce metabolic pathways that generate NADPH, such as the oxidized pentose phosphate (OPP) pathway, to facilitate scavenging of ROS [61].

5. Conclusions

Low-temperature wet stratification under 5 °C can release seed dormancy effectively. During the seed dormancy release, the seed germination rate was positively correlated with soluble sugar, GA_3, and IAA levels, as well as G-6-PDH, SOD, POD, CAT, and APX activity, but it was negatively correlated with MDH activity and ABA content. These imply that dormancy release might be attributed to the degradation of endogenous ABA and the oxidation of reactive oxygen species induced by low-temperature wet stratification. GA_3, IAA, and the metabolism of energy substrates may be correlated with the induction and promotion of germination.

Author Contributions: S.D. and J.M. conceived and designed the study. L.X., L.Z. and H.D. collected the samples and performed the experiments. J.L., Z.D. and S.D. analyzed the data and drafted the manuscript. Z.D. provided financial support, supervised the study, and revised the first draft of the manuscript. All authors have read and agreed to the published version of the manuscript.

Funding: This work was supported by the National Natural Sciences Foundation of China (31860073), the Science and Technology Research Project of Education Department of Hubei Province (B2021156) and Germplasm Engineering of Characteristic Plant Resources in Enshi Prefecture (2019–2021).

Data Availability Statement: Data are contained within the article.

Acknowledgments: We are grateful to Hua Xue (Beijing Forestry University) for help during the experiments and for his continuous support throughout this project.

Conflicts of Interest: The authors declare no conflict of interest.

References

1. Baskin, J.M.; Baskin, C.C. A classification system for seed dormancy. *Seed Sci. Res.* **2004**, *14*, 1–16. [CrossRef]
2. Baskin, C.C.; Baskin, J.M. *Seeds: Ecology, Biogeography, and Evolution of Dormancy and Germination*; Elsevier Academic Press Inc.: San Diego, CA, USA, 2014.
3. Finch-Savage, W.E.; Leubner-Metzger, G. Seed dormancy and the control of germination. *New Phytol.* **2006**, *171*, 501–523. [CrossRef] [PubMed]
4. Stevens, A.V.; Nicotra, A.B.; Godfree, R.C.; Guja, L.K. Polyploidy affects the seed, dormancy and seedling characteristics of a perennial grass, conferring an advantage in stressful climates. *Plant Biol.* **2020**, *22*, 500–513. [CrossRef] [PubMed]
5. Lee, S.Y.; Rhie, Y.H.; Kim, K.S. Dormancy breaking and germination requirements of seeds of Thalictrum uchiyamae (Ranunculaceae) with underdeveloped embryos. *Sci. Hortic.* **2018**, *231*, 82–88. [CrossRef]
6. Bewley, J.D. Seed germination and dormancy. *Plant Cell* **1997**, *9*, 1055–1066. [CrossRef]
7. Song, S.Q.; Liu, J.; Huang, H.; Wu, J.; Xu, H.; Zhang, Q.; Li, X.; Liang, J. Gibberellin metabolism and signaling and its molecular mechanism in regulating seed germina tion and dormancy. *Sci. China Life Sci.* **2020**, *50*, 599–615.
8. Song, S.Q.; Liu, J.; Xu, H.H.; Liu, X.; Huang, H. ABA metabolism and signaling and their molecular mechanism regulating seed dormancy and germination. *Sci. Agric. Sin.* **2020**, *50*, 599–615.
9. Song, S.Q.; Liu, J.; Tang, C.F.; Zhang, W.H.; Xu, H.H.; Zhang, Q.; Gao, J.D. Metabolism and signaling of auxins and their roles in regulating seed dormancy and germination. *Chin. Sci. Bull.* **2020**, *65*, 3924–3943. (In Chinese) [CrossRef]
10. Liu, X.; Wang, Z.; Xiang, Y.; Tong, X.; Wojtyla, Ł.; Wang, Y. Editorial: Molecular basis of seed germination and dormancy. *Front. Plant Sci.* **2023**, *14*, 1242428. [CrossRef]
11. Ali, F.; Qanmber, G.; Li, F.; Wang, Z. Updated role of ABA in seed maturation, dormancy, and germination. *J. Adv. Res.* **2022**, *35*, 199–214. [CrossRef]
12. Sano, N.; Marion-Poll, A. ABA Metabolism and Homeostasis in Seed Dormancy and Germination. *Int. J. Mol. Sci.* **2021**, *22*, 5069. [CrossRef]
13. Longo, C.; Holness, S.; De Angelis, V.; Lepri, A.; Occhigrossi, S.; Ruta, V.; Vittorioso, P. From the Outside to the Inside: New Insights on the Main Factors That Guide Seed Dormancy and Germination. *Genes* **2020**, *12*, 52. [CrossRef] [PubMed]
14. Bicalho, E.M.; Pintó-Marijuan, M.; Morales, M.; Müller, M.; Munné-Bosch, S.; Garcia, Q.S. Control of macaw palm seed germination by the gibberellin/abscisic acid balance. *Plant Biol.* **2015**, *17*, 990–996. [CrossRef] [PubMed]
15. Footitt, S.; Clewes, R.; Feeney, M.; Finch-Savage, W.E.; Frigerio, L. Aquaporins Influence seed dormancy and germination in response to stress. *Plant Cell Environ.* **2019**, *42*, 2325–2539. [CrossRef] [PubMed]

16. Xu, J.; Li, Q.; Yang, L.; Li, X.; Wang, Z.; Zhang, Y. Changes in carbohydrate metabolism and endogenous hormone regulation during bulblet initiation and development in Lycoris radiata. *BMC Plant Biol.* **2020**, *20*, 180. [CrossRef] [PubMed]
17. Wang, Z.; Chen, F.; Li, X.; Cao, H.; Ding, M.; Zhang, C.; Zuo, J.; Xu, C.; Xu, J.; Deng, X.; et al. Arabidopsis seed germination speed is controlled by SNL histone deacetylase-binding factor-mediated regulation of AUX1. *Nat. Commun.* **2016**, *7*, 13412. [CrossRef] [PubMed]
18. Anand, A.; Kumari, A.; Thakur, M.; Koul, A. Hydrogen peroxide signaling integrates with phytohormones during the germination of magnetoprimed tomato seeds. *Sci. Rep.* **2019**, *9*, 8814. [CrossRef] [PubMed]
19. Ma, L.; Hao, W.; Liu, D.; Wang, L.; Qiu, S. Effects of SA on the germination of seeds and the stability of cell membrane under cold stress. *J. Northwest AF Univ. (Nat. Sci. Ed.)* **2010**, *38*, 183–188.
20. Wang, Y.M.; Wang, L.J.; Yao, B.; Liu, Z.; Li, F. Changes in ABA, IAA, GA3, and ZR levels during seed dormancy release in Idesia polycarpa Maxim from Jiyuan. *Pol. J. Environ. Stud.* **2018**, *27*, 1833–1839.
21. Katsuya-Gaviria, K.; Caro, E.; Carrillo-Barral, N.; Iglesias-Fernández, R. Reactive Oxygen Species (ROS) and Nucleic Acid Modifications during Seed Dormancy. *Plants* **2020**, *9*, 679. [CrossRef]
22. Leymarie, J.; Vitkauskaité, G.; Hoang, H.H.; Gendreau, E.; Chazoule, V.; Meimoun, P.; Corbineau, F.; El-Maarouf-Bouteau, H.; Bailly, C. Role of reactive oxygen species in the regulation of Arabidopsis seed dormancy. *Plant Cell Physiol.* **2012**, *53*, 96–106. [CrossRef] [PubMed]
23. Luo, X.; Dai, Y.; Zheng, C.; Yang, Y.; Chen, W.; Wang, Q.; Chandrasekaran, U.; Du, J.; Liu, W.; Shu, K. The ABI4-RbohD/VTC2 regulatory module promotes reactive oxygen species (ROS) accumulation to decrease seed germination under salinity stress. *New Phytol.* **2021**, *229*, 950–962. [CrossRef]
24. Wang, W.B.; Kim, Y.H.; Lee, H.S.; Kim, K.Y.; Deng, X.P.; Kwak, S.S. Analysis of antioxidant enzyme activity during Germination of alfalfa under salt and drought stresses. *Plant Physiol. Biochem.* **2009**, *47*, 570–577. [CrossRef] [PubMed]
25. Oracz, K.; El-Maarouf-Bouteau, H.; Kranner, I.; Bogatek, R.; Corbineau, F.; Bailly, C. The mechanisms involved in seed dormancy alleviation by hydrogen cyanide unravel the role of reactive oxygen species as key factors of cellular signaling during germination. *Plant Physiol.* **2009**, *150*, 494–505. [CrossRef]
26. El-Maarouf-Bouteau, H.; Bailly, C. Oxidative signaling in seed germination and dormancy. *Plant Signal. Behav.* **2008**, *31*, 75–182. [CrossRef]
27. Zaynab, M.; Pan, D.; Chen, W. Transcriptomic approach to address low germination rate in Cyclobalnopsis gilva seeds. *S. Afr. J. Bot.* **2018**, *119*, 286–294. [CrossRef]
28. Roberts, E. Temperature and seed germination. *Symp. Soc. Exp. Biol.* **1988**, *42*, 109–132. [PubMed]
29. Zhi, L.M.; Zhang, Y.H.; Yu, F.Y. Biochemical and physiological changes of Euscaphis japonica seeds during the period of stratification. *J. Cent. South Univ. For. Technol.* **2016**, *36*, 36–40.
30. Wu, Z.Y.; Lu, A.M.; Tang, Y.C. *The Families and General of Angiosperm in China*; Science Press: Beijing, China, 2003; p. 702.
31. Zhang, J.R.; Cheng, G.F. National level to protect plants-bretschneidera sinensis Hemsl. *Bull. Biol.* **2009**, *44*, 7.
32. Li, T.H.; Zhou, Y.X.; Duan, X.P. A preliminary study of physiology dormancy character bretschneidera sinensis Hemsl Seeds. *J. Cent. South For. Univ.* **1997**, *17*, 41–44.
33. Li, X.Z.; Simpson, W.R.; Song, M.L.; Bao, G.S.; Niu, X.L.; Zhang, Z.H.; Xu, H.F.; Liu, X.; Li, Y.L.; Li, C.J. Effects of seed moisture content and Epichloe endophyte on germination and physiology of Achnatherum inebrians. *S. Afr. J. Bot.* **2022**, *134*, 407–414. [CrossRef]
34. Ma, L.Y.; Cheng, N.L.; Han, G.J.; Li, L. Effects of exogenous salicylic acid on seed germination and physiological characteristics of Coronilla varia under drought stress. *Chin. J. Appl. Ecol.* **2017**, *28*, 3274–3280.
35. Wu, M.J.; Wu, J.Y.; Gan, Y.B. The new insight of auxin functions: Transition from seed dormancy to germination and floral opening in plants. *Plant Growth Regul.* **2020**, *91*, 169–174. [CrossRef]
36. Boter, M.; Calleja-Cabrera, J.; Carrera-Castaño, G.; Wagner, G.; Hatzig, S.V.; Snowdon, R.J.; Legoahec, L.; Bianchetti, G.; Bouchereau, A.; Nesi, N.; et al. An integrative approach to analyze seed germination in Brassica napus. *Front. Plant Sci.* **2019**, *10*, 1342. [CrossRef] [PubMed]
37. Penfield, S. Seed dormancy and germination. *Curr. Biol.* **2017**, *27*, 874–878. [CrossRef] [PubMed]
38. Bian, F.; Su, J.; Liu, W.; Li, S. Dormancy release and germination of Taxus yunnanensis seeds during wet sand storage. *Sci. Rep.* **2018**, *8*, 3205. [CrossRef]
39. Lee, S.; Kim, S.G.; Park, C.M. Salicylic acid promotes seed germination under high salinity by modulating antioxidant activity in Arabidopsis. *New Phytol.* **2010**, *188*, 626–637. [CrossRef]
40. Zhang, J.J.; Wei, X.; Chai, S.F.; Wu, S.H.; Zou, R.; Qin, X.M.; Fu, R. Dormancy mechanism of the seeds of a rare and endangered plant, Garcinia paucinervis. *Chin. J. Ecol.* **2018**, *37*, 1371–1381.
41. Su, H.L.; Zhou, X.Z.; Li, X. Physicochemical changes of *Paris polyphylla* var. Chinensis seed during different stages of germination. *Chin. Tradit. Herb. Drugs* **2017**, *48*, 4755–4763.
42. Pluskota, W.E.; Pupel, P.; Głowacka, K.; Okorska, S.B.; Jerzmanowski, A.; Nonogaki, H.; Górecki, R.J. Jasmonic acid and ethylene are involved in the accumulation of osmotin in germinating tomato seeds. *J. Plant Physiol.* **2018**, *232*, 74–81. [CrossRef]
43. Xu, L.; Wang, P.; Ali, B.; Yang, N.; Chen, Y.; Wu, F.; Xu, X. Changes of the phenolic compounds and antioxidant activities in germinated adlay seeds. *J. Sci. Food Agric.* **2017**, *97*, 4227–4234. [CrossRef]

44. Oracz, K.; Bouteau, H.E.; Farrant, J.M.; Cooper, K.; Belghazi, M.; Job, C.; Job, D.; Corbineau, F.; Bailly, C. ROS production and protein oxidation as a novel mechanism for seed dormancy alleviation. *Plant J.* **2007**, *50*, 452–465. [CrossRef]
45. Huang, W.; Mayton, H.S.; Amirkhani, M.; Wang, D.; Taylor, A.G. Seed dormancy,germination and fungal infestation of eastern gamagrass seed. *Ind. Crops Prod.* **2017**, *99*, 109–116. [CrossRef]
46. Zhang, Y.; Chen, B.; Xu, Z.; Shi, Z.; Chen, S.; Huang, X.; Chen, J.; Wang, X. Involvement of reactive oxygen species in endosperm cap weakening and embryo elongation growth during lettuce seed germination. *J. Exp. Bot.* **2014**, *65*, 3189–3200. [CrossRef]
47. Jeevan Kumar, S.P.; Rajendra Prasad, S.; Banerjee, R.; Thammineni, C. Seed birth to death: Dual functions of reactive oxygen species in seed physiology. *Ann. Bot.* **2015**, *116*, 663–668. [CrossRef]
48. Amooaghaie, R. Triangular interplay between ROS, ABA and GA in dormancy alleviation of Bunium persicum seeds by cold stratification. *Russ. J. Plant Physiol.* **2017**, *64*, 588–599. [CrossRef]
49. Yang, Y.; Li, Y.X.; Xu, X. The activity of Principal antioxidant enzymes and the content of metabolites in dormancy breaking and germination of Davidia involucrate seeds. *Plant Divers. Resour.* **2015**, *37*, 779–787.
50. Marta, B.; Szafrańska, K.; Posmyk, M. Exogenous melatonin improves antioxidant defense in cucumber seeds (*Cucumis sativus* L.) germinated under chilling stress. *Front. Plant Sci.* **2016**, *7*, 575. [CrossRef]
51. Kazmi, R.H.; Willems, L.A.J.; Joosen, R.V.L.; Khan, N.; Ligterink, W.; Hilhorst, H.W.M. Metabolomic analysis of tomato seed germination. *Metabolomics* **2017**, *13*, 145. [CrossRef]
52. Alencar, N.L.; Innecco, R.; Gomes-Filho, E.; Gallão, M.I.; Alvarez-Pizarro, J.C.; Prisco, J.T.; De Oliveira, A.B. Seed reserve composition and mobilization during germination and early seedling establishment of *Cereus jamacaru* D.C. ssp. jamacaru (Cactaceae). *An. Acad. Bras. Ciênc.* **2012**, *84*, 823–832. [CrossRef]
53. Han, C.; Zhen, S.; Zhu, G.; Bian, Y.; Yan, Y. Comparative metabolome analysis of wheat embryo and endosperm reveals the dynamic changes of metabolites during seed germination. *Plant Physiol. Biochem.* **2017**, *115*, 320–327. [CrossRef] [PubMed]
54. Yang, Q.; Sang, S.; Chen, Y.; Wei, Z.; Wang, P. The role of Arabidopsis inositol polyphosphate kinase AtIPK2β in glucose suppression of seed germination and seedling development. *Plant Cell Physiol.* **2017**, *59*, 343–354. [CrossRef] [PubMed]
55. Zhao, M.; Zhang, H.; Yan, H.; Qiu, L.; Baskin, C.C. Mobilization and role of starch,protein, and fat reserves during seed germination of six wild grassland species. *Front. Plant Sci.* **2018**, *9*, 234. [CrossRef] [PubMed]
56. Sharma, S.; Sakshi, G.; Munshi, S. Changes in lipid and carbohydrate composition of germinating soybean seeds under different storage conditions. *Asian J. Plant Sci.* **2007**, *6*, 596.
57. Roberts, E.H.; Major, W. Dormancy in cereal seeds: II. The nature of gaseous exchange in imbibed barley and rice seeds. *J. Exp. Bot.* **1968**, *19*, 90–101.
58. Li, Z.L.; Tong, K.; Yan, S.; Yang, H.; Wang, Q.; Tang, Y.B.; Deng, M.S.; Tian, M.L. Physiological and biochemical change of Paris seed in after-ripening during variable temperature stratifications. *China J. Chin. Mater. Med.* **2015**, *40*, 629–633.
59. Smolikova, G.; Leonova, T.; Vashurina, N.; Frolov, A.; Medvedev, S. Desiccation Tolerance as the Basis of Long-Term Seed Viability. *Int. J. Mol. Sci.* **2021**, *22*, 101. [CrossRef]
60. Livingston, D.P.; Hincha, D.K.; Heyer, A.G. Fructan and its relationship to abiotic stress tolerance in plants. *Cell. Mol. Life Sci.* **2009**, *66*, 2007–2023. [CrossRef]
61. Couée, I.; Sulmon, C.; Gouesbet, G.; El Amrani, A. Involvement of soluble sugars in reactive oxygen species balance and responses to oxide tive stress in plants. *J. Exp. Bot.* **2006**, *57*, 449–459. [CrossRef]

Disclaimer/Publisher's Note: The statements, opinions and data contained in all publications are solely those of the individual author(s) and contributor(s) and not of MDPI and/or the editor(s). MDPI and/or the editor(s) disclaim responsibility for any injury to people or property resulting from any ideas, methods, instructions or products referred to in the content.

Article

Development of Commercial Eucalyptus Clone in Soil with Indaziflam Herbicide Residues

Josiane Costa Maciel [1,*], Tayna Sousa Duque [1], Aline Cristina Carvalho [2], Brenda Thaís Barbalho Alencar [2], Evander Alves Ferreira [3], José Cola Zanuncio [4], Bárbara Monteiro de Castro e Castro [4], Francisca Daniele da Silva [5], Daniel Valadão Silva [5] and José Barbosa dos Santos [1]

[1] Departamento de Agronomia, Universidade Federal dos Vales do Jequitinhonha e Mucuri, Diamantina 39100-000, Brazil; taynaduque24@gmail.com (T.S.D.); jbarbosa@ufvjm.edu.br (J.B.d.S.)
[2] Departamento de Engenharia Florestal, Universidade Federal dos Vales do Jequitinhonha e Mucuri, Diamantina 39100-000, Brazil; ninecarvalho87@gmail.com (A.C.C.); barbalhobrenda@gmail.com (B.T.B.A.)
[3] Instituto de Ciências Agrárias, Universidade Federal de Minas Gerais, Montes Claros 39404-547, Brazil; evanderalves@gmail.com
[4] Departamento de Entomologia/BIOAGRO, Universidade Federal de Viçosa, Viçosa 36570-900, Brazil; zanuncio@ufv.br (J.C.Z.); barbaramcastro@hotmail.com (B.M.d.C.e.C.)
[5] Departamento de Manejo Solo e Água, Universidade Federal Rural do Semi-Árido, Mossoró 59625-900, Brazil; danieleamancio20@gmail.com (F.D.d.S.); daniel.valadao@ufersa.edu.br (D.V.S.)
* Correspondence: josi-agronomia@hotmail.com; Tel.: +55-38-99171-6384

Citation: Maciel, J.C.; Duque, T.S.; Carvalho, A.C.; Alencar, B.T.B.; Ferreira, E.A.; Zanuncio, J.C.; de Castro e Castro, B.M.; da Silva, F.D.; Silva, D.V.; dos Santos, J.B. Development of Commercial Eucalyptus Clone in Soil with Indaziflam Herbicide Residues. *Forests* **2023**, *14*, 1923. https://doi.org/10.3390/f14091923

Academic Editors: Chunjian Zhao, Zhi-Chao Xia, Chunying Li and Jingle Zhu

Received: 24 July 2023
Revised: 6 September 2023
Accepted: 16 September 2023
Published: 21 September 2023

Copyright: © 2023 by the authors. Licensee MDPI, Basel, Switzerland. This article is an open access article distributed under the terms and conditions of the Creative Commons Attribution (CC BY) license (https://creativecommons.org/licenses/by/4.0/).

Abstract: The pre-emergent herbicide indaziflam is efficient in the management of weeds in eucalyptus crops, but this plant may develop less in soil contaminated with it. The objective was to evaluate the levels of chlorophylls a and b, the apparent electron transport rate (ETR), growth and dry mass of leaves, stems and roots of Clone I144, in clayey soil, contaminated with the herbicide indaziflam and the leaching potential of this herbicide. The design was completely randomized in a 3 × 5 factorial scheme, with four replications. The leaching of indaziflam in the clayey soil profile (69% clay) was evaluated in a bioassay with *Sorghum bicolor*, a plant with high sensitivity to this herbicide. The injury and height of this plant were evaluated at 28 days after sowing (DAS). We believe that this is the first work on *Eucalyptus* in soil with residues of the herbicide indaziflam. Chlorophyll a and b contents and ETR, and height and stem dry mass of Clone I144, were lower in soil contaminated with indaziflam residues. The doses of indaziflam necessary to cause 50% (C_{50}) of injury and the lowest height of sorghum plants were 4.65 and 1.71 g ha^{-1} and 0.40 and 0.27 g ha^{-1} in clayey soil and sand, respectively. The sorption ratio (SR) of this herbicide was 10.65 in clayey soil. The herbicide indaziflam leached up to 30 cm depth at doses of 37.5 and 75 g ha^{-1} and its residue in the soil reduced the levels of chlorophylls a and b, the apparent ETR and the growth of Clone I144.

Keywords: clonal eucalyptus; herbicide; indaziflam; leaching; soil profile

1. Introduction

The global demand for wood and wood products, a demand increasingly met by high-yield forest plantations, has been steadily growing [1]. These plantations have grown by an average of 4.4 million hectares annually, from 168 million hectares in 1990 to approximately 278 million hectares in 2015 [2]. To ensure high biomass production, significant quantities of agricultural materials are used [3]. Meeting these demands requires a dramatic increase in the global production and trade of forest products. This would imply a further increase in the global forest plantation area by about 25–67 million hectares to reach 303–345 million hectares by 2030, and there are predictions that the demand for roundwood supplied by forest plantations will increase by about 65% by 2070 [4].

Eucalyptus sp. is the most widely planted forest genus, with 25 million hectares [5,6] containing more than 110 species introduced in over 90 countries [7]. Brazil is the world leader in eucalyptus planted area, followed by China and India [8,9]. It has 9.93 million hectares of planted forest, of which 75.8% is eucalyptus plantation [10]. Furthermore, Brazil is a leader in productivity, with an average accumulated mass of 40 $m^3 ha^{-1}$ $year^{-1}$ [11], which has grown in recent years along tropical agricultural frontiers. Currently, the distribution and growth of eucalyptus plantation areas in Brazil is located in the Southeast, in Minas Gerais (30%) and São Paulo (13%), the Midwest, in Mato Grosso do Sul (14%), the Northeast, in Bahia (8%) and the South, in Rio Grande do Sul (8%) and Paraná (6%) [10]. Primary products, such as paper, pulp and wood, as well as secondary products, such as flooring and furniture, from Brazilian eucalyptus plantations are exported to many countries, highlighting the importance of Brazilian plantations for the international market [12].

Pure eucalyptus species, ranked in terms of importance [10], are mainly used in Brazilian plantations: *Eucalyptus grandis* (W. Hill ex Maiden), *Corymbia citriodora* (Hook.) KD Hill & LAS Johnson (formerly known as *E. citriodora*– basionym), *E. urophylla* (ST Blake), *E. saligna* (Sm.), *E. globulus* (Labill.), *E. camaldulensis* (Dehnh.), and hybrids *E. urophylla* × *E. grandis*, *E. urophylla* × *E. camaldulensis*, *E. grandis* × *E. camaldulensis* and *E. urophylla* × *E. globulus* [13,14]. These species or hybrids are selected for their characteristics, such as fast growth, wood quality, high productivity, profitability, strong adaptability to different soils and climatic conditions and ease of management [6,7]. We can also highlight an extensive history of investment in Brazil, and consolidated improvement techniques for silvicultural practices and forest genetic improvement.

Although the genetic improvement of this crop is at an advanced stage, another determining factor for the higher productivity of eucalyptus plantations is the control of diseases, pests and weeds [15,16]. Competition with weeds is a limiting factor for the development of most forest species [17]. Generally, weeds are considered the pest of greatest economic impact and phytosanitary risk in eucalyptus cultivation. Weeds seriously affect plant growth through interspecific competition for water, light and nutrients [18], causing serious damage to crop establishment, development and productivity. Although eucalyptus has potentially rapid growth rates, its tolerance to weed interference during establishment is low. Yield reduction due to weeds is greatest up to two years after eucalypt planting, when weed management in these crops is highly dependent on herbicides [19]. According to Silva et al. [20], specific plants can be controlled through the use of herbicides and their mechanisms of action.

Chemical control using herbicides is commonly employed for weed control. This weed control method in eucalyptus plantations is fast and efficient [21], with lower labor requirements and greater effectiveness. The development of a selective, broad-spectrum action herbicide, applied during the pre-emergence of weeds, would improve weed management for this crop and favor eucalyptus silviculture [22]. However, the number of herbicides used is reduced with few records for this crop [23] and most registered herbicides not being selective for eucalyptus [22].

Although there are some species that can be used as green manure to remove herbicides from the soil, as they have the ability to accumulate chemical compounds in tissues [24]. One of the areas with the greatest need for research development involves the use of chemical products for weed control in forest plantations, since application failures and herbicide drift can be harmful to the tree component and cause toxicity to plants, such that chemical controls must be used with caution. This situation is concerning given the low selectivity of herbicides to eucalyptus plantations, which can cause losses early on during tree development, leading to productivity losses [25]. The drift of glyphosate herbicide, non-selective to eucalyptus, can cause phytotoxicity, deformed apices, strongly developed necrosis along the leaf edges and marked leaf senescence [26]: nicosulfuron reduced stem diameter increment and fluazifop-p-butyl + fomesafen limited shoot dry mass accumulation [27].

The herbicide indaziflam N-[(1R,2S)-2,3-dihydro-2,6-dimethyl-1H-inden-1-yl]-6-[(1RS)-1-fluoroethyl]-1,3,5-triazine-2,4-diamine, an inhibitor of cellulose biosynthesis belonging to the alquilazinas group, is used during pre-emergence to manage weeds in coffee, citrus, sugarcane, pine and eucalyptus crops in Brazil [23]. The structural formula of indaziflam is described in Figure 1. This herbicide is safe for grape [28] and olive [29] crops with low solubility in water (0.0028 kg m^{-3} at 20 °C), o Koc < 1000 mL g^{-1} organic carbon, pKa = 3.5, log Kow at pH 4, 7 or 9 = 2.8, prolonged residual activity in the soil and half-life ($t_{1/2}$) greater than 150 days [30]. These features reduce the environmental impact from indaziflam leaching into the soil and contaminating the groundwater [31]. However, soil mobility in eucalyptus plantations and the tolerance of this plant to indaziflam are poorly understood, increasing the need to evaluate its residual effects, especially in planting rows [32]. Thus, we hypothesized that indaziflam soil residues would reduce eucalyptus development.

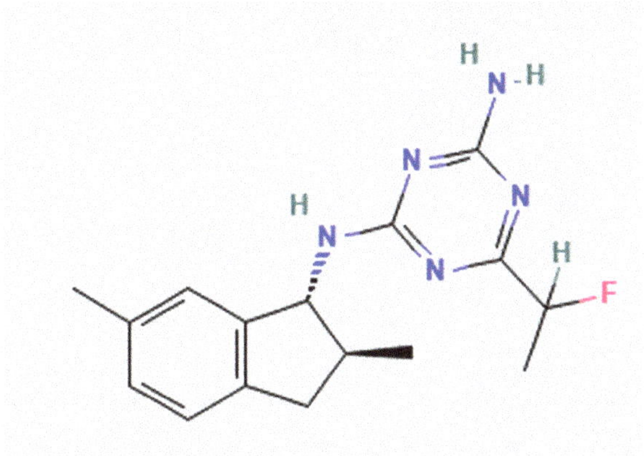

Source: PubChem (2023)

Figure 1. Representation of the chemical structure of indaziflam.

The objective was to evaluate the levels of chlorophyll a and b, ETR and the growth of Clone I144 in soil contaminated with indaziflam residues and the leaching potential of this herbicide.

2. Materials and Methods

The experiment was carried out in a greenhouse (minimum temperature of 25 °C and maximum temperature of 32 °C) at the Universidade Federal dos Vales do Jequitinhonha e Mucuri (UFVJM) in Diamantina, Minas Gerais, Brazil.

2.1. Experimental Design

The methodological design adopted in this study is outlined in Figure 2. The experiment had a completely randomized design, with treatments arranged in a 3 × 5 factorial scheme, with four replications. The first factor consisted of the control treatment (soil without herbicide) plus two doses, 35.7 and 75 g ha^{-1}, of Esplanade® herbicide (500 g a.i. L), with the doses corresponding to 25 and 50% of the commercially recommended dose for this product (150 g ha^{-1}). The second factor was the depth in soil profile: 0–10, 10–20, 20–30, 30–40 and 40–50 cm.

The eucalyptus clone used in the experiment was *Eucalyptus urograndis* (I144-*Eucalyptus urophylla* S.T. Blake × *Eucalyptus grandis* W. Hill ex Maiden). The eucalyptus clone was purchased in a nursery and was 45 days old. The clone was selected for its profitability, fast growth, high productivity and high-quality wood [15]. Each plot had a 150 mm PVC (polyvinyl chloride) tube, cut horizontally to form rings. The PVC columns were composed of five 10 cm high rings. Each one was filled with a sample of dystrophic red latosol (Table 1), previously fertilized as recommended for the crop.

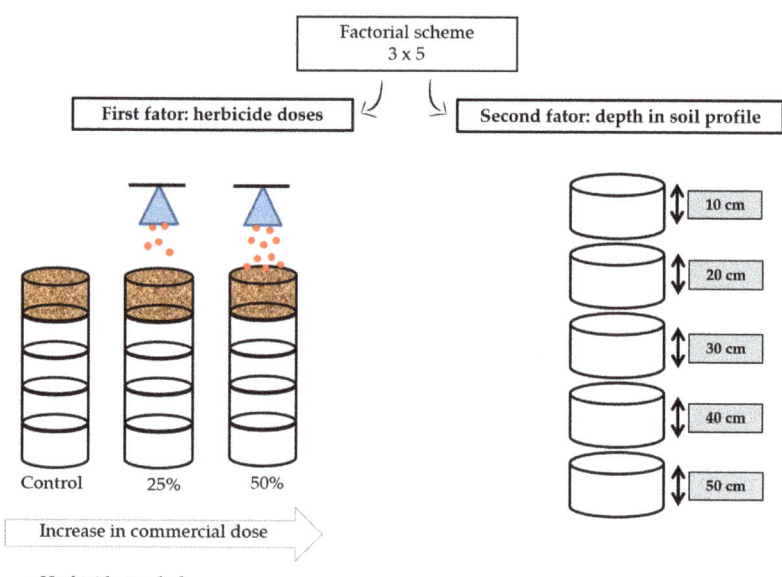

Figure 2. Diagram presenting the methodological design of the study.

Table 1. Physicochemical characteristics of the soil samples used in the experiment.

Physical Analysis											
Sand			Clay			Silt			Texture Class		
(dag kg^{-1})											
6			69			25			Very clayey		
Chemical analysis											
pH	P	K	Ca	Mg^{2+}	Al^{3+}	H+Al	SB	t	T	V	m	OM
(H$_2$O)	(mg dm^{-3})					(Cmol$_c$dm^{-3})				(%)		(dag kg^{-1})
5.00	0.54	31	0.18	0.13	0.80	4.62	0.39	1.19	5.01	7.8	67.2	1.88

P-K-Extractor Mehlich 1; Ca-Mg-Al-Extractor: KCl-1 mol/L; H + Al-Calcium Acetate Extractor 0.5 mol/L-pH 7.0; SB = Sum of Bases; t = Effective Cation Exchange Capacity; T = Cation Exchange Capacity at pH 7.0; V = Base Saturation Index; m = Aluminum Saturation Index; OM = Organic Matter (C.Org × 1724-Walkley–Black).

2.2. Application of Indaziflam

Irrigation was carried out before the herbicide was applied, keeping humidity between 70% and 80% of field capacity. Indaziflam was applied with an electric sprayer (Yamaho FT5®, 5 L capacity) in a solution with a spray volume of 120 L ha^{-1}. The eucalyptus seedlings were transplanted one day after herbicide application, with one plant remaining per experimental unit. Irrigation was carried out using sprinklers, without exceeding the daily simulation limit of 60 mm of rain.

2.3. Chlorophyll Index and Electron Transport Rate

The chlorophyll index was determined using a chlorophyll meter (ChlorofiLOG CFL 1030®) between 9 a.m. and 10 a.m. on fully expanded leaves, at 14 days after planting, and the chlorophyll fluorescence was measured with a portable fluorometer ((MINI model) -PAM II, Walz, Effeltrich, Germany), at 21 days after planting, in expanded and photosynthetically active leaves, using specific leaf support tweezers (model 2030-B). This evaluation was performed at night with at least 30 min of adaptation of the leaves to the dark.

2.4. Height and Dry Mass of Leaves, Stems and Roots

The height of eucalyptus plants was measured with a ruler graduated in centimeters 120 days after planting. Leaves, stems and roots of this plant were conditioned in paper bags and dried in a forced air circulation oven (65 °C) for 48 h. The dry mass was determined on a precision scale.

2.5. Sorghum Bicolor as a Bioindicator Plant

Sorghum bicolor (L.) Moench hybrid BRS 655 (sorghum) was used as a bioindicator plant [32]. This sorghum species was planted in soil samples with known herbicide concentrations (dose–response curve). Indaziflam was applied at doses of 0, 0.25, 0.5, 1, 2, 3, 5, 10, 20, 40 and 60 g ha^{-1}, established in the sorghum sensitivity test to this herbicide [32], in dystrophic red latosol soil samples. Dose–response curves were plotted to evaluate sorghum cultivated in the soil. Ten sorghum seeds were sown, one day after herbicide application, in transparent plastic pots with a volume of 250 cm^3, an area of 50 cm^2, a height of 6 cm and a diameter of 10 cm. The thinning was performed after emergence, leaving six seedlings per pot. Pots under the same cultivation conditions were filled with soil samples from the eucalyptus experiment in order to estimate the residue by comparison with the dose–response curve. The pots were kept in a greenhouse under minimum temperature conditions of 15 °C, maximum of 35 °C and 75% humidity.

Sorghum plant injuries were visually assessed 28 days after sowing (DAS) using a scale from 0 to 100%, with 0% being no symptoms and 100% being plant death [33]. Plant height was measured in centimeters with a ruler. The indaziflam residue adsorbed into the soil was evaluated, simultaneously, in washed sand. The sand (0.6 mm to 2.0 mm) was washed in running water to remove impurities, immersed in an acid solution (10% sulfuric acid) for 24 h, and washed again in running water until the acid residue was removed. The pH was corrected to neutral (7) with the addition of sodium hydroxide solution (NaOH). The sand was dried in the sun on plastic sheeting for 24 h. The indaziflam doses estimated for the sand were 0, 0.05, 0.1, 0.15, 0.25, 0.5, 1, 2, 3, 5, and 10 g ha^{-1} [32]. The sand volume and number of sorghum seeds were the same from the beginning to the end of the trial. The plants were irrigated with a nutrient solution (Table 2).

Table 2. Macro and micronutrients in the nutrient solution for irrigation of *Sorghum bicolor* in sand (CLARK 1975).

Element	Source	Molecular Formula	Amount (mg L^{-1})
N	Urea	CH_4N_2O	9.89
P	Phosphoric acid	H_3PO_4	0.15
K	Potassium chloride	KCl	5.36
Ca	Anhydrous calcium chloride	$CaCl_2$	11.56
Mg	Magnesium chloride	$MgCl_2(6H_2O)$	4.82
S	Sodium sulfate	Na_2SO_4	2.84
B	Boric acid	H_3BO_3	0.05
Cu	Copper chloride	CuCl	0.003
Fe	Iron chloride	$FeCl_3$	0.25
Mn	Manganese chloride	$MnCl_2(4H_2O)$	0.056
Zn	Zinc chloride	$ZnCl_2$	0.011
Mo	Sodium molybdate	Na_2MoO_4	0.0052
EDTA (5.44 g) + 0.824 g de NaOH			

2.6. Statistical Analysis

All analyses were carried out using the R Core Team software version 3.4.3 with the R Studio software. Analysis of variance (ANOVA) using the F test and the Tukey test were used using the ExpDes.pt packages version 1.2.2 [34]. Regression analysis and 3D response surfaces were performed for injury and sorghum plant height. The significance of the coefficients ($p < 0.05$) and the coefficient of determination were considered for the regression models. All statistical analyses were carried out at a 5% significance level.

The dose necessary to reduce the analyzed variable, injury or plant height, by 50% (C_{50}), was calculated for soil and sand, establishing a non-linear, log-logistic regression model with the equation of Seefeldt et al. [35]: $Y = C + D/1 + (X/C_{50})^{-b}$, where C = lower limit of the curve; D = difference between the upper and lower limits of the curve; b = slope of the curve; and C_{50} = curve inflection point corresponding to 50% response. Graphs and C_{50} were generated using SigmaPlot® (version 13.0, 2014, Systat Software, Inc., San Jose, CA, USA).

Indaziflam residue concentration by soil depth, was estimated by the percentage of visual injury of sorghum plants cultivated with soil depths of 0–10, 10–20, 20–30, 30–40 and 40–50 cm and with the C_{50} of the analyzed variable.

The sorption ratio (RS) for indaziflam was calculated from the data obtained from soil C_{50} in relation to sand, RS = C_{50}soil–C_{50}sand/C_{50}sand, which expresses the sorptive capacity of indaziflam into the soil, taking the soil and sand concentrations of the herbicide that inhibit 50% of the indicator plant's development as parameters.

3. Results

The results are outlined in Figure 3. Indaziflam reduced chlorophyll a and b levels, rate of electron transportation and height and dry mass of the eucalyptus plant stem. Indaziflam leached to a depth of 30 cm into clay soil (69% clay) at 121 days after application.

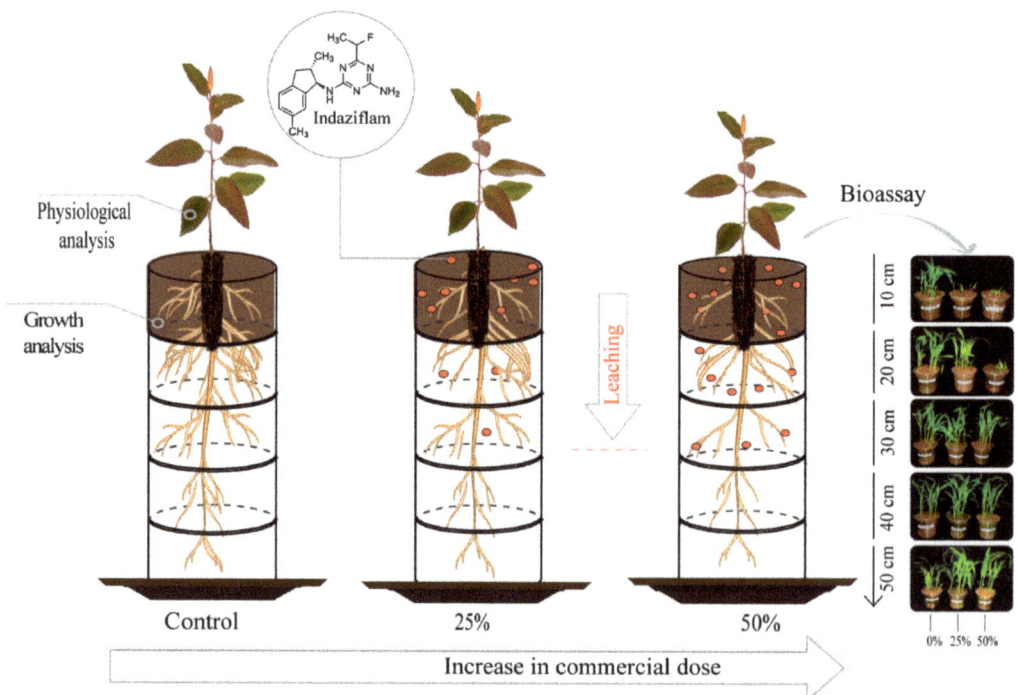

Figure 3. Representative scheme of the effect of indaziflam herbicide residues on the growth of Clone I144 (*Eucalyptus urophylla* × *Eucalyptus grandis*) at different soil depths.

3.1. Eucalyptus Plants

The chlorophyll a content of the Clone I144 was lower at 50% of the commercial indaziflam dose (Figure 4a), and the chlorophyll b content was lower at 25% and 50% of the commercial indaziflam dose, than in the control, 14 days after planting (Figure 4b). The electron transport rate (ETR) of Clone I144 exposed to herbicide residues, was lower at 25% and 50% of the commercial dose 21 days after planting (Figure 4c).

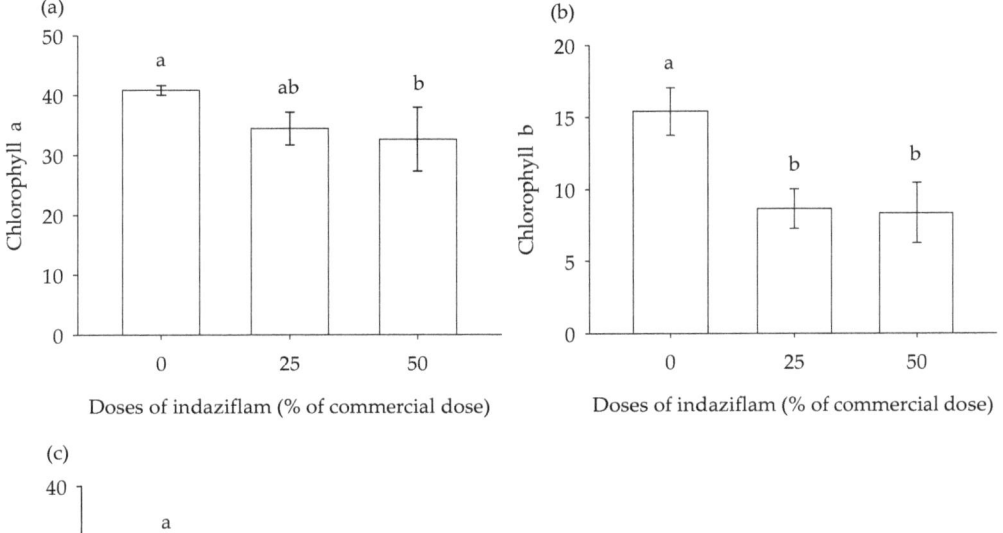

Figure 4. Chlorophyll (**a**,**b**) levels and electron transport rate (ETR) (**c**) of commercial eucalyptus clone I144 at 14 and 21 days after planting in soil contaminated with 25% and 50% of the commercial indaziflam dose (150 g ha^{-1}), respectively. Columns followed by the same lowercase letter, by parameter, do not differ by Tukey's test at 95% probability.

The height of Clone I144 was lower in soil contaminated with 25% and 50% of the commercial indaziflam dose, with a reduction of 12.46% under the effect of 50% of the commercial dose compared to the control (Figure 5).

Stem dry mass of Clone I144 was lower with 25% and 50% of the commercial indaziflam dose than in the control (Figure 6).

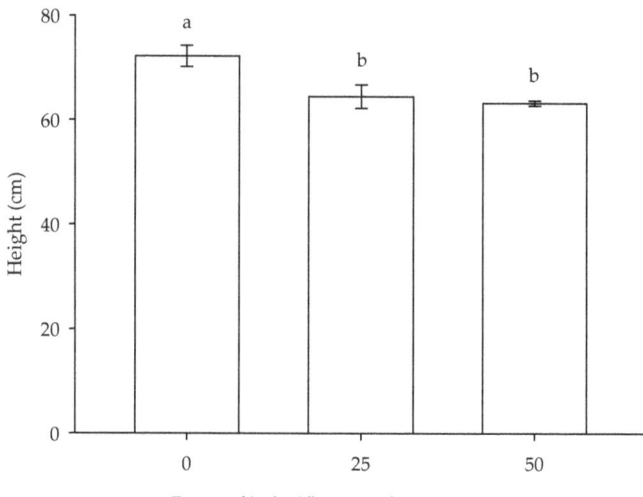

Figure 5. Height (cm) of commercial eucalyptus clone I144, 120 days after planting in soil contaminated with 25% and 50% of the commercial indaziflam dose (150 g ha^{-1}). Columns followed by the same lowercase letter do not differ according to Tukey's test at 95% probability.

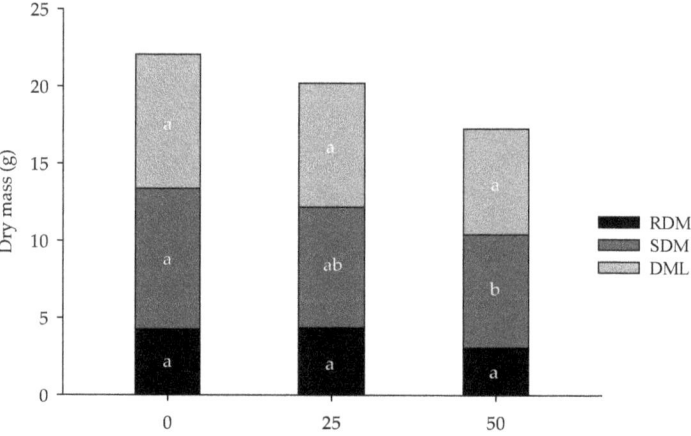

Figure 6. Dry mass of leaves (DML), stem (SDM) and roots (RDM) (g) of commercial eucalyptus clone I144, 120 days after planting in soil contaminated with 25% and 50% of the commercial indaziflam dose (150 g ha^{-1}). Columns followed by the same letter, per variable, do not differ by Tukey's test at 95% probability.

3.2. Sorghum Plants

The injury symptoms were maximal with soil removed at 15 cm and less than 10% with those at depths of 30–40 and 40–50 cm (Figure 7a). Sorghum plant height was lowest in soil contaminated with indaziflam up to 30 cm deep. Sorghum plant height variability was greater as a function of herbicide dose than of soil contamination depth, with the shortest height observed in soil contaminated up to 30 cm after being contaminated with the largest herbicide dose (Figure 7b). Initial symptoms observed in sorghum plants with increasing herbicide dose were leaf tissue reddening, leaf blade chlorosis and reduced growth.

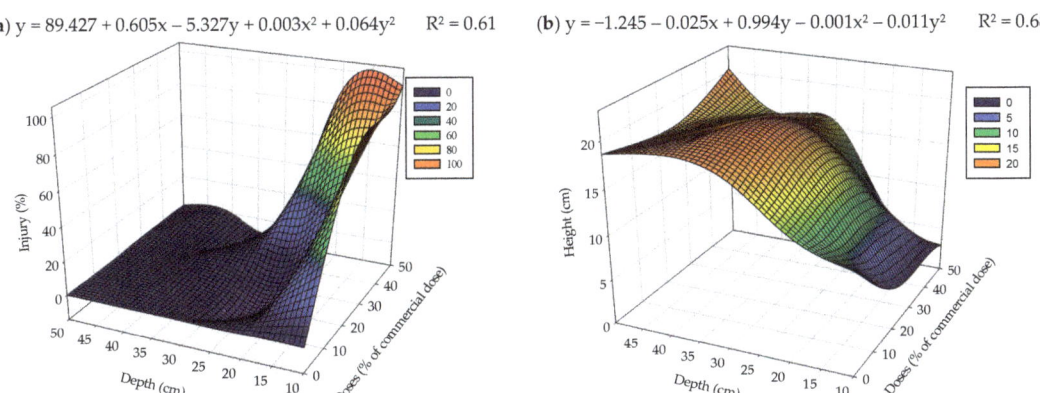

Figure 7. Injury (**a**) and height (**b**) of sorghum plants at 28 DAS as a function of indaziflam dose and soil depth cultivated with eucalyptus for 120 days.

The indaziflam doses necessary to cause 50% injury and reduced sorghum plant height, were 4.65 and 1.71 g ha^{-1} in soil and 0.40 and 0.27 g ha^{-1} in sand, respectively (Figure 8a,b).

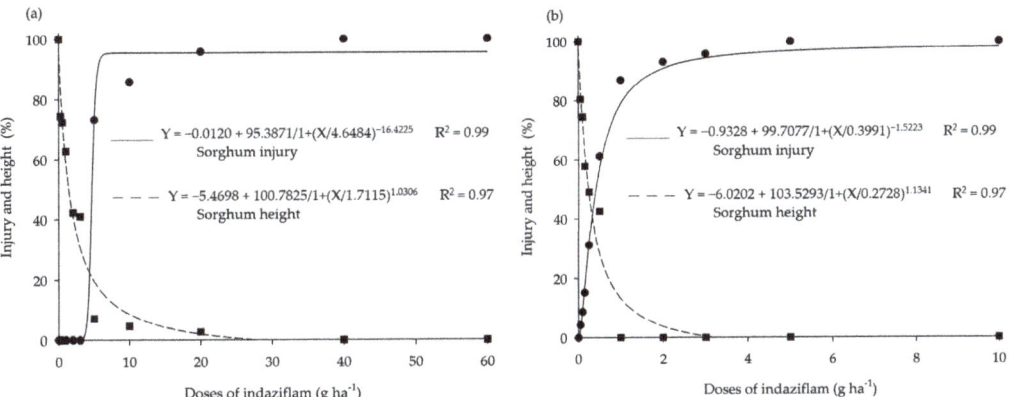

Figure 8. The dose–response curve for sorghum plant injury and height at 28 DAS grown in soil (**a**) and sand (**b**) with indaziflam doses of 0, 0.25, 0.5, 1, 2, 3, 5, 10, 20, 40 and 60 g ha^{-1} and 0, 0.05, 0.1, 0.15, 0.25, 0.5, 1, 2, 3, 5, and 10 g ha^{-1}.

3.3. Indaziflam Soil Residues

The indaziflam soil residues, collected at 0–10, 10–20, 20–30, 30–40 and 40–50 cm depth, with 25% and 50% of the commercial herbicide dose, were 7.79 and 8.72 g ha^{-1}, 5.12 and 7.44 g ha^{-1}, 2.33 and 2.79 g ha^{-1}, 0 and 0 g ha^{-1}, and 0 and 0 g ha^{-1}, respectively (Figure 9).

The sorption ratio (SR) of the herbicide from the data obtained from soil C$_{50}$ in relation to sand was 10.65 in clayey soil.

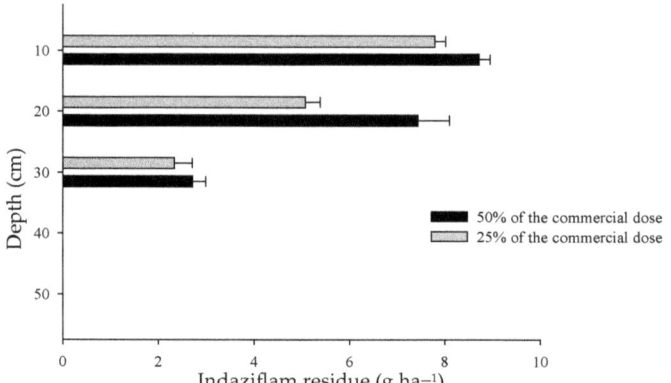

Figure 9. Indaziflam residue in soil samples cultivated with commercial eucalyptus clone I144 120 days after application of 25% and 50% of the recommended commercial dose (150 g ha^{-1}).

4. Discussion

Chemical control is a weed management alternative for forest plantations. Injuries caused by inadequate application, drift or herbicide soil residues are among the main problems reported when chemical control is employed [36]. Some herbicide use for weed control has a residual effect in the soil and can reduce the physiological and growth characteristics of the crop, affecting productivity.

The lower chlorophyll a and b levels at 50%, and 25 and 50%, respectively, of the commercial indaziflam dose can be explained by the indirect interference of the herbicide, affecting the translocation of divalent cations, such as magnesium (Mg) and manganese (Mn) [37], to the meristematic tissues, thereby inhibiting photosynthetic activity. Mg and Mn are essential for photosynthetic light reactions. Previous studies reported that the absorption bands of both chlorophylls a and b were directly related to the emission spectra of Mg [38]. The lower ETR of eucalyptus in soil contaminated with indaziflam is the result of the indirect effect of this herbicide on chlorophyll, reducing the emission of fluorescence signals by plant leaves [37]. ETR is a variable closely correlated with chlorophyll content [39]. Thus, lower chlorophyll levels reduce photon absorption and, consequently, lower ETR to the photosystem II binding site. Although the phytotoxic effects of indaziflam do not require light [40], it has been proposed that, as a cellulose biosynthesis inhibitor, it inhibits photosystem II [41].

The lower height of Clone I144 in soil contaminated with 25% and 50% of the commercial indaziflam dose is related to the action mechanism of this herbicide [32]. Indaziflam inhibits cellulose biosynthesis in plants [42,43], which is considered to be the main source of rigidity and structural support for plant cell walls [44,45]. Several accessory proteins are necessary for cellulose production and deposition, including Cellulose Synthase A (CESA), Korrigan, Cobra and Cellulose Synthase Interacting1 [46]. Loss of function in any of the necessary cellulose synthase subunit proteins causes complete or partial loss of anisotropic growth in expanding cells [40]. Interestingly, all these proteins are potential action sites for herbicides that inhibit cellulose biosynthesis [47]. These effects are seen not only in grasses [48], but symptoms have also been reported in perennial species such as macauba [49], sweet potato [43], *Coffea arabica* cultivar IBC12 [50] and pecan [51]. The lower height of Clone I144 at the highest dose (50% of the commercial dose) demonstrates the plant's susceptibility to lower than commercially recommended doses.

One potential explanation for the lower stem dry mass in the treatment with 50% of the commercial herbicide dose compared to the control is due to the action of the product, which inhibits cellulose biosynthesis, thereby leading to loss of integrity of the primary and secondary cell wall, formed by thin and thick layers of cellulose microfibrils [42]. It has already been reported that indaziflam inhibits the cellulose microfibril cross-linking stage [50], reducing cell formation and consequently, plant dry mass. In trees, the secondary vascular tissues come from the activity of the secondary meristems that promote secondary growth in stem thickness. However, the effect of the herbicide on cellulose microfibril cross-linking and the inhibition of crystal deposition in the cell wall affect cell formation, division and elongation [52], leading to reduced stem dry mass. This may explain why the stem was the element most affected by indaziflam. The results of this study agree with previous research that shows varying lesions on the trunk of pecan plants between three and four months after indaziflam application [51].

Indaziflam leached presenting a residual effect up to a depth of 30 cm, and the symptoms observed in the bioindicator plant demonstrate the presence of the herbicide in numerous soil layers and their high sensitivity to indaziflam. The symptoms observed in sorghum plants, such as chlorosis of the young tissues, reddening of leaf tissue, necrosis and plant death, are characteristic of sensitive species exposed to herbicides that inhibit cellulose biosynthesis [32]. The injury and lower height of the sorghum plants at higher herbicide doses at a depth of up to 30 cm is due to direct herbicide action that inhibits cellulose biosynthesis [42], which can promote polymerization of cellulose from the UDP-glucose substrate by glucosyltransferase and also by inhibiting cell multiplication of other polysaccharides due to nitric acid accumulation [52]. Furthermore, cell division inhibition in meristematic tissue has also been suggested as a secondary mode of action that reduces cell formation and, consequently, plant height [53].

The value of the C_{50} dose of 0.40 g ha^{-1} for damage to sorghum plants grown in sand was also observed in a bioassay study [32]. This is the result of an inert substrate, in which the physical and chemical characteristics, such as the absence of organic matter, surface loads and clay, make it impossible for the herbicide to sorb, resulting in availability of the substrate and absorption by the plant roots [54]. This explains the greater injury and lower height of sorghum plants in sand than in soil.

The sorption ratio (SR) of 10.65 indicates a high amount of adsorbed indaziflam residue. Thus, the sorption ratio evaluated may be directly relate to the high clay content (69 dag kg^{-1}), organic matter (1.88 dag kg^{-1}) and pH (5.00) of the soil used in the study, which are similar to those reported in an experiment with Red-Yellow Latosol with pH (5.1) and Cambissolo with (SR) equal to 10 [32]. Physical and chemical soil characteristics generate different sorption capacities for herbicides, especially mineralogy and organic matter content, which are attributes that are directly involved in the sorption process of these products, as they have three-dimensional sites responsible for the sorption of ionic and non-ionic herbicides that form hydrogen bonds with the herbicides [55]. Herbicides applied pre-emergence and those derived from weak acids, such as indaziflam, are more adsorbed in the soil solution at low pH [56,57].

In this context, soils with a high sand content cause the herbicide to move downwards through the soil profile, due to the greater number of macropores as well as the low clay and organic matter levels [56]. Understanding the behavior and destination of this herbicide in the soil, as well as potential contamination risks stemming from variable soil properties, is important when explaining the possible presence of indaziflam in the planting lines affecting the crop.

5. Conclusions

Eucalyptus Clone I144 was sensitive to the herbicide indaziflam. The herbicide reduced chlorophyll a and b levels, the electron transport rate, and the height and dry mass of the stem of the clone evaluated. It leached to a depth of 30 cm at doses of 37.5 and 75 g ha^{-1}. This is the first report of the effects of indaziflam residue on the physiological and growth characteristics of a eucalyptus clone.

Author Contributions: J.C.M.: conceptualization, formal analysis, investigation, writing—original draft, writing—review and editing. T.S.D.: formal analysis, investigation, writing—original draft, writing—review and editing. A.C.C.: formal analysis, investigation, writing—original draft. B.T.B.A.: formal analysis, investigation, writing—original draft. E.A.F.: conceptualization, methodology, resources, writing—review and editing. J.C.Z.: resources, writing—original draft, writing—review and editing, supervision. B.M.d.C.e.C.: resources, writing—original draft, writing—review and editing, supervision. F.D.d.S.: writing—review and editing. D.V.S.: writing—review and editing, resources. J.B.d.S.: conceptualization, methodology, resources, writing—original draft, writing—review and editing. All authors have read and agreed to the published version of the manuscript.

Funding: This research was funded by Conselho Nacional de Desenvolvimento Científico e Tecnológico (CNPq), Programa MAI DAI UFVJM, Coordenação de Aperfeiçoamento de Pessoal de Nível Superior (CAPES)—Código Financeiro 001 and Fundação de Amparo à Pesquisa do Estado de Minas Gerais (FAPEMIG).

Data Availability Statement: Data is contained within the article.

Acknowledgments: To the "Conselho Nacional de Desenvolvimento Científico e Tecnológico (CNPq)", "Coordenação de Aperfeiçoamento de Pessoal de Nível Superior (CAPES)—Código Financeiro 001" and "Fundação de Amparo à Pesquisa do Estado de Minas Gerais (FAPEMIG)" and "Programa Cooperativo sobre Proteção Florestal (PROTEF) do Instituto de Pesquisas e Estudos Florestais (IPEF)" for financial support. Phillip John Villani (University of Melbourne, Australia), a professional editor and proofreader and native English speaker, has reviewed and edited this article for structure, grammar, punctuation, spelling, word choice, and readability.

Conflicts of Interest: The authors declare that they have no conflict of interest.

References

1. Lock, P.; Legg, P.; Whittle, L.; Black, S. *Global Outlook for Wood Markets to 2030: Projections of Future Production, Consumption and Trade Balance*; Australian Bureau of Agricultural and Resource Economics and Sciences: Canberra, Australia, 2021.
2. Payn, T.; Carnus, J.M.; Freer-Smith, P.; Kimberley, M.; Kollert, W.; Liu, S.; Wingfield, M.J. Changes in planted forests and future global implications. *For. Ecol. Manag.* **2015**, *352*, 57–67. [CrossRef]
3. Fernandes, B.C.C.; Mendes, K.F.; Júnior, A.F.D.; Caldeira, V.P.S.; Teófilo, T.M.S.; Silva, T.S.; Mendonça, V.; Souza, M.F.; Silva, D.V. Impact of pyrolysis temperature on the properties of eucalyptus wood-derived biochar. *Materials* **2020**, *13*, 5841. [CrossRef] [PubMed]
4. Nepal, P.; Korhonen, J.; Prestemon, J.P.; Cubbage, F.W. Projecting global planted forest area developments and the associated impacts on global forest product markets. *J. Environ. Manag.* **2019**, *240*, 421–430. [CrossRef] [PubMed]
5. Ferreto, D.O.C.; Reichert, J.M.; Cavalcante, R.B.L.; Srinivasan, R. Water budget fluxes in catchments under grassland and Eucalyptus plantations of different ages. *Can. J. For. Res.* **2021**, *51*, 513–523. [CrossRef]
6. Martins, F.B.; Benassi, R.B.; Torres, R.R.; de Brito Neto, F.A. Impacts of 1.5 C and 2 C global warming on Eucalyptus plantations in South America. *Sci. Total Environ.* **2022**, *825*, 153820. [CrossRef]
7. Elli, E.F.; Sentelhas, P.C.; Huth, N.; Carneiro, R.L.; Alvares, C.A. Gauging the effects of climate variability on Eucalyptus plantations productivity across Brazil: A process-based modelling approach. *Ecol. Indic.* **2020**, *114*, 106325. [CrossRef]
8. Wen, Y.; Zhou, X.; Yu, S.; Zhu, H. The predicament and countermeasures of development of global Eucalyptus plantations. *Guangxi Sci.* **2018**, *25*, 107–116. [CrossRef]
9. Zhang, C.; Xiao, X.; Zhao, L.; Qin, Y.; Doughty, R.; Wang, X.; Yang, X. Mapping Eucalyptus plantation in Guangxi, China by using knowledge-based algorithms and PALSAR-2, Sentinel-2, and Landsat images in 2020. *Int. J. Appl. Earth Obs. Geoinf.* **2023**, *120*, 103348. [CrossRef]
10. Ibá. Indústria Brasileira de Árvores. 2022. Available online: https://www.iba.org/publicacoes (accessed on 5 July 2023).
11. Avisar, D.; Azulay, S.; Bombonato, L.; Carvalho, D.; Dallapicolla, H.; de Souza, C.; Silva, W. Safety Assessment of the CP4 EPSPS and NPTII Proteins in Eucalyptus. *GM Crops Food* **2023**, *14*, 1–14. [CrossRef]
12. Florêncio, G.W.L.; Martins, F.B.; Fagundes, F.F.A. Climate change on Eucalyptus plantations and adaptive measures for sustainable forestry development across Brazil. *Ind. Crops Prod.* **2022**, *188*, 115538. [CrossRef]
13. Hakamada, R.E.; Hubbard, R.M.; Stape, J.L.; de Paula Lima, W.; Moreira, G.G.; de Barros Ferraz, S.F. Stocking effects on seasonal tree transpiration and ecosystem water balance in a fast-growing Eucalyptus plantation in Brazil. *For. Ecol. Manag.* **2020**, *466*, 118149. [CrossRef]
14. Manzato, B.L.; Manzato, C.L.; Dos Santos, P.L.; Passos, J.D.S.; Da Silva Junior, T.A.F. Diversity of macroscopic basidiomycetes in reforestation areas of *Eucalyptus* spp. *Sci. For.* **2020**, *48*, e3305. [CrossRef]
15. Hakamada, R.; da Silva, R.M.L.; Moreira, G.G.; Teixeira, J.D.S.; Takahashi, S.; Masson, M.V.; Martins, S.D.S. Growth and canopy traits affected by myrtle rust (*Austropuccinia psidii* Winter) in *Eucalyptus grandis* x *Eucalyptus urophylla*. *For. Pathol.* **2022**, *52*, e12736. [CrossRef]

16. Hutapea, F.J.; Weston, C.J.; Mendham, D.; Volkova, L. Sustainable management of *Eucalyptus pellita* plantations: A review. *For. Ecol. Manag.* **2023**, *537*, 120941. [CrossRef]
17. Braga, A.F.; Barroso, A.A.M.; Amaral, C.L.; Nepomuceno, M.P.; Alves, P.L.C.A. Population interference of glyphosate resistant and susceptible ryegrass on eucalyptus initial development. *Planta Daninha* **2018**, *36*, e018170148. [CrossRef]
18. Deng, Y.; Yang, G.; Xie, Z.; Yu, J.; Jiang, D.; Huang, Z. Effects of different weeding methods on the biomass of vegetation and soil evaporation in eucalyptus plantations. *Sustainability* **2020**, *12*, 3669. [CrossRef]
19. Smethurst, P.J.; Valadares, R.V.; Huth, N.I.; Almeida, A.C.; Elli, E.F.; Neves, J.C. Generalized model for plantation production of Eucalyptus grandis and hybrids for genotype-site-management applications. *For. Ecol. Manag.* **2020**, *469*, 118164. [CrossRef]
20. Silva, T.S.; Freitas, S.M.; Teófilo, T.M.S.; Santos, M.S.; Porto, M.A.F.; Souza, C.M.M.; Santos, J.B.; Silva, D.V. Use of neural networks to estimate the sorption and desorption coefficients of herbicides: A case study of diuron, hexazinone, and sulfometuron-methyl in Brazil. *Chemosphere* **2019**, *236*, 124333. [CrossRef]
21. De Carvalho, L.B.; Duke, S.O.; Alves, P.D.C. Physiological responses of *Eucalyptus* × *urograndis* to glyphosate are dependent on the genotype. *Sci. For.* **2018**, *46*, 177–187. [CrossRef]
22. Minogue, P.J.; Osiecka, A.; Lauer, D.K. Selective herbicides for establishment of *Eucalyptus benthamii* plantations. *New For.* **2018**, *49*, 529–550. [CrossRef]
23. Agrofit-Sistema de Agrotóxicos Fitossanitários. 2023. Available online: http://agrofit.agricultura.gov.br/agrofit_cons/principal_agrofit_cons (accessed on 13 February 2023).
24. Teófilo, T.M.S.; Mendes, K.F.; Fernandes, B.C.C.; Oliveira, F.S.; Silva, T.S.; Takeshita, V.; Souza, M.F.; Tornisielo, V.L.; Silva, D.V. Phytoextraction of diuron, hexazinone, and sulfometuron-methyl from the soil by green manure species. *Chemosphere* **2020**, *256*, 127059. [CrossRef]
25. Agostinetto, D.; Tarouco, C.P.; Markus, C.; Oliveira, E.D.; Da Silva, J.M.B.V.; Tironi, S.P. Selectivity of eucalyptus genotypes to herbicides rates. *Semin. Ciências Agrárias* **2010**, *31*, 585–598. [CrossRef]
26. De Abreu, K.M.; de Castro Santos, D.; Pennacchi, J.P.; Calil, F.N.; Moura, T.M.; Alves, E.M.; de Souza, S.O. Differential tolerance of four tree species to glyphosate and mesotrione used in agrosilvopastoral systems. *New For.* **2022**, *53*, 831–850. [CrossRef]
27. Tiburcio, R.A.S.; Ferreira, F.A.; Paes, F.A.S.V.; Melo, C.A.D.; Medeiros, W.N. Growth of eucalyptus clones seedlings submitted to simulated drift of different herbicides. *Rev. Árvore* **2012**, *36*, 65–73. [CrossRef]
28. Basinger, N.T.; Jennings, K.M.; Monks, D.W.; Mitchem, W.E. Effect of rate and timing of indaziflam on 'Sunbelt' and muscadine grape. *Weed Technol.* **2019**, *33*, 380–385. [CrossRef]
29. Grey, T.L.; Rucker, K.; Webster, T.M.; Luo, X. High-density plantings of olive trees are tolerant to repeated applications of indaziflam. *Weed Sci.* **2016**, *64*, 766–771. [CrossRef]
30. Brosnan, J.T.; Breeden, G.K.; McCullough, P.E.; Henry, G.M. Pre and post control of annual bluegrass (*Poa annua*) with indaziflam. *Weed Technol.* **2012**, *26*, 48–53. [CrossRef]
31. Alonso, D.G.; Oliveira, R.S.D.; Koskinen, W.C.; Hall, K.; Constantin, J.; Mislankar, S. Sorption and desorption of indaziflam degradates in several agricultural soils. *Sci. Agrícola* **2016**, *73*, 169–176. [CrossRef]
32. Gonçalves, V.A.; Ferreira, L.R.; Teixeira, M.F.F.; De Freitas, F.C.L.; D'Antonino, L. Sorption of indaziflam in brazilian soils with different pH values. *Rev. Caatinga* **2021**, *34*, 494–504. [CrossRef]
33. SBCPD-Sociedade Brasileira da Ciência das Plantas Daninhas. *Procedimentos Para Instalação, Avaliação e Análise de Experimentos com Herbicidas*; SBCPD-Sociedade Brasileira da Ciência das Plantas Daninhas: Londrina, PR, Brazil, 1995.
34. Ferreira, E.B.; Cavalcanti, P.P.; Nogueira, D.A. ExpDes.pt: Experimental Designs Package. 2013. Available online: http://cran.r-project.org/package=ExpDes.pt (accessed on 25 July 2023).
35. Seefeldt, S.S.; Jensen, J.E.; Fuerst, E.P. Log-Logistic analysis of herbicide dose-response relationships. *Weed Technol.* **1995**, *9*, 218–227. [CrossRef]
36. Rabelo, J.S.; Dos Santos, E.A.; de Melo, E.I.; Vaz, M.G.M.V.; de Oliveira Mendes, G. Tolerance of microorganisms to residual herbicides found in eucalyptus plantations. *Chemosphere* **2023**, *329*, 138630. [CrossRef]
37. Jones, P.A.; Brosnan, J.T.; Kopsell, D.A.; Armel, G.R.; Breeden, G.K. Preemergence herbicides affect hybrid bermudagrass nutrient content. *J. Plant Nutr.* **2015**, *38*, 177–188. [CrossRef]
38. Levitt, L.S. The role of magnesium in photosynthesis. *Science* **1954**, *120*, 33–135. [CrossRef] [PubMed]
39. Najafpour, M.M.; Zaharieva, I.; Zand, Z.; Hosseini, S.M.; Kouzmanova, M.; Hołyńska, M.; Allakhverdiev, S.I. Water-oxidizing complex in Photosystem II: Its structure and relation to manganese-oxide based catalysts. *Coord. Chem. Rev.* **2020**, *409*, 213183. [CrossRef]
40. Brabham, C.; Lei, L.; Gu, Y.; Stork, J.; Barrett, M.; DeBolt, S. Indaziflam herbicidal action: A potent cellulose biosynthesis inhibitor. *Plant Physiol.* **2014**, *166*, 1177–1185. [CrossRef] [PubMed]
41. Meyer, D.F. Indaziflam-A New Herbicide for Pre-Emergent Control of Grasses and Broadleaf Weeds for Turf and Ornamentals. Meeting Abstracts. 2019. Available online: http://wssa.net/meeting/ (accessed on 20 June 2023).
42. Sebastian, D.J.; Fleming, M.B.; Patterson, E.L.; Sebastian, J.R.; Nissen, S.J. Indaziflam: A new cellulose-biosynthesis-inhibiting herbicide provides long-term control of invasive winter annual grasses. *Pest Manag. Sci.* **2017**, *73*, 2149–2162. [CrossRef]
43. Smith, S.C.; Jennings, K.M.; Monks, D.W.; Jordan, D.L.; Reberg-Horton, S.C.; Schwarz, M.R. Sweetpotato tolerance and Palmer Amaranth control with indaziflam. *Weed Technol.* **2022**, *36*, 202–206. [CrossRef]
44. Jarvis, M.C. Cellulose biosynthesis: Counting the chains. *Plant Physiol.* **2013**, *163*, 1485–1486. [CrossRef]

45. Hu, Z.; Zhang, T.; Rombaut, D.; Decaestecker, W.; Xing, A.; D'Haeyer, S.; De Veylder, L. Genome editing-based engineering of CESA3 dual cellulose-inhibitor-resistant plants. *Plant Physiol.* **2019**, *180*, 827–836. [CrossRef]
46. Gu, Y.; Kaplinsky, N.; Bringmann, M.; Cobb, A.; Carroll, A.; Sampathkumar, A.; Somerville, C.R. Identification of a cellulose synthase-associated protein required for cellulose biosynthesis. *Proc. Natl. Acad. Sci. USA* **2010**, *107*, 12866–12871. [CrossRef]
47. Tateno, M.; Brabham, C.; DeBolt, S. Cellulose biosynthesis inhibitors—A multifunctional toolbox. *J. Exp. Bot.* **2016**, *67*, 533–542. [CrossRef]
48. Davies, K.W.; Boyd, C.S.; Baughman, O.W.; Clenet, D.R. Effects of Using Indaziflam and Activated Carbon Seed Technology in Efforts to Increase Perennials in *Ventenata dubia*–Invaded Rangelands. *Rangel. Ecol. Manag.* **2023**, *88*, 70–76. [CrossRef]
49. Da Costa, Y.K.S.; De Freitas, F.C.L.; Da Silveira, H.M.; Nascimento, R.S.M.; Sediyama, C.S.; Alcantara-de la Cruz, R. Herbicide selectivity on Macauba seedlings and weed control efficiency. *Ind. Crops Prod.* **2020**, *154*, 112725. [CrossRef]
50. Gomes, C.A.; Pucci, L.F.; Alves, D.P.; Leandro, V.A.; Pereira, G.A.M.; Reis, M.R.D. Indaziflam application in newly transplanted arabica coffee seedlings. *Coffee Sci.* **2019**, *14*, 373–381. Available online: http://www.sbicafe.ufv.br/handle/123456789/12516 (accessed on 5 July 2023). [CrossRef]
51. González-Delgado, A.M.; Ashigh, J.; Shukla, M.K.; Perkins, R. Mobility of indaziflam influenced by soil properties in a semi-arid area. *PLoS ONE* **2015**, *10*, e0126100. [CrossRef]
52. Diasa, R.C.; Gomes, D.M.; Anunciato, V.M.; Bianchi, L.; Simões, P.S.; Carbonari, C.A.; Velini, E.D. Selection of bioindicator species for the herbicide indaziflam. *Rev. Bras. Herbic.* **2019**, *18*, 1–11. [CrossRef]
53. U.S. EPA. Pesticide Fact Sheet for Indaziflam. 2010. Available online: http://www.epa.gov/opprd001/factsheets/indaziflam.pdf (accessed on 22 July 2023).
54. Jhala, A.J.; Singh, M. Leaching of indaziflam compared with residual herbicides commonly used in Florida citrus. *Weed Technol.* **2012**, *26*, 602–607. [CrossRef]
55. Schneider, J.G.; Haguewood, J.B.; Song, E.; Pan, X.; Rutledge, J.M.; Monke, B.J.; Xiong, X. Indaziflam effect on bermudagrass (*Cynodon dactylon* L. Pers.) shoot growth and root initiation as influenced by soil texture and organic matter. *Crop Sci.* **2015**, *55*, 429–436. [CrossRef]
56. Alonso, D.G.; Oliveira Jr, R.S.; Hall, K.E.; Koskinen, W.C.; Constantin, J.; Mislankar, S. Changes in sorption of indaziflam and three transformation products in soil with aging. *Geoderma* **2015**, *239*, 250–256. [CrossRef]
57. Mendes, K.F.; Furtado, I.F.; Sousa, R.N.D.; Lima, A.D.C.; Mielke, K.C.; Brochado, M.G.D.S. Cow bonechar decreases indaziflam pre-emergence herbicidal activity in tropical soil. *J. Environ. Sci. Health Part B* **2021**, *56*, 532–539. [CrossRef]

Disclaimer/Publisher's Note: The statements, opinions and data contained in all publications are solely those of the individual author(s) and contributor(s) and not of MDPI and/or the editor(s). MDPI and/or the editor(s) disclaim responsibility for any injury to people or property resulting from any ideas, methods, instructions or products referred to in the content.

Article

Effects of the Larch–Ashtree Mixed Forest on Contents of Secondary Metabolites in *Larix olgensis*

Hong Jiang [1,2,3], Shanchun Yan [1,2], Zhaojun Meng [1,2,*], Shen Zhao [3], Dun Jiang [1,2] and Peng Li [3,*]

1 School of Forestry, Northeast Forestry University, Harbin 150040, China; yanshanchun@nefu.edu.cn (S.Y.)
2 Key Laboratory of Sustainable Forest Ecosystem Management-Ministry of Education, Northeast Forestry University, Harbin 150040, China
3 Heilongjiang Academy of Land Reclamation Sciences, Harbin150038, China
* Correspondence: mengzj2018@nefu.edu.cn (Z.M.); liuzhihua@chinabdh.com (P.L.)

Abstract: To understand the insect resistance mechanism of the larch, *Larix olgensis*, in a mixed forest, larch (*Larix olgensis*) seedlings and ashtree (*Fraxinus mandshurica*) seedlings were planted with mixed banding forests in the proportion of 1:1 ($BMF_{1:1}$), 3:3 ($BMF_{3:3}$) and 5:5 ($BMF_{5:5}$), in pots and in the field. One year later, the content of secondary metabolites in the needles of each larch treatment were tested with an ultraviolet spectrophotometer. The results showed that the allelopathic effect of *F. mandshuricas* (ashtree) on *L. olgensis* (larch) could increase the content of secondary metabolites in larch needles. It was found that the flavonoid content in the needles of $BMF_{5:5}$ was higher than that in the needles of $BMF_{1:1}$ and $BMF_{3:3}$ ($p < 0.05$). The tannin content in the needles of $FBMF_{3:3}$ and $FBMF_{5:5}$ was significantly higher than that of $FBMF_{1:1}$, whereas the tannin content in the needles of $PBMF_{3:3}$ reached 1.27 mg/g, which was the highest ($p < 0.05$). The lignin content in the needles of $FBMF_{3:3}$ reached 2.27 mg/g, which was significantly more increased than that in the control group in a dose-dependent manner, while that in the needles of $PBMF_{3:3}$ and $PBMF_{5:5}$ was higher than that in the needles of $PBMF_{1:1}$ ($p < 0.05$). The tannin and lignin content in the needles of FBMF was higher than that of PBMF. However, there was no difference in the content of flavonoids in the needles of FBMF and PBMF. These results suggest that banding mixed larches and ashtrees can significantly increase the content of secondary metabolites (phenolic compounds) in the needles of *L. olgensis* and improve its chemical defense, and the allelopathic effect of ashtrees on larches is related to the mixed proportion. Thus, the effect of mixed banding forests in the proportion of 3:3 and 5:5 is better.

Keywords: *Larix olgensis*; *Fraxinus mandshurica*; banding mixed forest; secondary metabolites; allelopathy

Citation: Jiang, H.; Yan, S.; Meng, Z.; Zhao, S.; Jiang, D.; Li, P. Effects of the Larch–Ashtree Mixed Forest on Contents of Secondary Metabolites in *Larix olgensis*. Forests 2023, 14, 871. https://doi.org/10.3390/f14050871

Academic Editor: Ido Izhaki

Received: 31 January 2023
Revised: 29 March 2023
Accepted: 30 March 2023
Published: 24 April 2023

Copyright: © 2023 by the authors. Licensee MDPI, Basel, Switzerland. This article is an open access article distributed under the terms and conditions of the Creative Commons Attribution (CC BY) license (https:// creativecommons.org/licenses/by/ 4.0/).

1. Introduction

Plant interaction is one of the fundamental scientific problems in ecological research [1], in which the chemical interactions among and within plants have been widely and deeply studied. Plant allelopathy is a natural ecological phenomenon, in which a plant releases chemicals to affect another plant. It is a chemical response strategy of a plant to the same or different plants that coexist with it [2]. In the face of animal feeding, microbial infection and other plant competition, plants often respond by synthesizing and releasing secondary metabolites. They adjust their biomass distribution by identifying the information of adjacent species, thus deciding whether to adopt chemical defense strategies or not [3,4]. Secondary metabolites result from the interaction between plants and their living environment during long-term evolution. They are products of complex branching metabolic pathways, which determine the color, smell and taste of plants [5]. Although many secondary metabolites do not participate in the metabolism of plants, they can suppress the digestion and utilization of food by herbivorous insects and then interfere with their mating behavior, attracting natural enemies. As an important physiological indicator of performance, they play an essential role in the process of plants resisting insect invasion [6,7].

As the content of secondary metabolites in plant tissue cell walls increases, the number of nutrients, such as proteins and sugars, obtained by phytophagous insects decreases [8,9]. Phenolic compounds in plant secondary metabolites are essential chemical defense substances in plant insect resistance [10], and there is a close relationship between the composition and the content of such phenolic compounds, which are a class of substances with complex genes synthesized by the shikimate acid pathway and the malonate acid pathway, including flavonoids, tannins, lignin and other substances [11,12]. Tannin is a highly polymerized polyphenol compound in polyphenols that is usually divided into condensed tannins and hydrolyzable tannins. It can combine with proteins and digestive enzymes to form a complex compound insoluble in water that interferes with insect food utilization [13,14]. Liu XX et al. [15] added different mass fractions of tannins to an artificial diet to feed *Hyphantria cunea* larvae and found that tannins had a significant inhibitory effect on the food utilization of *Hyphantria cunea*. Flavonoids are also important phenolic compounds in the process of plant insect resistance. They exist in plants through the shikimic acid–phenylpropane metabolic pathway. They not only have antioxidant activity, but also increase the metabolic burden of insects and affect the normal life activities of insects, e.g., soybean, which can synthesize flavonoids to inhibit the feeding harm of lepidopteran larvae [16]. Lignin is a complex phenolic polymer filled in the cellulose framework and formed of three alcohol monomers. Its metabolic pathways intersect with those of other plant secondary metabolites. Its metabolism is closely related to plant disease resistance, insect resistance, waterlogging resistance and cold resistance, and other forms of stress resistance physiology have certain correlations [17]. After being ingested by insects, lignin can reduce the efficiency of insects' food utilization [18].

At present, pure artificial forests with a single tree species in forestry production systems are considered to be typical representatives of simplified ecosystems, which are usually sensitive, easily disturbed, and even have outbreaks of insect pests. Therefore, increasing the diversity of tree species can improve the resistance of trees to pests [19]. In mixed forests, the interactions among the same or different trees affect the function of the forest ecosystem through competition, predation, parasitism and mutualism [20]. According to research, *Ostryopsis davidiana* can promote the growth of *Pinus tabulaeformis* [21]. The content of phenolic acid in the rhizosphere soil of a mixed forest of *P. tabulaeformis* and *O. davidiana* is significantly lower than that of pure forest *P. tabulaeformis*, which reduces the autotoxic effect caused by phenolic acid content that is too high in the soil of a pure forest of *P. tabulaeformis*. The author's previous research found the 30-year or 20-year growth of *L. olgensis* (larch)-*F. mandshurica* (ashtree) banding mixed forests can significantly enhance the activity of the defense proteins in larch needles, thus improving the resistance of larch to phytophagous insects and their chemical defense [22]. In this study, the allelopathy of *L. olgensis* (larch)-*F. mandshurica* (ashtree) banding mixed forests on larch young trees was studied by measuring the content of the primary secondary metabolites (tannin, lignin and flavonoids) in larch needles, and the different allelopathy of banding mixed ashtree on larch young trees was compared by the pot experiment and the field experiment, which illuminated the allelopathic mechanism of the mixed banding forest in improving the insect resistance of larches. The results lay a theoretical foundation for applying forest management measures to control forest defoliators, such as *Dendrolimus superans*.

2. Materials and Methods

2.1. Mixed Mode Setting of L. olgensis and F. mandshurica

In mid-April 2015, two-year-old larch seedlings and one-year-old ashtree seedlings were planted in pots (23 × 23 × 25 cm) and in the field at the Maoershan experimental forestry farm of Northeast Forestry University, Shangzhi, Heilongjiang Province, P.R. China. Larch and ashtree seedlings were planted with mixed banding forests in the proportion of 1:1, 3:3 and 5:5, in pots and in the field, and then larch pure forest was used as the control (Table 1).

The distance between one mixed mode and the other was 4 m, and the distance between the two lines of seedlings was 25 cm (Figure 1). The seedlings were covered with gauze to avoid the occurrence of diseases, insect pests and human-caused mechanical damage.

Table 1. Two planting modes and different banding mixed proportions (four samples each, $n = 4$).

Planting Modes	Banding Mixed Modes	Abbreviation Code
planting in the field	larch pure forests	FLPF
	larch–ashtree banding mixed forests in the proportion of 1:1	$FBMF_{1:1}$
	larch–ashtree banding mixed forests in the proportion of 3:3	$FBMF_{3:3}$
	larch–ashtree banding mixed forests in the proportion of 5:5	$FBMF_{5:5}$
planting in pots	larch pure forests	PLPF
	larch–ashtree banding mixed forests in the proportion of 1:1	$PBMF_{1:1}$
	larch–ashtree banding mixed forests in the proportion of 3:3	$PBMF_{3:3}$
	larch–ashtree banding mixed forests in the proportion of 5:5	$PBMF_{5:5}$

Figure 1. Larch and ashtree young trees planted with mixed banding forests in the field and in pots.

2.2. Collection of Larch Needles

Larch needles were collected on 22 July, 1 August, 12 August, 22 August and 1 September. There were 4 repetitions for each treatment and about 30 g of needles for each repetition. Then, those needles were stored at $-40\ °C$ in a freezer for the sample testing.

2.3. Determination of Secondary Metabolites Content

2.3.1. Determination of Tannin Content

Tannin in needles was extracted and identified according to the method described by Yan S.C. et al. [23]. Then, 5 g of frozen needles of each treatment were frozen to a constant weight by a freeze dryer. Next, 1 g of dry needles were homogenized using a mortar and pestle, and then the powder was placed into a 20 mL screw-cap centrifuge tube with 10 mL of 95% ethanol and extracted for 24 h at $-20\ °C$. The mixture was centrifuged for 10 min at 10,000 rpm and at $4\ °C$. Then, 1 mL of supernatant was added to a test tube containing 9 mL of 70% ethanol for the mixture solution, 0.5 mL mixture solution was added to an aluminum foil-covered test tube with 3 mL 4% vanillin of ethanol and 1.5 mL concentrated hydrochloric acid, and the tubes were heated at 20 °C for 20 min in the water bath. The absorbance was measured at 510 nm, and 70% ethanol was the blank control.

2.3.2. Determination of Flavonoid Content

The extraction and identification of flavonoid were performed according to the method described by Jiang et al. [24], with modifications. Needles were extracted with 50 mL of 95% aqueous methanol on a shaker for 24 h and then extracted by ultrasonic extraction for 2 h. The extracted liquid was centrifuged for 15 min at 14,000 rpm. Next, 1 mL of supernatant was added to a test tube containing 4 mL of water, and then 0.5 mL of 50% $NaNO_2$ was added to the test tube. After setting for 6 min, 0.5 mL of 10% $AlCl_3$ was added to the tube. The mixture in the tube was set for 6 min again, and 4 mL of 4% NaOH was added to the tube. The absorbance was measured at 510 nm, and 95% aqueous was the blank control. The flavonoid content was expressed with milligrams of rutin equivalents per gram of fresh leaf weight.

2.3.3. Determination of Lignin Content

Lignin in the needles was extracted and identified according to the method described by Ren et al. [25,26]. First, 0.5 g of needles in 2 mL of 95% ethanol were homogenized using a mortar and pestle and then were centrifuged for 10 min at 4500 rpm. The residues were washed by successive stirring and centrifugation—five times with 95% methanol and three times with 1:2 mixtures of ethanol and n-hexane (v/v)—and then were dried in an oven at 60 °C overnight. All dry residues were placed into a 10 mL screw-cap centrifuge tube with 1 mL of glacial acetic acid containing 25% acetyl bromide. While the tubes were heated at 70 °C for 30 min in the water bath, the reaction was immediately stopped by adding 0.9 mL of 2 M NaOH solution, and then 5 mL of glacial acetic acid and 0.1 mL 7.5 M hydroxylammonium chloride were added. The mixture was centrifuged for 5 min at 4500 rpm, and the absorption of the supernatant after dilution with glacial acetic acid was determined at 280 nm, and 95% ethanol was the blank control. The lignin content was expressed as OD/g (optical density, OD) at the weight of the fresh needles.

2.4. Statistical Analysis

All the data were analyzed with SPSS 19.0 for Windows. The contents data of secondary metabolites in the pot and field treatments were analyzed by one-way analysis of variance (ANOVA), followed by least significant difference (Bonferroni), multiple comparisons at $\alpha = 0.05$. Two-tailed t-tests of independent samples were used for comparing the contents data of tannin, flavonoid and lignin in the same mixed proportion treatments in planting pots and the field, whose data were log-transformed to achieve variance homogeneity and normal distribution.

3. Results

3.1. Effect of the Banding Mixed Forest in the Field on Content of Secondary Metabolites in Needles of L. olgensis

Except for 22 July (F = 0.97, df_1 = 3, df_2 = 8, $p > 0.05$), the tannin content of needles in FBMF was significantly higher than that in FLPF. On 1 August and 12 August the tannin content in needles of $FBMF_{3:3}$ was significantly higher than that of $FBMF_{1:1}$ (F = 27.10, df_1 = 3, df_2 = 8, $p < 0.05$ on 1 August; F = 39.73, df_1 = 3, df_2 = 8, $p < 0.05$ on 12 August), but there was no difference between $FBMF_{3:3}$ and $FBMF_{5:5}$. On 22 August, the tannin content in needles of $FBMF_{3:3}$ was significantly higher than that of $FBMF_{1:1}$ and $FBMF_{5:5}$, while that of $FBMF_{5:5}$ was significantly higher than that of $FBMF_{1:1}$ (F = 243.26, df_1 = 3, df_2 = 8, $p < 0.05$). On 1 September, the tannin content in needles of $FBMF_{3:3}$ was significantly higher than that of $FBMF_{1:1}$ (F = 33.06, df_1 = 3, df_2 = 8, $p < 0.05$), but that of $FBMF_{5:5}$ was not different from $FBMF_{3:3}$ and $FBMF_{1:1}$ (Figure 2A). The results indicated that the allelopathic effect of ashtrees on larches in FBMF could increase the tannin content in needles, and then the tannin content in needles of $FBMF_{3:3}$ and $FBMF_{5:5}$ significantly increased.

The flavonoids content of needles in FBMF was higher than that in FLPF. On 1 August, the flavonoids content of needles in FBMF$_{5:5}$ was significantly higher than that in FBMF$_{1:1}$, FBMF$_{3:3}$ and FLPF (F = 4.76, df$_1$ = 3, df$_2$ = 8, p < 0.05), but there was no difference in FBMF$_{1:1}$, FBMF$_{3:3}$ and FLPF. On 12 August and 22 August, the flavonoids content of needles in FBMF$_{5:5}$ and FBMF$_{3:3}$ was significantly higher than that in FLPF (F = 13.40, df$_1$ = 3, df$_2$ = 8, p < 0.05 on 12 August; F = 22.16, df$_1$ = 3, df$_2$ = 8, p < 0.05 on 22 August), while that in FBMF$_{5:5}$ was significantly higher than that in FBMF$_{1:1}$ on 22 August, but there was no difference between that in FBMF$_{3:3}$ and FBMF$_{1:1}$. On 1 September, the flavonoids content of needles in FBMF$_{5:5}$ was significantly higher than that in and FLPF (F = 7.41, df$_1$ = 3, df$_2$ = 8, p < 0.05), but there was no difference between that in FBMF5:5 and FBMF1:1 or FBMF3:3 (Figure 2B). The results showed that ashtree promoted the flavonoids compound in needles, and the flavonoids content of needles in FBMF$_{5:5}$ was higher than that in FBMF$_{1:1}$ and FBMF$_{3:3}$.

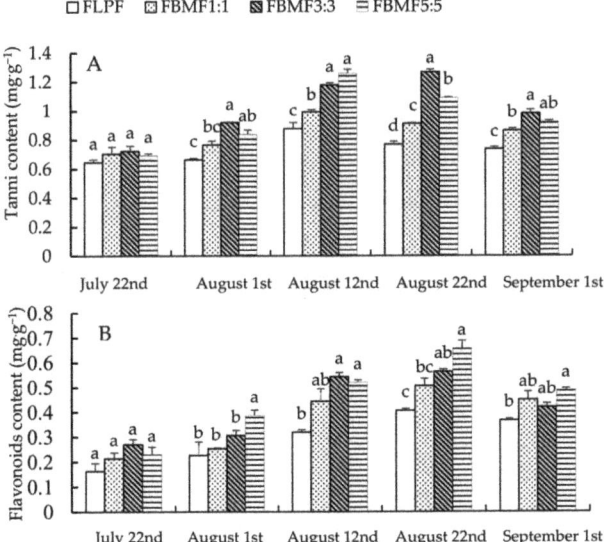

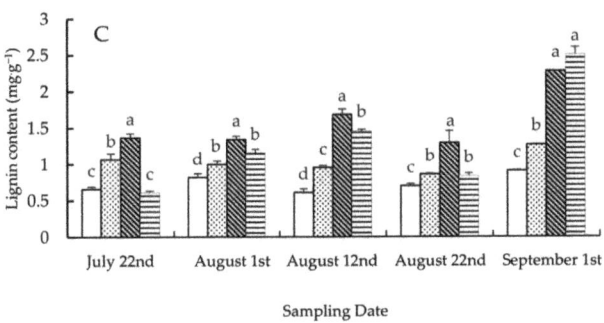

Figure 2. Changes of contents of tannin, flavonoids and lignin in needles of FBMF. (**A**) Tannin content, (**B**) Flavonoids content, (**C**) Lignin content. The values presented in the figure are means ± standard deviation (n = 3). Different lowercase letters in the same age histogram mean there is a significant difference between different treatments at the same time in the same mixed forest (ANOVA followed by Bonferroni multiple comparisons, p < 0.05). The same holds for Figures 2 and 3 below.

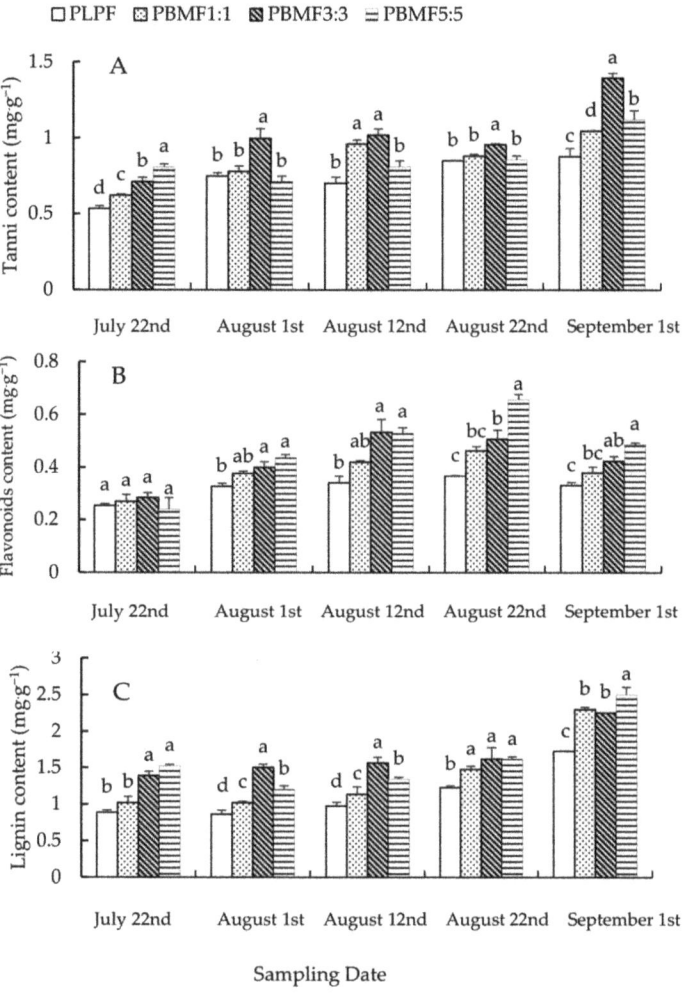

Figure 3. Changes of contents of tannin, flavonoids and lignin in needles of PBMF. (**A**) Tannin content, (**B**) Flavonoids content, (**C**) Lignin content.

Except for 22 July and 1 August, the lignin content in needles treated with FBMF was significantly higher than that of needles treated with FLPF ($p < 0.05$). On 22 July, the lignin content in needles of FBMF$_{1:1}$ and FBMF$_{3:3}$ was significantly higher than that of FLPF (F = 32.63, df$_1$ = 3, df$_2$ = 8, $p < 0.05$), but there was no difference between them. On 1 August and 12 August, the lignin content in needles of larch–ashtree FBMF was, from high to low, FBMF$_{3:3}$ > FBMF$_{5:5}$ > FBMF$_{1:1}$ (F = 6.84, df$_1$ = 3, df$_2$ = 8, $p < 0.05$ on 1 August; F = 11.63, df$_1$ = 3, df$_2$ = 8, $p < 0.05$ on 12 August). On 22 August, the lignin content of needles in FBMF$_{3:3}$ was significantly higher than that in FBMF$_{1:1}$ and FBMF$_{5:5}$ (F = 31.60, df$_1$ = 3, df$_2$ = 8, $p < 0.05$), but there was no difference between that in FBMF$_{1:1}$ and FBMF$_{5:5}$. On 1 September, the lignin content in needles of FBMF$_{3:3}$ and FBMF$_{5:5}$ was significantly higher than that of FBMF$_{1:1}$ (F = 87.68, df$_1$ = 3, df$_2$ = 8, $p < 0.05$), but there was no significant difference between FBMF$_{3:3}$ and FBMF$_{5:5}$ (Figure 2C). These results indicated that ashtree in FBMF had a significant effect on the lignin content of needles, and then the lignin content in the needles of FBMF$_{3:3}$ was higher than that of other treatments.

3.2. Effects of the Banding Mixed Forest in Pots on Content of Secondary Metabolites in Needles of L. olgensis

The tannin content of needles in $PBMF_{3:3}$ was always significantly higher than that in PLPF, and the tannin content of needles in $PBMF_{5:5}$ was significantly higher than that in PLPF on 22 July and 1 September. On 22 July, the tannin content in needles of PBMF was, from high to low, $PBMF_{5:5} > PBMF_{3:3} > PBMF_{1:1}$ ($F = 32.38$, $df_1 = 3$, $df_2 = 8$, $p < 0.05$). On 1 August, 22 August and 1 September, the tannin content of needles in $PBMF_{3:3}$ was significantly higher than in $PBMF_{1:1}$ and $PBMF_{5:5}$ ($F = 9.13$, $df_1 = 3$, $df_2 = 8$, $p < 0.05$ on 1 August; $F = 11.52$, $df_1 = 3$, $df_2 = 8$, $p < 0.05$ on 22 August; $F = 25.50$, $df_1 = 3$, $df_2 = 8$, $p < 0.05$ on 1 September), but there was no difference in $PBMF_{1:1}$ and $PBMF_{5:5}$ on 1 August and 22 August. On 12 August, the tannin content in needles of $PBMF_{1:1}$ and $PBMF_{3:3}$ was significantly higher than that of $PBMF_{5:5}$ ($F = 15.11$, $df_1 = 3$, $df_2 = 8$, $p < 0.05$), but there was no difference between them (Figure 3A). These results showed that the allelopathy of ashtree in PBMF could increase the content of tannin in larch needles, and the tannin content in the needles of $PBMF_{3:3}$ was higher than that of $PBMF_{1:1}$ and $PBMF_{5:5}$.

Except for 22 July, the flavonoids content in needles treated with $PBMF_{3:3}$ and $PBMF_{5:5}$ was significantly higher than that of needles treated with PLPF ($p < 0.05$). On 1 August and 12 August, the content of flavonoids in needles of $PBMF_{5:5}$ was significantly higher than that of $PBMF_{1:1}$ ($F = 10.61$, $df_1 = 3$, $df_2 = 8$, $p < 0.05$ on 1 August; $F = 9.95$, $df_1 = 3$, $df_2 = 8$, $p < 0.05$ on 12 August), and there was no difference between that of $PBMF_{3:3}$ and $PBMF_{1:1}$. On 22 August, the content of flavonoids in needles of $PBMF_{5:5}$ was significantly higher than that of $PBMF_{1:1}$ and $PBMF_{3:3}$, but there was no difference between that of $PBMF_{1:1}$ and $PBMF_{3:3}$ ($F = 30.49$, $df_1 = 3$, $df_2 = 8$, $p > 0.05$). On 1 September, the content of flavonoids in needles of $PBMF_{5:5}$ was significantly higher than that of $PBMF_{1:1}$ ($F = 11.79$, $df_1 = 3$, $df_2 = 8$, $p < 0.05$), but there was no difference between that of $PBMF_{3:3}$ and $PBMF_{1:1}$ or $PBMF_{5:5}$. (Figure 3B). These results showed that the allelopathy of ashtree in PBMF could increase the content of flavonoids in needles, and the content of flavonoids in the needles of $PBMF_{5:5}$ was higher than that of $PBMF_{1:1}$ and $PBMF_{3:3}$, which is consistent with the results of the field experiment.

Except for 22 July, the lignin content in needles treated with PBMF was significantly higher than that of needles treated with PLPF ($p < 0.05$). On 22 July, the lignin content of needles in $PBMF_{3:3}$ and $PBMF_{5:5}$ were significantly higher than that of PLPF and $PBMF_{1:1}$ ($F = 33.87$, $df_1 = 3$, $df_2 = 8$, $p < 0.05$), but there was no difference between them. On 1 August and 12 August, the lignin content in needles treated with PBMF was, from high to low, $PBMF_{3:3} > PBMF_{5:5} > PBMF_{1:1}$. On 1 September, the lignin content in needles of $PBMF_{5:5}$ was significantly higher than that of $PBMF_{1:1}$ and $PBMF_{3:3}$. However, there was no difference between that of $PBMF_{1:1}$ and $PBMF_{3:3}$ ($F = 4.69$, $df_1 = 3$, $df_2 = 8$, $p > 0.05$) (Figure 3C). These results showed that the allelopathy of ashtree in needles of PBMF could increase the lignin content, and the lignin content in needles of $PBMF_{3:3}$ and $PBMF_{5:5}$ was significantly higher than that of $PBMF_{1:1}$ ($p < 0.05$).

3.3. Effects of Two Planting Methods on Content of Secondary Metabolites in Needles of L. olgensis

The difference of the tannin content in needles was irregular in FBMF and PBMF. On 12 August, the tannin content in needles of $FBMF_{1:1}$ was significantly lower than that of $PBMF_{1:1}$ ($t = -5.89$, $df = 4$, $p < 0.01$), and there was no difference between them at other times. On 22 July and 1 September, the tannin content in needles of $FBMF_{3:3}$ was significantly lower than that of $PBMF_{3:3}$ ($t = -3.07$, $df = 4$, $p < 0.05$ on 22 July; $t = -4.90$, $df = 4$, $p < 0.05$ on 1 September), and on 22 August, the tannin content in needles of $FBMF_{3:3}$ was significantly higher than that of $PBMF_{3:3}$ ($t = 4.95$, $df = 4$, $p < 0.05$). On 1 August, 12 and 22, the tannin content in needles of $FBMF_{5:5}$ was significantly higher than that of $PBMF_{5:5}$ ($t = 3.55$, $df = 4$, $p < 0.05$ on 1 August; $t = 4.5$, $df = 4$, $p < 0.05$ on 12 August; $t = 6.33$, $df = 2.02$, $p < 0.05$ on 22 August), while the tannin content in needles of $FBMF_{5:5}$ was significantly lower than that of $PBMF_{5:5}$ on 22 July ($t = -10.78$, $df = 4$, $p < 0.01$). These results showed that the allelopathy of ashtree in needles of $FBMF_{5:5}$ was more beneficial to improving the tannin content than that of $PBMF_{5:5}$ (Table 2).

Table 2. Comparative analysis of content of secondary metabolites in larch needles by two planting modes.

Secondary Metabolites	Month/Date	Ratio of Secondary Metabolites Content to Control								
		$BMF_{1:1}$			$BMF_{3:3}$			$BMF_{5:5}$		
		Planting in Field	Planting in Pots	Sig.	Planting in Field	Planting in Pots	Sig.	Planting in Field	Planting In Pots	Sig.
tannin	7/22	1.09 ± 0.07	1.18 ± 0.02	ns	1.11 ± 0.05	1.35 ± 0.06	*	1.07 ± 0.02	1.53 ± 0.04	**
	8/01	1.16 ± 0.04	1.04 ± 0.05	ns	1.40 ± 0.01	1.33 ± 0.09	ns	1.27 ± 0.04	1.04 ± 0.05	*
	8/12	1.13 ± 0.02	1.37 ± 0.04	**	1.34 ± 0.02	1.46 ± 0.06	ns	1.44 ± 0.03	1.16 ± 0.05	*
	8/22	1.18 ± 0.00	1.28 ± 0.04	ns	1.65 ± 0.03	1.36 ± 0.05	*	1.42 ± 0.00	1.08 ± 0.05	*
	9/01	1.17 ± 0.02	1.19 ± 0.00	ns	1.33 ± 0.04	1.58 ± 0.04	**	1.25 ± 0.01	1.28 ± 0.07	ns
flavonoid	7/22	1.34 ± 0.15	1.08 ± 0.10	ns	1.69 ± 0.13	1.13 ± 0.07	*	1.44 ± 0.50	0.96 ± 0.17	ns
	8/01	1.10 ± 0.01	1.14 ± 0.03	ns	1.33 ± 0.39	1.21 ± 0.06	ns	1.68 ± 0.09	1.32 ± 0.04	*
	8/12	1.39 ± 0.16	1.24 ± 0.01	ns	1.70 ± 0.05	1.57 ± 0.14	ns	1.63 ± 0.03	1.56 ± 0.06	ns
	8/22	1.24 ± 0.07	1.25 ± 0.05	ns	1.37 ± 0.02	1.37 ± 0.09	ns	1.60 ± 0.08	1.77 ± 0.05	ns
	9/01	1.21 ± 0.09	1.15 ± 0.08	ns	1.14 ± 0.04	1.28 ± 0.06	ns	1.31 ± 0.02	1.47 ± 0.02	*
lignin	7/22	1.62 ± 0.08	1.15 ± 0.07	**	2.07 ± 0.05	1.56 ± 0.01	**	0.93 ± 0.04	1.71 ± 0.06	**
	8/01	1.22 ± 0.05	1.19 ± 0.09	ns	1.64 ± 0.02	1.75 ± 0.03	ns	1.41 ± 0.03	1.40 ± 0.01	ns
	8/12	1.56 ± 0.14	1.16 ± 0.03	*	2.76 ± 0.00	1.60 ± 0.00	**	2.39 ± 0.00	1.37 ± 0.02	**
	8/22	1.23 ± 0.09	1.20 ± 0.13	ns	1.85 ± 0.02	1.31 ± 0.02	**	1.20 ± 0.03	1.31 ± 0.01	ns
	9/01	1.40 ± 0.09	1.33 ± 0.11	ns	2.50 ± 0.02	1.30 ± 0.01	**	2.75 ± 0.02	1.45 ± 0.03	**

Note: * = $p < 0.05$; ** = $p < 0.01$; ns = $p > 0.05$ (independent sample t-test).

The content of flavonoids in needles of $FBMF_{3:3}$ was significantly higher than that of $PBMF_{3:3}$ on 22 July ($t = 3.71$, df = 4, $p < 0.05$); the content of flavonoids in needles of $FBMF_{5:5}$ was significantly higher than that of $PBMF_{5:5}$ on 1 August ($t = 3.61$, df = 4, $p < 0.05$); and the flavonoids content of needles in $PBMF_{5:5}$ was significantly higher than that of $FBMF_{5:5}$ on 1 September ($t = -4.85$, df = 4, $p < 0.05$). Except for those results, there was no difference between FBMF and PBMF at other times or treatments. (Table 2). These results showed that the effects of the allelopathy of ashtree on needles in FBMF and PBMF were similar for improving the content of flavonoids.

The lignin content of needles in FBMF was higher than that in PBMF. Except for that result, the lignin content in needles of $FBMF_{5:5}$ was significantly lower than that of $PBMF_{5:5}$ on 22 July ($t = -9.675$, df = 4, $p < 0.01$), while the lignin content in needles of $FBMF_{1:1}$ ($t = 5.76$, df = 4, $p < 0.01$) and $FBMF_{3:3}$ ($t=6.25$, df=4, $p < 0.01$) on 22 July; that of $FBMF_{1:1}$ ($t = 6.93$, df = 4, $p < 0.01$), $FBMF_{3:3}$ ($t=13.96$, df = 4, $p < 0.01$) and $FBMF_{5:5}$ ($t = 17.49$, df = 4, $p < 0.01$) on 12 August; and that of $FBMF_{3:3}$ ($t = 20.96$, df = 4, $p < 0.01$) and $FBMF_{5:5}$ ($t = 16.04$, df = 4, $p < 0.01$) on 1 September were all significantly higher than that of PBMF (Table 2). These results showed that compared with PBMF, FBMF was more beneficial in increasing the lignin content in larch needles.

4. Discussion

Plant allelopathy is a way of transmitting information between plants. It is also a natural chemical regulation phenomenon in the ecosystem and an ecological mechanism for plants to adapt to the environment [27]. Plants can sense and recognize the information substances that coexist within the same species or alien plants. Different plants regulate plant growth and development through different types of allelochemicals; change plants' physiological, biochemical and compositional levels; and then improve their adaptability [28]. When *Mallotus japonicus* grows with its related species, the nectar secretion will be reduced, and their own chemical defense capabilities will be changed [29]. Among secondary metabolites, flavonoids and lignin are important phenylpropanoid pathway metabolites that can increase the metabolic burden of herbivores and inhibit their food consumption, digestibility and assimilation rates [16] (Matthias and Danie, 2020). Our study results showed that the allelopathy of ashtree can increase the content of tannin, flavonoids and lignin in needles, whether planting occurred in pots or the field. The reason may be due to the induction of some chemical substances released by the ashtree volatilization, rain and fog leaching, plant residue decomposition, root exudation and other substances infiltrating the soil to reach the larch [30]. The results are also consistent with the research results, in which the growth of larch–ashtree banded mixed forest for twenty and thirty years significantly enhanced larch's resistance to insect pests [22]. This shows that a mixed forest of two-year-old larch–ashtree can promote a level of secondary metabolism in larches and the synthesis of phenols to enhance larch's resistance to phytophagous insects and improve its chemical defenses.

The stable stand structure of mixed plantations gives full play to forest ecological function and benefit, compared with monocultural *L. olgensis* plantations with large-scale and successive planting [31]. It has many obvious advantages in enhancing system stability, resisting diseases and pests and increasing biological diversity. At the same time, increasing the richness of stand species can improve the trees' resistance to pests [32]. In a broad-leaved forest, the overall damage of forest pests to broad-leaved tree species is significantly reduced with an increased number of tree species [33]. Cao B [34] used 3-year-old *Ailanthus altissima* and *Populus bolleana* to establish a mixed forest for controlling *Anoplophora glabripennis*. The results showed that the insect resistance of *P. bolleana* was better when *P. bolleana* and *A. altissima* were banding mixed in a ratio of 3:2 and 2:3. Our study showed that $BMF_{3:3}$ and $BMF_{5:5}$ of larch–ashtree had the most obvious effect on improving the content of secondary metabolites in needles, which was consistent with the previous study, which demonstrated that the content of secondary metabolites in the needles of 20-year-old larch–ashtree mixed by strips of 4:4 was significantly stronger than that in needles of 20-year-old larch–ashtree mixed by strips of 2:10 [22]. This shows that the allelopathy intensity of *F. mandshurica* on larch is different in different mixed modes. There are differences in stand density, population structure, interspecific competition, nutrient absorption, the content of allelochemicals released by ashtrees and the allelochemicals secreted by its roots among the three banding mixed modes of larch–ashtree—$BMF_{1:1}$, $BMF_{3:3}$ and $BMF_{5:5}$—which all affect the synthesis of chemical defense substances in the needles. Therefore, when constructing larch–ashtree banding mixed forest, three or five rows of banded mixed forest are more conducive to improving the chemical defense ability of the trees.

Allelopathy among plants is mainly concentrated in the aboveground and underground parts [35]. Plants release specific secondary metabolites (allelochemicals) to the environment through aboveground parts' (stems, leaves, flowers, fruits or seeds) volatilization, leaching and root exudation, which directly or indirectly affect the growth and development of neighboring plants, allowing neighboring plants to adjust their biomass allocation in order to determine whether to employ strategies such as chemical defense [36,37]. The present study showed that compared with the pot experiment, larches in the field experiment were more conducive to increasing their content of tannin and lignin. This shows that the synergistic chemical action of the aboveground volatiles and underground compounds of *F. mandshurica* is more conducive to enhancing the synthesis of insect-resistant secondary metabolites in needles. However, there is no significant difference between the effects of the two planting methods on the content of flavonoids in needles. It may be that some allelochemicals secreted by the roots of *F. mandshurica* in the ground-planting method have an antagonistic effect on the synthesis of secondary metabolites of flavonoids [38,39], which leads to the fact that the allelopathic effect of *F. mandshurica* in the field experiment on flavonoids is not stronger than that of *F. mandshurica* in the pot experiment. The mechanism of this phenomenon and the allelochemicals involved in enhancing the chemical defense of larch should be further studied.

With the improvement of people's awareness of ecological and environmental protection, pest control has developed from comprehensive control measures based on chemical control to natural control based on forestry measures. The establishment of a mixed forest is an essential means of forest pest control and a measure of ecological pest control, too. Furthermore, allelopathy is an ecological factor that cannot be ignored in diverse forest ecosystems. It not only exists universally, but also greatly impacts the structural layout, function, benefit and development of forest communities [34,40]. Therefore, we should fully use the current healthy forest ecosystem and combine the underground and aboveground chemical links and their ecological synergy to explore the corresponding mechanisms and further construct natural chemical regulation in the future [21,32].

5. Conclusions

In summary, banding mixed larch and ashtree can significantly increase the content of secondary metabolites (phenolic compounds) in *L. olgensis* and improve its chemical defense. FBMF treatments showed more significant effects on the tannin and lignin content in needles than PBMF treatments. However, the allelopathic intensity of *F. mandshurica* on larches is different in different mixed modes, and then the effect of banding mixed forests by the proportion of 3:3 and 5:5 is better. These findings lay a theoretical foundation for applying forest management measures to control forest defoliators, such as *Dendrolimus superans*. Yet, the underground and aboveground chemical links and their ecological synergy mechanism for larch–ashtree banding mixed forests are still not clear, which needs further research.

Author Contributions: H.J. and Z.M. designed the study, performed the experiments and participated in developing, drafting and finalizing the manuscript. S.Y. formulated the overarching research goals, commentary and revision of the manuscript. S.Z. and D.J. helped analyze the data. P.L. helped analyze the data, finalized the manuscript and provided helpful advice on the revision. All authors have read and agreed to the published version of the manuscript.

Funding: This research was supported by the Natural Science Foundation of Heilongjiang Province of China (LH2021C086) and the Special Fund Project for Basic Scientific Research Business Expenses of Central Universities (DL13BAX31).

Institutional Review Board Statement: Not applicable.

Informed Consent Statement: Not applicable.

Data Availability Statement: All the data supporting the conclusions of this study are included in the manuscript.

Conflicts of Interest: The authors declare no conflict of interest.

References

1. Kong, C. Inter-specific and intra-specific chemical interactions among plants. *Chin. J. Appl. Ecol.* **2020**, *31*, 2141–2150.
2. Li, Y.; Xia, Z.; Kong, C. Allelobiosis in the interference of allelopathic wheat with weeds. *Pest Manag. Sci.* **2016**, *72*, 2146–2153. [CrossRef]
3. Zhang, Y.B.; Lou, Y.G. Research progresses in chemical interactions between plants and phytophagous Insects. *Chin. J. Appl. Ecol.* **2020**, *31*, 2151–2160.
4. Wu, J.S. The "chemical defense" of plants against pathogenic microbes: Phytoa-lexins biosynthesis and molecular regulations. *Chin. J. Appl. Ecol.* **2020**, *31*, 2161–2167.
5. Zhang, W.H.; Liu, G.J. A review on plant secondary substances in plant resistance to insect pests. *Chin. Bull. Bot.* **2003**, *20*, 522–530.
6. Jiang, D.; Yan, S. Effects of Cd, Zn or Pb stress in Populus alba berolinensis on the development and reproduction of Lymantria dispar. *Ecotoxicology* **2017**, *26*, 1305–1313. [CrossRef]
7. Zhang, K.; Wang, Q.; He, D. Temporal changes of phenolicacids in Phellodendron amurense Rupr. Leaves and its resistance to insects. *J. Northeast. For. Univ.* **2014**, *42*, 126–130.
8. Jiang, D.; Wang, Y.Y.; Yan, S.C. Effects of Zn stress on growth development and chemical defense of Populus alba'berolinensis' seedlings. *J. Beijing For. Univ.* **2018**, *40*, 42–48.
9. Guo, Y.; Zhang, P.; Guo, M. Secondary metabolites and plant defence against pathogenic disease. *Plant Physiol. Jounal.* **2012**, *48*, 429–434.
10. Weston, L. Mechanisms for cellular transport and release of allelochemicals from plant roots into the rhizosphere. *J. Exp. Bot.* **2012**, *63*, 3445–3454. [CrossRef]
11. Li, M.; Zeng, R.; Luo, S. Secondary metabolites related with plant resistance against pathogenic microorganisms and insect pests. *Chin. J. Biol. Control.* **2007**, *23*, 269–273.
12. Rice, E.L. *Allelopathy*, 2nd ed.; Academic Press: Orlando, FL, USA, 1984.
13. Shi, X.P.; Chen, Y.P.; Yan, Z.Q. Research progress on plant allelopathy. *Biotechnol. Bull.* **2020**, *36*, 215–222.

14. Yuan, H.; Yan, S.; Tong, L. Content differences of condensed tannin in needles of Larix gmelinii by cutting needles and insect feeding. *Acta Ecol. Sin.* **2009**, *29*, 1415–1420.
15. Liu, X.X.; Jiang, D.; Meng, Z.J. Effects of secondary substances on food Utilization by Hyphantria cunea larvae. *J. Northeast. For. Univ.* **2020**, *48*, 99–103.
16. Dučaiová, Z.; Sajko, M.; Mihaličová, S.; Repčák, M. Dynamics of accumulation of coumarin-related compounds in leaves of Matricaria chamomilla after methyl jasmonate elicitation. *Plant Growth Regul.* **2016**, *79*, 81–94. [CrossRef]
17. Shahabinejad, M.; Shojaaddini, M.; Maserti, B. Exogenous application of methyl jasmonate and salicylic acid increases antioxidant activity in the leaves of pistachio (*Pistacia vera* L. cv. Fandoughi) trees and reduces the performance of the phloem-feeding psyllid Agonoscena pistaciae. *Arthropod Plant Interact.* **2014**, *8*, 525–530. [CrossRef]
18. Mitsuda, N.; Seki, M.; Shinozaki, K.; Ohmetakagi, M. The NAC Transcription Factors NST1 and NST2 of Arabidopsis Regulate Secondary Wall Thickenings and Are Required for Anther Dehiscence. *Plant Cell* **2005**, *17*, 2993–3006. [CrossRef]
19. Klapwijk, M.J.; Björkman, C. Mixed forests to mitigate risk of insect outbreaks. *Scand. J. For. Res.* **2018**, *33*, 772–780. [CrossRef]
20. Jactel, H.; Bauhus, J.; Boberg, J. Tree diversity drives forest stand resistance to natural disturbances. *Curr. For. Rep.* **2017**, *3*, 223–243. [CrossRef]
21. Zhang, Y.J. Allelopathic Effects of Pinus tabulaeformis Carr.Littles Extract on Castanea mollissima Bl. and Quercus variabilis Bl. Seedling Growth. *Diss. Beijing For. Univ.* 2009.
22. Jiang, H.; Yan, S.; Xue, Y. Effects of forest type on activity of several defense proteins and contents of secondary metabolites in Larch needles. *For. Res.* **2018**, *31*, 24–28.
23. Wu, Y.Q.; Guo, Y.Y. Determination of tannin in cotton plant. *J. Appl. Ecol.* **2000**, *11*, 243–245.
24. Jiang, D.; Wang, Y.Y.; Dong, X.W.; Yan, S.C. Inducible defense responses in populus alba berolinensis to pb stress. *S. Afr. J. Bot.* **2018**, *119*, 295–300. [CrossRef]
25. Ren, Q.; Hu, Y.J.; Li, Z.Y. Content variation of lignin and peroxidase activities from damaged Pinus massioniana. *ActaEcologicaSinica* **2007**, *27*, 4895–4899.
26. Fukushima, R.; Hatfield, R. Extraction and isolation of lignin for utilization as a standard to determine lignin concentration using the acetyl bromide spectrophotometric method. *J. Agric. Food Chem.* **2001**, *49*, 3133–3139. [CrossRef]
27. Hisashi, F.; Fukiko, K.; Osamu, O. Involvement of allelopathy in inhibition of understory growth in red pine forests. *J. Plant Physiol.* **2017**, *218*, 66–73.
28. Kong, C. Chemical interactions between plant and othe rorganisms: A potential strategy for pest mana-gement. *Sci. Agric. Sin.* **2007**, *40*, 712–720.
29. Yamawo, A. Relatedness of Neighboring Plants Alters the Expression of Indirect Defense Traits in an Extrafloral Nectary-Bearing Plant. *Evol. Biol.* **2015**, *42*, 12–19. [CrossRef]
30. Zhang, W.; Zhang, X.; Jiang, Y. Allelochemicals from root exudates and their effects on soil biota. *Adv. Earth Sci.* **2005**, *20*, 330–337.
31. Pretzsch, P.H. Tree species mixing can increase maximum stand density. *Can. J. For. Res.* **2016**, *46*, 45–52. [CrossRef]
32. Chen, B.; During, H.; Anten, N. Detect thy neighbor: Identity recognition at the root level in plants. *Plant Sci.* **2012**, *195*, 157–167. [CrossRef] [PubMed]
33. Guyot, V.; Castagneyrol, B.; Vialatte, A. Tree diversity reduces pest damage in mature forests across Europe. *Biol. Lett.* **2016**, *4*, 12–17. [CrossRef] [PubMed]
34. Cao, B.; Xu, X. Effects of mixed forest of Ailanthus altissima and Populus bolleana on host choice of Anoplophora glabripennis. *Sci. Silvae Sin.* **2006**, *42*, 56–60.
35. Zhang, Y.; Chang, S.; Song, Y. Application of plant allelopathy in agro-ecosystems. *Chin. Agric. Sci. Bull.* **2018**, *34*, 61–68. [CrossRef]
36. Semchenko, M.; Saar, S.; Lepik, A. Plant root exudates mediate neighbour recognition and trigger complex behavioural changes. *New Phytol.* **2014**, *204*, 631–637. [CrossRef]
37. Semchenko, M.; John, E.A.; Hutchings, M.J. Effects of physical connection and genetic identity of neighbouring ramets on root-placement patterns in two clonal species. *New Phytol.* **2007**, *176*, 644–654. [CrossRef]
38. Santonja, M.; Bousquet-Mélou, A.; Greff, S. Allelopathic effects of volatile organic compounds released from Pinus halepensis needles and roots. *Ecol. Evol.* **2019**, *9*, 8201–8213. [CrossRef]
39. Qian, C.Y.; Tang, F.H.; Li, C.C. Review on allelopathic effect of forest tree. *J. Northwest For. Univ.* **2019**, *34*, 79–85.
40. Erb, M.; Kliebenstein, D.J. Plant Secondary Metabolites as Defenses, Regulators, and Primary Metabolites: The Blurred Functional Trichotomy. *Plant Physiol.* **2020**, *184*, 39–52. [CrossRef]

Disclaimer/Publisher's Note: The statements, opinions and data contained in all publications are solely those of the individual author(s) and contributor(s) and not of MDPI and/or the editor(s). MDPI and/or the editor(s) disclaim responsibility for any injury to people or property resulting from any ideas, methods, instructions or products referred to in the content.

Article

Patterns of Needle Nutrient Resorption and Ecological Stoichiometry Homeostasis along a Chronosequence of *Pinus massoniana* Plantations

Qiqiang Guo [1,2,*], Huie Li [3], Xueguang Sun [1], Zhengfeng An [2] and Guijie Ding [1]

[1] Key Laboratory of Forest Cultivation in Plateau Mountain of Guizhou Province, Institute for Forest Resources and Environment of Guizhou, College of Forestry, Guizhou University, Guiyang 550025, China
[2] Department of Renewable Resources, University of Alberta, Edmonton, AB T6G2E3, Canada
[3] College of Agriculture, Guizhou University, Guiyang 550025, China
* Correspondence: qqguo@gzu.edu.cn; Tel.: +86-085-18385-3069

Abstract: Nutrient resorption and stoichiometry ratios are vital indicators to explore nutrient transfer and use efficiency for plants, particularly under the condition of nutrient limitation. However, the changing rules about nutrient resorption and ecological stoichiometry homeostasis are still unclear with the development of plantations. We determined carbon (C), nitrogen (N), and phosphorus (P) concentrations in soil and in fresh and senesced needles along a chronosequence of *Pinus massoniana* plantations (10, 20, 30, and 36 years old) in Guizhou Province, China. We also calculated the N and P resorption efficiency (NRE and PRE, respectively) and the homeostasis coefficient. The results showed that fresh and senesced needles' C and N concentrations maintained an increasing trend, whereas their P concentrations decreased initially and subsequently increased as the plantations' ages increased. Fresh needles' N:P ratios indicated that N limitation existed before 20 years old, while P limitation appeared in the 30-year-old plantations. The NRE and PRE showed patterns of increasing initially and decreasing subsequently along the chronosequence of *P. massoniana* plantations, which was coupled with weak stoichiometric homeostasis to reduce nutrient deficiency. Therefore, the appropriate nutrient management measurements should be induced to promote tree growth and the sustainable development of *P. massoniana* plantations.

Keywords: nutrient resorption; nutrient limitation; ecological stoichiometry homeostasis; *Pinus massoniana* plantations

Citation: Guo, Q.; Li, H.; Sun, X.; An, Z.; Ding, G. Patterns of Needle Nutrient Resorption and Ecological Stoichiometry Homeostasis along a Chronosequence of *Pinus massoniana* Plantations. *Forests* **2023**, *14*, 607. https://doi.org/10.3390/f14030607

Academic Editor: Sune Linder

Received: 14 February 2023
Revised: 16 March 2023
Accepted: 17 March 2023
Published: 18 March 2023

Copyright: © 2023 by the authors. Licensee MDPI, Basel, Switzerland. This article is an open access article distributed under the terms and conditions of the Creative Commons Attribution (CC BY) license (https://creativecommons.org/licenses/by/4.0/).

1. Introduction

Carbon (C), nitrogen (N), and phosphorus (P) are vital indicators of nutrient utilisation in trees. Their ecological stoichiometry ratios are closely related to the life histories and adaptation strategies of tree species [1,2]. Generally, nutrient accumulation and the transformation of tree and soil nutrient release often change because of the biotic and abiotic influence during the development of the plantation [3,4]. These variations in N and P concentrations in trees and soil often lead to the nutrient limitation of tree individuals within the background of the long growth cycle and the dynamic understory habitats [5,6]. According to previous reports, N and P limitations often happen because of insufficient nutrient supply in young and ageing forests, respectively [7–9]. Inevitably, these situations would have substantial effects on tree biomass accumulation and timber yield [10]. Hence, clarifying these effects will further strengthen the understanding of the adaptability of plantations under nutrient changes.

Ecological stoichiometry focuses on analysing the balance of chemical elements in ecological interactions [11]. To date, this related theory has been widely used to explain the relationship and feedback between plants and the environmental factors in the ecological system [12,13]. Leaf N:P ratios are acknowledged to be indicators of N or P limitation for plant growth. For example, N or P limitation occurred at the N:P < 14 or >16 for plant leaves, respectively [14]. Simultaneously, soil nutrient state and fertiliser supply capacity affect plant nutrient and growth characteristics. For one plant species, the values of its stoichiometry usually maintain a relatively constant composition regardless of the change in environmental factors, which is called stoichiometric homeostasis [15,16]. At present, these patterns have been discovered in many kinds of woody plants [17]. However, stoichiometric homeostasis may have different thresholds in various environments, including different growth stages in plants [18,19].

Nutrient resorption from ageing tissues is an important approach for plants to accumulate nutrient elements and optimise their use efficiency, which make them less dependent on soil nutrients to maintain stoichiometric homeostasis [15]. Hence, plants have greater N or P resorption efficiency (NRE/PRE, respectively) once N/P limitation occurs [20]. In addition to soil nutrient status, studies have reported that nutrient resorption efficiency (NuRE) could also be affected by fresh leaf and litter nutrients [21], leading to various patterns in response to different-age stands [17]. Many researchers found that NuRE presented a negative correlation with soil nutrients [22,23]. On the contrary, some studies reported that plants growing in infertile soil did not always show higher NuRE [24,25]. These results indicate that NuRE with plantation development and the correlations with stoichiometry and homeostasis are still uncertain. Therefore, clarifying the pattern of nutrient resorption and ecological stoichiometry homeostasis along a chronosequence of plantations is necessary.

Pinus massoniana is a pioneer afforestation species and the main timber tree species in South China, covering about one million hm^2 of afforestation area. *P. massoniana* plays an extremely important role in China's timber production and forest ecosystem service function [26]. Usually, the mature age of *P. massoniana* plantations is about 40 years in south of China. However, their rotation period could be from 20 to 25 years based on the different purpose of utilization and to maximize economic benefit. In recent years, the productivity and economic value of *P. massoniana* plantations have attracted much attention. For example, the earlier, pure *P. massoniana* plantations have begun to degenerate because of the simple community structure, the growth rate and timber productivity, and soil fertility reduction and acidification [27–29]. Previously, the majority of experiments on the ecological stoichiometry of *P. massoniana* plantations were carried out only at one growth stage [30], which made the nutrient dynamic and limitation status along a chronosequence of *P. massoniana* plantations unclear. Therefore, we measured the needle and soil total organic C, N, and P concentrations and calculated their stoichiometric ratios, including needle nutrient resorption in *P. massoniana* plantations across a chronosequence of 10- (young), 20- (middle-aged), 30- (near-mature), and 36-year-old (mature) stands in Guizhou Province, China. The aim was to illustrate the patterns of nutrient resorption and stoichiometric homeostasis along a chronosequence of *P. massoniana* plantations. Based on this, we want to verify two hypotheses: (i) NuRE would increase along the chronosequence of the *P. massoniana* plantation because nutrient demand may increase with tree growth, and (ii) NuRE would present a negative relationship between soil and needle nutrients to conserve nutrients for trees.

2. Materials and Methods

2.1. Study Site

A description of this study site (Figure 1) can be found in the study by Pan et al., 2021 [26]. The soil belongs to the alfisols type (the acidic siliceous yellow soil) and has moderate fertility, with a soil depth of 60–80 cm. The earliest local vegetation was a *P. massoniana*, *Pinus armandii*, and broadleaf tree (*Populus davidiana* and *Quercus fabri*) artificial forest, which was established as a pure forest of *P. massoniana* in the 1980s.

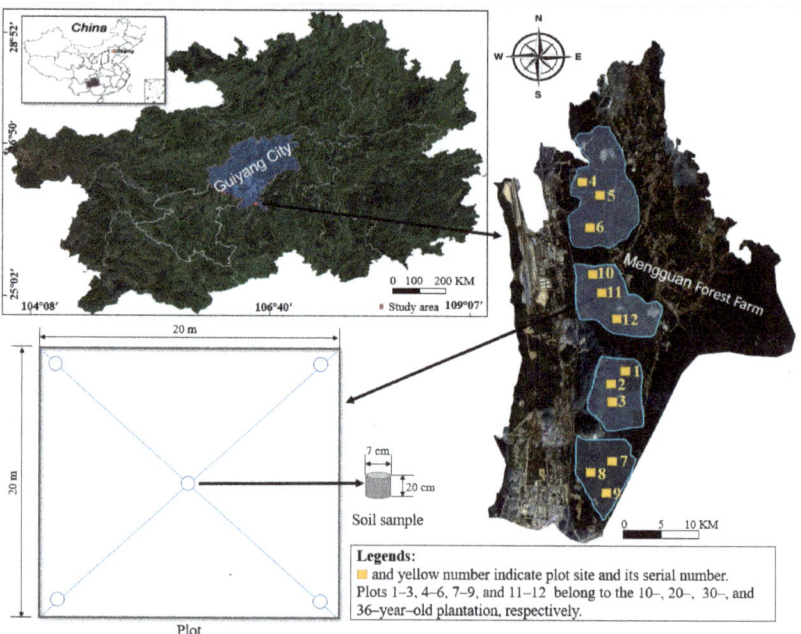

Figure 1. Location of Mengguan Forest Farm study area in Guizhou, China and soil sample distribution in each sample plot.

2.2. Experimental Design and Sample Collection

We selected four age classes (10, 20, 30, and 36 years old) of *P. massoniana* plantations as the research objects and carried out the preliminary investigation in August 2017. The four aged plantations were established in 2007, 1997, 1987, and 1981, respectively, with a plant/row distance of 1 m × 2 m. They were tended (irrigation and weeding) and managed (insect control and grazing prohibition) only in the first 3 years after afforestation. The first and the second thinning were conducted with 30% of trees being removed 10 and 20 years after establishment, respectively, and a density of approximately 1500 trees·ha^{-1} was kept. Three typical 20 m × 20 m replicated sample plots were established within each of the same age plantations. All plots were on relatively flat topography as well as similar soil type, altitude, and climate conditions as much as possible (Table 1). Within each plot, the tree height, diameter at breast height (DBH), canopy, and others were measured and recorded. Meanwhile, three well-growing individuals were selected according to the average DBH, with a >10 m distance between any two sample trees, and were labelled.

Table 1. Characteristics of *Pinus massoniana* plantations along a chronosequence.

Plantation Age	10 Years Old	20 Years Old	30 Years Old	36 Years Old
Altitude/m	1194 ± 6.4	1175 ± 8.3	1206 ± 8.5	1214 ± 9.2
Canopy density	0.90 ± 0.07	0.85 ± 0.10	0.75 ± 0.09	0.80 ± 0.08
Stand density/(trees·ha^{-1})	4675 ± 256	2812 ± 135	1356 ± 103	1083 ± 94
Mean diameter at breast height (DBH)/cm	8.53 ± 1.56	12.66 ± 2.55	18.46 ± 3.01	21.26 ± 5.26
Mean tree height/m	7.43 ± 1.83	15.64 ± 2.74	18.94 ± 2.43	19.87 ± 3.26

Note: each value denotes the mean (±SD).

Fresh (1- and 2-year-old matured leaves) and senesced needles from the same tree were sampled in August and September 2017, respectively, during vigorous growth and peak abscission. Twelve branches in the middle sun-exposed crown from the labelled individual were cut down and sampled in approximately equal quantities of 1- and 2-year-old-matured fresh needles, respectively, and mixed to form a single sample. The senesced needles were also collected from the same branches and mixed to form a litter needle sample. In the laboratory, all the samples were dried for 48 h at 65 °C and powderised using a mechanical grinder for chemical measurement.

Belonging to the shallow-rooted trees, the majority of *P. massoniana* fine roots were distributed in the topsoil layer [31]. Thus, soil samples were taken from the 0–20 cm soil layer in each sample plot (Figure 1). After clearing the surface plant litter, five soil samples were collected using the five-point method. A total of 60 soil samples were obtained from the 4 age stands.

2.3. Chemical and Physical Measurements

Fresh and senesced needles were ground with ball mill equipment (HM2L, Shandong, China) for chemical determination. Soil samples were also air-dried indoors and sieved (2 mm mesh) for chemical measurements. The total C of all the samples was determined using the $K_2Cr_2O_7$ oxidation method, and their total N concentrations were determined with the Kjeldahl method using a Kjeldahl autoanalyser (GK–600P, Shandong, China) [32]. The total P concentrations of the needles and soil samples were determined with the colourimetry method using a spectrophotometer (V–5600, Shanghai, China). Soil available N (AN) and available P (AP) concentrations were analysed using the alkali diffusion and colourimetric methods, respectively [33]. Each needle and soil sample was divided into three parts equally as three duplicate samples for chemical determination.

2.4. Calculations

NuRE was calculated using the formula of Vergutz et al. [34] as follows:

$$\mathrm{NuRE} = \left(\frac{(Nu_g - Nu_s) \times \mathrm{MLCV}}{Nu_g} \right) \times 10 \quad (1)$$

where the mass loss correction value (MLCV) was 0.745 for conifers [34], and Nu_g and Nu_s represented the nutrient concentrations in all age classes of fresh needles and senesced needles, respectively.

The homeostatic coefficient (H) was calculated using the method of Sterner and Elser [35], which was derived from the following model:

$$y = \frac{1}{H} \times \log x + B \quad (2)$$

where y is the concentration of N, P, or N:P ratio in the fresh needles of *P. massoniana* samples, respectively, whereas x is the corresponding value in soil. Mean N and P concentrations and the N:P ratio of the 0–20 cm soil layers were calculated, which were used as soil values. B and H values were calculated through a linear regression analysis. For the convenience of statistics, the slope (1/H with a value ranging from 0 to 1) was used to measure the homeostasis degree. According to Person et al. [15], homeostasis was divided into four patterns based on the values of 1/H, which were stability (the value ranges from 0 to 0.25), weak stability (the value ranges from 0.25 to 0.5), weak sensitivity (the value ranges from 0.5 to 0.75), and sensitivity (the value ranges from 0.75 to 1), respectively.

2.5. Statistical Analysis

C, N, and P stoichiometric ratios of all the samples were calculated using the mass ratios. One-way analysis of variance was used to compare the significant differences in soil total organic C, N, and P concentrations and NuRE along a chronosequence of *P. massoniana* plantations. A two-sample *t*-test was used to analyse the differences between NRE and PRE under the same-age plantation. Linear or quadratic regression was conducted to determine the relationships between plantation age and needle C, N, and P concentrations and their stoichiometric ratios, including NRE:PRE ratios. Covariance analysis was used to test the significant differences in the slopes of stoichiometry between fresh and senesced needles, including the slopes of stoichiometric homeostasis in all-aged plantations. Pearson's correlations were used to test the relationships between NuRE and nutrient concentrations of the needles and soil. All the statistical analyses were conducted with SPSS 24.0 (SPSS Inc., Chicago, IL, USA) for Windows, and all the figures were plotted with Origin 2022b (Microcal Inc., Northampton, MA, USA).

3. Results

3.1. Soil C, N, and P Concentrations and Their Stoichiometric Ratios

Soil C, N, and P concentrations and their stoichiometric ratios presented different change trends along a chronosequence of *P. massoniana* plantations (Table 2). C and AP concentrations and the AN:AP ratios all tended to increase with increasing plantation age. N and AN concentrations and the C:P ratio decreased initially and subsequently increased, with their lowest values appearing in the 20-year-old plantation. On the contrary, the P concentration and C:N ratio increased initially and then decreased, and both reached the highest values in the 20-year-old plantation.

Table 2. Soil main element concentrations and their stoichiometric ratios along a chronosequence of *P. massoniana* plantations.

Nutrition Indexes	10 Years Old	20 Years Old	30 Years Old	36 Years Old
C (g·kg^{-1})	9.98 ± 1.89 b	9.48 ± 1.24 b	11.58 ± 1.50 a	11.85 ± 1.14 a
N (g·kg^{-1})	1.95 ± 0.34 b	1.51 ± 0.22 c	2.32 ± 0.28 a	2.37 ± 0.24 a
P (g·kg^{-1})	0.38 ± 0.05 b	0.49 ± 0.04 a	0.24 ± 0.04 c	0.21 ± 0.05 c
C:N	5.87 ± 0.34 b	6.63 ± 0.25 a	5.87 ± 0.26 b	5.45 ± 0.31 c
C:P	28.88 ± 7.36 b	20.09 ± 5.20 c	48.09 ± 5.09 a	43.89 ± 5.87 a
N:P	5.86 ± 1.41 c	2.24 ± 1.11 d	9.45 ± 1.43 a	7.93 ± 1.26 b
AN (mg·kg^{-1})	12.96 ± 1.76 b	11.04 ± 1.90 c	13.62 ± 1.73 b	15.35 ± 2.38 a
AP (mg·kg^{-1})	7.14 ± 0.93 a	1.41 ± 0.22 b	1.79 ± 0.13 b	1.59 ± 0.14 b
AN:AP	4.55 ± 0.78 c	8.81 ± 1.20 b	8.52 ± 0.76 b	10.03 ± 0.79 a

Note: Parameters include organic carbon (C); nitrogen (N); phosphorus (P); available nitrogen (AN); and available phosphorus (AP). Each value denotes the mean (±SD). Different lowercase letters represent differences among different stand ages under the same nutrition indexes ($p < 0.05$).

3.2. Needle C, N, and P Concentrations and Their Stoichiometric Ratios

The fresh- and senesced-needle C concentrations revealed a linear increasing trend along the chronosequence of *P. massoniana* plantations, ranging from 411.43 to 472.21 mg·g^{-1} and from 368.02 to 451.13 mg·g^{-1}, respectively (Figure 2a). Meanwhile, both N concentrations increased quadratically from 11.42 to 18.45 mg·g^{-1} and from 7.44 to 11.23 mg·g^{-1}, respectively (Figure 2b). In addition, both P concentrations tended to drop initially and subsequently rise along the chronosequence (Figure 2c). Fresh needle C:N ratios tended to decrease with increasing plantation age, whereas they increased initially and decreased subsequently in the senesced needles (Figure 2d). Generally, the fresh needle C:P ratios showed a trend of first decreasing and then increasing, which was opposite to the senesced needles (Figure 2e). Synchronously, the fresh- and senesced-needle N:P ratios increased along the chronosequence, whereas the fresh needle N:P ratios maintained the higher values before the approximately 23-year-old plantation, and then were exceeded by the senesced needles (Figure 2f).

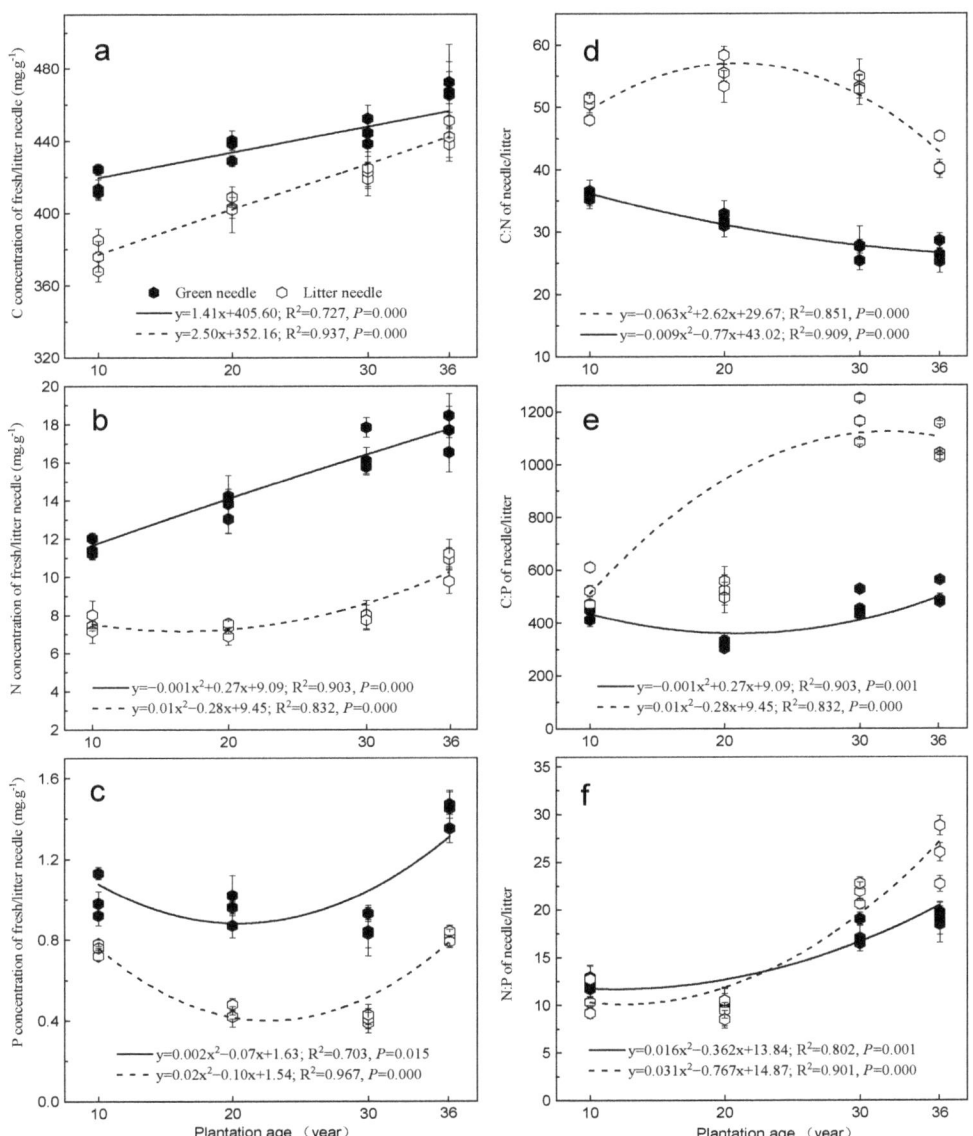

Figure 2. Fresh and senesced needles' C, N, and P concentrations (**a–c**) and their stoichiometry ratios (**d–f**) along the chronosequence of *P. massoniana* plantations.

3.3. Needle NuRE and Their Relationships with Soil Stoichiometry

As shown in Figure 3a, the NRE and PRE exhibited the same change trend, which continually increased from the 10- to 30-year-old stage and then decreased in the 36-year-old stage. However, the NRE was higher than the PRE before the 20-year-old plantations, whereas it was the opposite after 30-year-old stages. NRE:PRE ratios showed a rapid decrease along the chronosequence of *P. massoniana* plantations (Figure 3b). The NRE was significantly negatively correlated with the senesced-needle N concentration (Table 3). However, the PRE was positively correlated with C concentrations and C:N ratios in the fresh needles, the ratios of C:P and N:P in the senesced needles, and the AN:AP ratio in the soil. Undoubtedly, the PRE was negatively correlated with the P concentration in the senesced needles, including the soil AP concentration (Table 3).

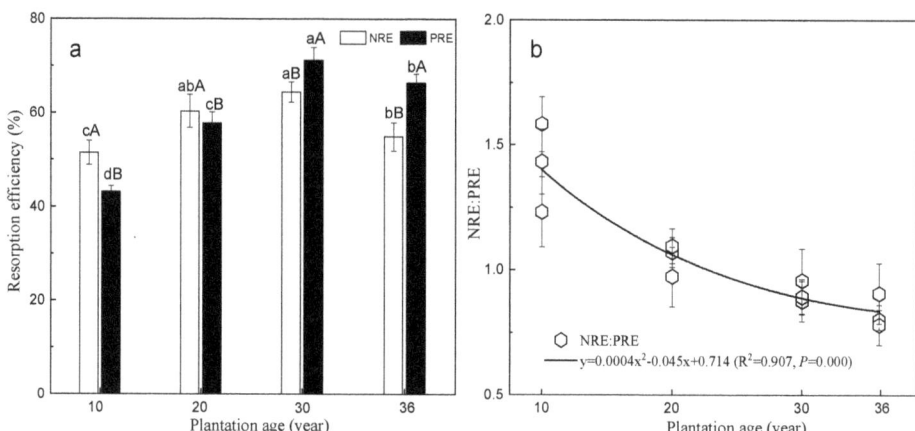

Figure 3. Resorption efficiency of N and P (**a**) and their ratios (**b**) in fresh and senesced needles along the chronosequence of *P. massoniana* plantations. Different capital letters represent differences between fresh and senesced needles in the same-age plantations, respectively ($p < 0.05$). Different lowercase letters represent differences amongst fresh or senesced needles along the chronosequence, respectively ($p < 0.05$).

Table 3. Pearson correlations between nitrogen/phosphorus resorption efficiency and element concentrations, and their stoichiometric ratios in needles and soil.

Factors	Items	Pearson Correlations	
		NRE (p Value)	PRE (p Value)
Fresh needle	C		0.866 ** (0.000)
	C:N		0.921 ** (0.000)
Senesced needle	P		−0.790 ** (0.002)
	N	−0.615 * (0.033)	
	C:P		0.834 ** (0.001)
	N:P		0.756 ** (0.004)
Soil	AP	−0.665 * (0.018)	−0.777 * (0.003)
	AN:AP		0.769 ** (0.003)

Note: Parameters include organic carbon (C); nitrogen (N); phosphorus (P); available nitrogen (AN); available phosphorus (AP); NRE (nitrogen resorption efficiency); and PRE (phosphorus resorption efficiency). Only the items with significant correlations are shown in this table. The value shows the correlation coefficient; positive values represent positive correlation and negative values represent negative correlation. * $p < 0.05$, ** $p < 0.01$.

3.4. Pattern of Needle Stoichiometric Homeostasis

The slopes (1/H) of linear regression relationships between the fresh needle and soil for N, P, and N:P ratios were 0.019, 0301, and 0.287, respectively (Figure 4). These results indicated that the homeostatic adjustment existed in N and P concentrations along the chronosequence of *P. massoniana* plantations. Comparatively, the N concentration had a stronger ability of homeostatic adjustment than the P concentration and N:P ratio during the development of *P. massoniana* plantations (Figure 3a).

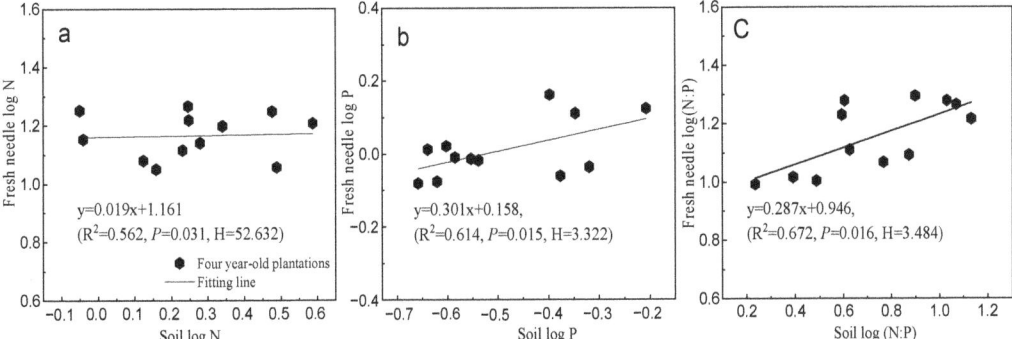

Figure 4. Line regression relationships between fresh needle and soil for N concentrations (**a**), P concentrations (**b**), and N:P ratios (**c**) along the chronosequence of *P. massoniana* plantations.

4. Discussion

4.1. Changes in Soil C, N, and P Concentrations along a Chronosequence

In this research, soil C, N, and P concentrations showed significant fluctuation along the chronosequence of *P. massoniana* plantations. The previous researcher indicated that soil C concentration mainly depended on the quantity of organic matter and the size of the soil humus [36]. We observed the increasing C concentration in the soil, and this pattern was in agreement with some previous results [18,23,37]. The researchers also found that decaying plant litter would provide approximately 58% of the source of N in plantation ecosystems [38]. As we know, 30% of trees were removed during the 10 and 20 years of the plantations in this study area, which made the amount of litter decrease dramatically. This case greatly limited the input of topsoil nutrients, particularly the N element [39]. As is known, soil P transformation is mainly regulated by biochemical mineralisation under abiotic factors [40,41]. Here, we also found that the soil P concentrations quickly increased (Table 2) in plantations before year 20. Thus, the thinning was responsible for this change. After the thinning, more sunshine reaching the understory made the surface soil temperatures increase and accelerated the rate of P mineralisation. Eventually, this led to the P accumulation and concentration increasing in the soil.

4.2. Changes in Nutrient Limitation and Nutrient Resorption Efficiency along a Chronosequence

The C, N, and P concentrations in fresh and senesced needles also showed significant correlations along the chronosequence of *P. massoniana* plantations. The C and N concentrations in the two kinds of needle also maintained the increasing trend along the chronosequence. Similar results have been reported in *C. lanceolata* [18] and *Larix kaempferi* [42]. Obviously, the senesced needle would transfer the N element to the fresh needle for the nutrient resorption strategy. The increasing NRE before the age of 30 years old also supported our opinion. The decreasing soil N concentration portended the N limitation appearing during the development process of the plantation, which was demonstrated further by the lower N:P ratio of 14 in the fresh needles (Figure 2f).

The P concentrations in the fresh needles showed that the changing trend decreased initially and then increased along the chronosequence of *P. massoniana* plantations. Lower P concentrations of the fresh and senesced needles were observed in the 20-year-old plantation. This case may be caused by the high demand for N in the early stage of plantations, which inhibited the uptake of the P element to some degree. After the 20-year-old stage, the demand for P increased rapidly with the further development of plantations. Generally, we found a higher N:P ratio of 16 after the age of 30 years old in plantations (Figure 2f), which indicated the appearance of a P limitation during this period. Based on the theory of the plant nutrient utilisation hypothesis [19], we also observed that *P. massoniana* continually maintained the increasing PRE to relieve the P deficiency until the age of 30 years old. However, the PRE began to decrease after the age 30 years old (Figure 3a).

Evidently, this study indicated that the NRE and PRE were contrary to our first hypothesis. This phenomenon could be caused by the growth variations in *P. massoniana* individuals at different ages. That is, NRE and PRE maintained the increasing trend until the plantation age reached 30 years old, which also illustrated that the increasing nutrient element resorption significantly improved the N and P concentrations of plants to promote tree growth. After 30 years, the *P. massoniana* plantations had rich understory soil and a developed root system, which could obtain more nutrients from the soil and reduce the nutrient utilization from senesced needles (Table 1). Meanwhile, as an effect of the natural closure of the canopy as the trees grew, the light intensity reaching the interior of the tree canopy was reduced. This shade condition sped up the senescence of needles, which shed from the tree earlier, shortening the time for nutrient transfer from the senesced leaves. Hence, the reducing NRE and PRE appeared after 30 years in the *P. massoniana* plantations.

This case was confirmed by the growth rhythm of *P. massoniana* plantations. According to Table 1, the mean DBH of *P. massoniana* individuals increased quickly before the 30-year-old plantations, and then slowly. Meanwhile, the DBH increment quickly increased until the 20-year-old plantation and tended to decrease gradually. Similar growth patterns have been reported about *P. sylvestris* var. *mongolica* [23], *Pinus tabulaeformis* [37], and *Robinia pseudoacacia* [43] along the chronosequence. Subsequently, the NRE and PRE decreased with plantation development after the 30-year-old stage, and a growing number of N and P elements in senesced needle were returned to the soil.

4.3. Interrelationship of Needle Nutrient Utilization, Stoichiometry, and Homeostasis

The C and N concentrations in fresh and senesced needles all increased along a chronosequence (Figure 2a,b), which may have been caused by the accumulation of C structural compounds as the trees grew. Generally, a higher C concentration in the needles had a stronger self-protection ability [44], indicating that *P. massoniana* individuals obtained a strong resistance with increasing stand age in the plantation systems. Simultaneously, the increasing C concentration and C:N ratio in fresh needles could significantly improve the vegetative growth of *P. massoniana* and the PRE (Table 3). Generally, higher N concentrations and N:P in the senesced needles improve litter decomposition rates and nutrient release, which is helpful in enhancing soil nutrient availability and easing plant nutrient limitation [21]. As shown in Figure 2f and Table 3, the N limitation disappeared after the 20-year-old plantations, and soil N concentrations also kept an increasing trend. Evidently, the increasing NRE had a significant negative correlation with the senesced-needle N concentration to conserve nutrients for tree growth with the development of *P. massoniana* plantations, which was consistent with our second hypothesis.

The rapid growth of the young-aged trees requires more P for the composition of the genetic material [45]. Once the soil P concentration is deficient, the plant has to transfer the senesced needle P to a fresh needle [46]. Clearly, the above condition was consistent with our results, which eventually led to an increase in the PRE (Figure 3a). In most cases, the greater NuRE for plants is caused by soil nutrient deficiency [47]. In this study, the same result was found, supported by the negative correlations between the PRE and soil P and AP concentrations (Table 3). The change in the greater NRE and PRE before and after the 20-year-old stage in *P. massoniana* plantations led to the decrease in NRE:PRE ratio, respectively (Figure 3b), which also indicated that relatively more P from the senesced needles was reused along the chronosequence. The above findings were in line with the relative nutrient resorption hypothesis [48,49]. We also observed that *P. massoniana* individuals, in fact, resorbed more N or P elements during the N− or P−limited stage (Figure 2b,c). Generally, the increase in NRE and PRE in the fresh needle and the decreased P and AP concentrations in the soil confirmed our second hypothesis.

In this research, $(1/H)N:P$ and $(1/H)P$ ranged from 0.25 to 0.75 across all the ages of *P. massoniana* plantations (Figure 4c). Based on the homeostatic patterns, *P. massoniana* showed weak homeostasis in P concentration and N:P ratio across all the plantation ages, indicating it had the same stoichiometric characteristic as most subtropical tree species [50]. However, a lower value of (1/H)N ranging from 0 to 0.25 was observed (Figure 4a), which showed that the needle N concentration in *P. massoniana* plantations had stable nitrogenous homeostasis. According to scholars' previous reports, plants with stable stoichiometric homeostasis have an efficient nutrient regulation ability when the soil nutrients do not meet their growth needs, and they are obliged to regulate nutrient composition to store nutrients [51,52]. *P. massoniana* is a fast-growing species in South China, and the rapid biomass and volume increase must be supported by a stable nitrogen supply. The presence of the lower N concentration in soil and the N limitation in fresh needles made plants have to increase the NRE from the senesced needles, which was also consistent with the increasing trend in the NRE before 30-year-old *P. massoniana* plantations. As a whole, the relatively stable stoichiometric homeostasis is one of the adaptive strategies to cope with heterogeneous habitats for plants [53], which could be realised by regulating nutrient resorption from leaves [54].

4.4. Inspiration for P. massoniana Plantation Management

The analysis of the N:P and NRE:PRE ratios in needles and soil along the chronosequence of *P. massoniana* plantations indicated that tree growth might improve the transfer from N to P limitation. Although the P concentration in soil decreased, the PRE maintained an increasing trend along the chronosequence. These results also implied that *P. massoniana* had a certain tolerance to low-P habitats in the earlier stage of the plantation, and a similar characteristic has been reported previously [55]. The soil nutrient concentration was low owing to the thin solum and serious leaching in most of Southern China, particularly N and P deficiency. Therefore, N and P limitations usually occur in artificial forest ecosystems. Similar results have been found in other forests or plantations, such as *C. lanceolata* plantations limited by N and P [2], *Metasequoia glyptostroboides* forests limited by P [56], and Eucalyptus plantations limited by P [57]. Evidently, N and P supplements should also be vital for not only younger but also older plantations. Therefore, introducing mixed tree species will be an effective way to alleviate nutrient demand conflict with the increasing age of artificial plantations. For example, introducing N-fixing tree species improved nitrogen absorption on Eucalyptus plantations [58], and the mixed-species tree plantations could significantly ease or eliminate P limitation in some forest ecosystems [59]. Certainly, the suitable tree species for mixing with *P. massoniana* still require further research in the future. Certainly, efficient fertilisation management strategies could also supply the appropriate means to alleviate nutrient limitations in timber forests.

5. Conclusions

In summary, our results found that a synergic relationship between leaf litter and soil in *P. massoniana* plantations. The significant differences in tree nutrient components at different ages suggested that their ecological stoichiometry and nutrient resorption may be flexible. The soil organic C concentration increased with the increasing age of *P. massoniana* plantations, whereas their N and P concentrations presented the opposite fluctuation trend. Generally, the nutrient element accumulation of fresh and senesced needles increased along the chronosequence. N and P limitation existed in *P. massoniana* plantations before the age of 20 years old and after 30 years old, respectively. The patterns of NRE and PRE initially increased and then decreased along the chronosequence of *P. massoniana* plantations, indicating the increasing nutrient element resorption could promote tree growth. The weak stoichiometric homeostasis would improve nutrient utilization during the development process of *P. massoniana* plantations. To ease nutrient limitation, introducing suitable tree species for mixing with *P. massoniana* should be advocated to realise the more effective cultivation of plantations.

Author Contributions: Q.G. and X.S. conceived and designed the study. Q.G. and H.L. performed the experiments. Q.G. wrote the paper. H.L., X.S. and G.D. reviewed and edited the manuscript. Z.A. performed the constructive comments and writing—review and editing. All authors have read and agreed to the published version of the manuscript.

Funding: This research was funded by the National Natural Science Foundation of China (31971572), China Scholarship Council ([2021]15), and Technological Projects of Guizhou Province, China ([2018]5261).

Data Availability Statement: All data used in this research are publicly available for download. Please contact corresponding author with further inquiries about accessing data.

Conflicts of Interest: The authors declare no conflict of interest.

References

1. Tang, Z.; Xu, W.; Zhou, G.; Bai, Y.; Li, J.; Tang, X.; Chen, D.; Liu, Q.; Ma, W.; Xiong, G.; et al. Patterns of plant carbon, nitrogen, and phosphorus concentration in relation to productivity in China's terrestrial ecosystems. *Proc. Natl. Acad. Sci. USA* **2018**, *115*, 4033–4038. [CrossRef]
2. Chang, Y.; Zhong, Q.; Yang, H.; Xu, C.; Hua, W.; Li, B. Patterns and driving factors of leaf C, N, and P stoichiometry in two forest types with different stand ages in a mid-subtropical zone. *For. Ecosyst.* **2022**, *9*, 100005. [CrossRef]
3. Erb, M.; Lu, J. Soil abiotic factors influence interactions between belowground herbivores and plant roots. *J. Exp. Bot.* **2013**, *64*, 1295–1303. [CrossRef]
4. Chen, R.; Ran, J.; Hu, W.; Dong, L.; Ji, M.; Jia, X.; Lu, J.; Gong, H.; Aqeel, M.; Yao, S.; et al. Effects of biotic and abiotic factors on forest biomass fractions. *Natl. Sci. Rev.* **2021**, *8*, nwab025. [CrossRef]
5. Fisher, J.B.; Malhi, Y.; Torres, I.C.; Metcalfe, D.B.; van de Weg, M.J.; Meir, P.; Silva-Espejo, J.E.; Huasco, W.H. Nutrient limitation in rainforests and cloud forests along a 3,000–m elevation gradient in the Peruvian Andes. *Oecologia* **2012**, *172*, 889–902. [CrossRef]
6. Zhang, P.; Lü, X.T.; Li, M.H.; Wu, T.; Jin, G. N limitation increases along a temperate forest succession: Evidences from leaf stoichiometry and nutrient resorption. *J. Plant Ecol.* **2022**, *15*, 1021–1035. [CrossRef]
7. Vitousek, P.M.; Farrington, H. Nutrient limitation and soil development: Experimental test of a biogeochemical theory. *Biogeochemistry* **1997**, *37*, 63–75. [CrossRef]
8. Ågren, G.I.; Wetterstedt, J.Å.M.; Billberger, M.F.K. Nutrient limitation on terrestrial plant growth—Modeling the interaction between nitrogen and phosphorus. *New Phytol.* **2012**, *194*, 953–960. [CrossRef]
9. Hou, E.; Luo, Y.; Kuang, Y.; Chen, C.; Lu, X.; Jiang, L.; Luo, X.; Wen, D. Global meta–analysis shows pervasive phosphorus limitation of aboveground plant production in natural terrestrial ecosystems. *Nat. Commun.* **2020**, *11*, 637. [CrossRef]
10. Sponseller, R.A.; Gundale, M.J.; Futter, M.; Ring, E.; Nordin, A.; Näsholm, T.; Laudon, H. Nitrogen dynamics in managed boreal forests: Recent advances and future research directions. *Ambio* **2016**, *45*, 175–187. [CrossRef]
11. Elser, J.J.; Fagan, W.F.; Kerkhoff, A.J.; Swenson, N.G.; Enquist, B.J. Biological stoichiometry of plant production: Metabolism, scaling and ecological response to global change. *New Phytol.* **2010**, *186*, 593–608. [CrossRef] [PubMed]
12. Allen, A.P.; Gillooly, J.F. Towards an integration of ecological stoichiometry and the metabolic theory of ecology to better understand nutrient cycling. *Ecol. Lett.* **2009**, *12*, 369–384. [CrossRef]
13. Zechmeister–Boltenstern, S.; Keiblinger, K.M.; Mooshammer, M.; Peñuelas, J.; Richter, A.; Sardans, J.; Wanek, W. The application of ecological stoichiometry to plant–microbial–soil organic matter transformations. *Ecol. Monogr.* **2015**, *85*, 133–155. [CrossRef]
14. Reich, P.B.; Oleksyn, J. Global patterns of plant leaf N and P in relation to temperature and latitude. *Proc. Natl. Acad. Sci. USA* **2004**, *101*, 11001–11006. [CrossRef] [PubMed]
15. Persson, J.; Fink, P.; Goto, A.; Hood, J.M.; Jonas, J.; Kato, S. To be or not to be what you eat: Regulation of stoichiometric homeostasis among autotrophs and heterotrophs. *Oikos* **2010**, *119*, 741–751. [CrossRef]
16. Wang, J.; Wang, J.; Wang, L.; Zhang, H.; Guo, Z.; Wang, G.G.; Smithd, W.K.; Wu, T. Does stoichiometric homeostasis differ among tree organs and with tree age? *Forest Ecol. Manag.* **2019**, *453*, 117637. [CrossRef]
17. Fanin, N.; Fromin, N.; Buatois, B.; Hättenschwiler, S. An experimental test of the hypothesis of non–homeostatic consumer stoichiometry in a plant litter–microbe system. *Ecol. Lett.* **2013**, *16*, 764–772. [CrossRef]
18. Selvaraj, S.; Duraisamy, V.; Huang, Z.; Guo, F.; Ma, X. Influence of long–term successive rotations and stand age of Chinese fir (*Cunninghamia lanceolata*) plantations on soil properties. *Geoderma* **2017**, *306*, 127–134. [CrossRef]
19. Sardans, J.; Rivas-Ubach, A.; Peñuelas, J. The C:N:P stoichiometry of organisms and ecosystems in a changing world: A review and perspectives. *Perspect. Plant Ecol.* **2012**, *14*, 33–47. [CrossRef]
20. Côté, B.; Fyles, J.W.; Djalilvand, H. Increasing N and P resorption efficiency and proficiency in northern deciduous hardwoods with decreasing foliar N and P concentrations. *Ann. For. Sci.* **2002**, *59*, 275–281. [CrossRef]
21. Krishna, M.; Mohan, M. Litter decomposition in forest ecosystems: A review. *Energy Ecol. Environ.* **2017**, *2*, 236–249. [CrossRef]
22. See, C.R.; Yanai, R.D.; Fisk, M.C.; Vadeboncoeur, M.A.; Quintero, B.A.; Fahey, T.J. Soil nitrogen affects phosphorus recycling: Foliar resorption and plant–soil feedbacks in a northern hardwood forest. *Ecology* **2015**, *96*, 2488–2498. [CrossRef]

23. Wang, K.; Wang, G.G.; Song, L.; Zhang, R.; Yan, T.; Li, Y. Linkages between nutrient resorption and ecological stoichiometry and homeostasis along a chronosequence of Mongolian pine plantations. *Front. Plant Sci.* **2021**, *12*, 692683. [CrossRef] [PubMed]
24. Lal, C.; Annapurna, C.; Raghubanshi, A.; Singh, J. Effect of leaf habit and soil type on nutrient resorption and conservation in woody species of a dry tropical environment. *Can. J. Bot.* **2001**, *79*, 1066–1075.
25. Gerdol, R.; Iacumin, P.; Brancaleoni, L. Differential effects of soil chemistry on the foliar resorption of nitrogen and phosphorus across altitudinal gradients. *Funct. Ecol.* **2019**, *33*, 1351–1361. [CrossRef]
26. Pan, J.; Guo, Q.; Li, H.; Luo, S.; Zhang, Y.; Yao, S.; Fan, X.; Sun, X.; Qi, Y. Dynamics of soil nutrients, microbial community structure, enzymatic activity, and their relationships along a chronosequence of *Pinus massoniana* plantations. *Forests* **2021**, *12*, 376. [CrossRef]
27. Xue, L.; Li, Q.; Chen, H. Effects of a wildfire on selected physical, chemical and biochemical soil properties in a *Pinus massoniana* forest in South China. *Forests* **2014**, *5*, 2947–2966. [CrossRef]
28. Liang, H.; Huang, J.G.; Ma, Q.; Li, J.; Wang, Z.; Guo, X.; Zhu, H.; Jiang, S.; Zhou, P.; Yu, B.; et al. Contributions of competition and climate on radial growth of *Pinus massoniana* in subtropics of China. *Agric. For. Meteorol.* **2019**, *274*, 7–17. [CrossRef]
29. Yin, X.; Zhao, L.; Fang, Q.; Ding, G. Differences in soil physicochemical properties in different–aged *Pinus massoniana* plantations in Southwest China. *Forests* **2021**, *12*, 987. [CrossRef]
30. Wang, B.; Chen, J.; Huang, G.; Zhao, S.; Dong, F.; Zhang, Y.; He, W.; Wang, P.; Yan, Z. Growth and nutrient stoichiometry responses to N and P fertilization of 8–year–old Masson pines (*Pinus massoniana*) in subtropical China. *Plant Soil* **2022**, *477*, 343–356. [CrossRef]
31. Li, X.; Su, Y.; Yin, H.; Liu, S.; Chen, G.; Fan, C.; Feng, M.; Li, X. The effects of crop tree management on the fine root traits of *Pinus massoniana* in Sichuan Province, China. *Forests* **2020**, *11*, 351. [CrossRef]
32. Bremner, J. Determination of nitrogen in soil by the Kjeldahl method. *J. Agric. Sci.* **1960**, *55*, 11–33. [CrossRef]
33. Stone, W. A new colorimetric reagent for micro determination of ammonia. *Proc. Soc. Exp. Biol. Med.* **1956**, *93*, 589–591. [CrossRef]
34. Vergutz, L.; Manzoni, S.; Porporato, A.; Novais, R.F.; Jackson, R.B. Global resorption efficiencies and concentrations of carbon and nutrients in leaves of terrestrial plants. *Ecol. Monogr.* **2012**, *82*, 205–220. [CrossRef]
35. Sterner, R.W.; Elser, J.J. *Ecological Stoichiometry: Biology of Elements from Molecules to the Biosphere*; Princeton University Press: Princeton, NJ, USA, 2003; pp. 53–62.
36. Müller, T.; Höper, H. Soil organic matter turnover as a function of the soil clay content: Consequences for model applications. *Soil Biol. Biochem.* **2004**, *36*, 877–888. [CrossRef]
37. Zhang, W.; Qiao, W.; Gao, D.; Dai, Y.; Deng, J.; Yang, G. Relationship between soil nutrient properties and biological activities along a restoration chronosequence of *Pinus tabulaeformis* plantation forests in the Ziwuling Mountains, China. *Catena* **2018**, *161*, 85–95. [CrossRef]
38. Zheng, Y.; Hu, Z.; Pan, X.; Chen, X.; Derrien, D.; Hu, F.; Liu, M.; Hättenschwiler, S. Carbon and nitrogen transfer from litter to soil is higher in slow than rapid decomposing plant litter: A synthesis of stable isotope studies. *Soil Biol. Biochem.* **2021**, *156*, 108196. [CrossRef]
39. Ni, X.; Lin, C.; Chen, G.; Xie, J.; Yang, Z.; Liu, X.; Xiong, D.; Xu, C.; Yue, K.; Wu, F.; et al. Decline in nutrient inputs from litterfall following forest plantation in subtropical China. *Forest Ecol. Manag.* **2021**, *496*, 119445. [CrossRef]
40. Deng, J.; Wang, S.; Ren, C.; Zhang, W.; Zhao, F.; Li, X.; Zhang, D.; Han, X.; Yang, G. Nitrogen and phosphorus resorption in relation to nutrition limitation along the chronosequence of black locust (*Robinia pseudoacacia* L.) plantation. *Forests* **2019**, *10*, 261. [CrossRef]
41. Baker, T.R.; Burslem, D.F.; Swaine, M.D. Associations between tree growth, soil fertility and water availability at local and regional scales in Ghanaian tropical rain forest. *J. Trop. Ecol.* **2003**, *19*, 109–125. [CrossRef]
42. Yan, T.; Lü, X.T.; Zhu, J.J.; Yang, K.; Yu, L.Z.; Gao, T. Changes in nitrogen and phosphorus cycling suggest a transition to phosphorus limitation with the stand development of larch plantations. *Plant Soil* **2018**, *422*, 385–396. [CrossRef]
43. Zhang, W.; Liu, W.; Xu, M.; Deng, J.; Han, X.; Yang, G.; Feng, Y.; Ren, G. Response of forest growth to C:N:P stoichiometry in plants and soils during *Robinia pseudoacacia* afforestation on the Loess Plateau, China. *Geoderma* **2019**, *337*, 280–289. [CrossRef]
44. Waterman, J.M.; Hall, C.R.; Mikhael, M.; Cazzonelli, C.I.; Hartley, S.E.; Johnson, S.N. Short–term resistance that persists: Rapidly induced silicon anti–herbivore defence affects carbon–based plant defences. *Funct. Ecol.* **2021**, *35*, 82–92. [CrossRef]
45. Kobe, R.K.; Lepczyk, C.A.; Iyer, M. Resorption efficiency decreases with increasing green leaf nutrients in a global data set. *Ecology* **2005**, *86*, 2780–2792. [CrossRef]
46. Beroueg, A.; Lecompte, F.; Mollier, A.; Pagès, L. Genetic variation in root architectural traits in *Lactuca* and their roles in increasing phosphorus–use–efficiency in response to low phosphorus availability. *Front. Plant Sci.* **2021**, *12*, 658321. [CrossRef]
47. Ullah, H.; Santiago–Arenas, R.; Ferdous, Z.; Attia, A.; Datta, A. Improving water use efficiency, nitrogen use efficiency, and radiation use efficiency in field crops under drought stress: A review. *Adv. Agron.* **2019**, *156*, 109–157. [CrossRef]
48. Tong, R.; Zhou, B.; Jiang, L.; Ge, X.; Cao, Y. Spatial patterns of leaf carbon, nitrogen, and phosphorus stoichiometry and nutrient resorption in Chinese fir across subtropical China. *Catena* **2021**, *201*, 105221. [CrossRef]
49. Sartori, K.; Violle, C.; Vile, D.; Vasseur, F.; de Villemereuil, P.; Bresson, J.; Gillespie, L.; Fletcher, L.R.; Sack, L.; Kazakou, E. Do leaf nitrogen resorption dynamics align with the slow-fast continuum? A test at the intraspecific level. *Funct. Ecol.* **2022**, *36*, 1315–1328. [CrossRef]

50. Yan, J.; Li, K.; Peng, X.; Huang, Z.; Liu, S.; Zhang, Q. The mechanism for exclusion of *Pinus massoniana* during the succession in subtropical forest ecosystems: Light competition or stoichiometric homoeostasis? *Sci. Rep.* **2015**, *5*, 10994. [CrossRef]
51. Güsewell, S. N: P ratios in terrestrial plants: Variation and functional significance. *New Phytol.* **2004**, *164*, 243–266. [CrossRef]
52. Pan, Y.; Fang, F.; Tang, H. Patterns and internal stability of carbon, nitrogen, and phosphorus in soils and soil microbial biomass in terrestrial ecosystems in China: A data synthesis. *Forests* **2021**, *12*, 1544. [CrossRef]
53. Ci, H.; Guo, C.; Tuo, B.; Zheng, L.T.; Xu, M.S.; Sai, B.L.; Yang, B.Y.; Yang, Y.C.; You, W.H.; Yan, E.R.; et al. Tree species with conservative foliar nutrient status and strong phosphorus homeostasis are regionally abundant in subtropical forests. *J. Ecol.* **2022**, *110*, 1497–1507. [CrossRef]
54. Freschet, G.T.; Cornelissen, J.H.; van Logtestijn, R.S.; Aerts, R. Substantial nutrient resorption from leaves, stems and roots in a subarctic flora: What is the link with other resource economics traits? *New Phytol.* **2010**, *186*, 879–889. [CrossRef] [PubMed]
55. Fan, F.; Wang, Q.; Li, H.; Ding, G.; Wen, X. Transcriptome–wide identification and expression profiles of Masson pine WRKY transcription factors in response to low phosphorus stress. *Plant Mol. Biol. Rep.* **2021**, *39*, 1–9. [CrossRef]
56. Zhang, H.; Wang, J.; Wang, J.; Guo, Z.; Wang, G.G.; Zeng, D.; Wu, T. Tree stoichiometry and nutrient resorption along a chronosequence of *Metasequoia glyptostroboides* forests in coastal China. *Forest Ecol. Manag.* **2018**, *430*, 445–450. [CrossRef]
57. Fan, H.; Wu, J.; Liu, W.; Yuan, Y.; Hu, L.; Cai, Q. Linkages of plant and soil C:N:P stoichiometry and their relationships to forest growth in subtropical plantations. *Plant Soil* **2015**, *392*, 127–138. [CrossRef]
58. Yao, X.; Zhang, Q.; Zhou, H.; Nong, Z.; Ye, S.; Deng, Q. Introduction of *Dalbergia odorifera* enhances nitrogen absorption on *Eucalyptus* through stimulating microbially mediated soil nitrogen–cycling. *For. Ecosyst.* **2021**, *8*, 59. [CrossRef]
59. Forrester, D.I.; Bauhus, J.; Cowie, A.L. On the success and failure of mixed–species tree plantations: Lessons learned from a model system of *Eucalyptus globulus* and *Acacia mearnsii*. *Forest Ecol. Manag.* **2005**, *209*, 147–155. [CrossRef]

Disclaimer/Publisher's Note: The statements, opinions and data contained in all publications are solely those of the individual author(s) and contributor(s) and not of MDPI and/or the editor(s). MDPI and/or the editor(s) disclaim responsibility for any injury to people or property resulting from any ideas, methods, instructions or products referred to in the content.

Article

Relationship between Leaf Scorch Occurrence and Nutrient Elements and Their Effects on Fruit Qualities in Chinese Chestnut Orchards

Rongrong Chen [1,†], Jingle Zhu [2,†], Jiabing Zhao [1,3,†], Xinru Shi [1], Wenshi Shi [4,*], Yue Zhao [1], Jiawei Yan [1], Lu Pei [1], Yunxia Jia [5], Yanyan Wu [1], Haitao Liu [6], Zeping Jiang [7], Changming Ma [3,*] and Shengqing Shi [1,*]

1. State Key Laboratory of Tree Genetics and Breeding, Research Institute of Forestry, Chinese Academy of Forestry, Beijing 100091, China
2. Key Laboratory of Non-Timber Forest Germplasm Enhancement & Utilization of National Forestry and Grassland Administration, Research Institute of Non-Timber Forestry, Chinese Academy of Forestry, Zhengzhou 450003, China
3. Department of Forest Cultivation, College of Forestry, Hebei Agricultural University, Baoding 071000, China
4. Department of Achievement Transformation and Industrial Development, Chinese Academy of Forestry, Beijing 100091, China
5. Zunhua Natural Resources and Planning Bureau, Zunhua 064200, China
6. Huairou Chestnut Experiment Technology and Extension Station, Beijing 102206, China
7. Key Laboratory of Forest Ecology and Environment of National Forestry and Grassland Administration, Ecology and Nature Conservation Institute, Chinese Academy of Forestry, Beijing 100091, China
* Correspondence: shiwens@caf.ac.cn (W.S.); machangming@126.com (C.M.); shi.shengqing@caf.ac.cn (S.S.)
† These authors contributed equally to this work.

Citation: Chen, R.; Zhu, J.; Zhao, J.; Shi, X.; Shi, W.; Zhao, Y.; Yan, J.; Pei, L.; Jia, Y.; Wu, Y.; et al. Relationship between Leaf Scorch Occurrence and Nutrient Elements and Their Effects on Fruit Qualities in Chinese Chestnut Orchards. *Forests* 2023, 14, 71. https://doi.org/10.3390/f14010071

Academic Editor: Wenjie Liu

Received: 10 November 2022
Revised: 22 December 2022
Accepted: 27 December 2022
Published: 30 December 2022

Copyright: © 2022 by the authors. Licensee MDPI, Basel, Switzerland. This article is an open access article distributed under the terms and conditions of the Creative Commons Attribution (CC BY) license (https://creativecommons.org/licenses/by/4.0/).

Abstract: Chinese chestnut (*Castanea mollissima*) is a multipurpose tree providing nuts and timbers, which holds an important position in the mountainous villages in China. However, leaf scorch disease is becoming more and more serious in the chestnut orchards of Yanshan Mountain areas, but the cause of occurrence is still unclear. In this study, the nutrient elements were analyzed from the leaves, roots, and surrounding soils of roots as well as the nut qualities in the healthy and scorched trees from two adjacent chestnut orchards. The results showed that the elements of nitrogen (N), iron (Fe), boron (B), and zinc (Zn) in leaves significantly increased in the scorched trees as well as N and B in roots, and potassium (K), and available potassium (AP) in soils, but leaf magnesium (Mg), root manganese (Mn), and soil Mg, copper (Cu), Fe, and B significantly decreased. Correlation analysis demonstrated that B, Zn, Mg, and Fe had a greater influence on the status of leaf health, and soil AK, K, Fe, B, and Cu had an impact on leaf B concentration. In addition, the occurrence of leaf scorch affected the nut sizes, contents of total soluble proteins and ascorbic acid as well as the catalase activity in the nuts. These results indicated that the disruption of soil-element balance may be one of the main causes resulting in the occurrence of leaf scorch, which would provide a theoretical basis and practical guidance for the prevention of chestnut leaf scorch disease.

Keywords: chestnut; leaf scorch; soil elements; leaf elements; nut quality

1. Introduction

Chestnut trees (*Castanea*), belonging to the family Fagaceae, are economically important fruit trees known as woody grain plants in Asia, North America, and Europe [1]. In China, chestnut cultivation has a long growing history of over 3000 years, which is particularly widespread, being distributed in arid, semi-arid mountainous areas. Its nuts have high contents of nutrients, such as carbohydrates, proteins, and vitamin C, so it is also called a source of nutritional health care [2,3]. In recent years, however, leaf scorch disease is becoming severe and has spread in the chestnut orchards from the Yanshan Mountains, which has threatened the sustainable development of the chestnut industry.

Leaf scorch is a common type of leaf disease in herbaceous or woody plants. Its etiology is very complex, resulting not only from pathogens, but also from environmental factors such as nutrient abundance, temperature, and drought [4–6]. At present, most studies on leaf scorch disease focus on the identification of pathogenic bacteria, while there has not been a systematic investigation of the aspects of cultivation management and nutrient elements [7]. Leaf scorch can be caused by *Neoscytalidium dimidiatum*, and then dieback of twigs, branches, and even whole trees of olive [8], as well as by *Xylella fastidiosa* in olive [9] and almond [10]. Maydis leaf blight, caused by *Cochliobolus heterostrophus*, appears as small lesions with a dark brown margin and a straw to light brown colored center to absolute foliage blight [11]. A serious leaf disease in Chinese chestnut (*C. mollissima*), referred to as the brown margin leaf blight with irregular dry rot in a large area, is caused by *Phomopsis castaneae-mollissimae* [12] coordinated with *Ophiognomonia castaneae* [13], which is different from the yellow or brown wilted leaves caused by *Cryphonectria parasitica* in chestnut blight in Iran [14]. In addition, leaf scorch is also caused by a combination of environmental factors. For example, zinc (Zn) deficiency causes leaf mottled yellowing in citrus trees, which is more serious when citrus trees grow in the sun [15], and the effective Zn content in soils is affected by the joint influence of pH, organic matter, and soil structure [16]. The low potassium (K) content in leaves is related to leaf scorch of *Ginkgo biloba*, and comprehensive environmental factors such as high temperature and drought also cause its physiological scorching phenomena [17], while excessive sodium content in soils may lead to the scorched death of apple leaf margins [18]. High temperature and low rainfall lead to an increase in the content of available phosphorus (P) in soils, which in turn leads to a significant increase in Cl^- and Ca^{2+} in leaves, and a significant decrease in Mg^{2+}, resulting in scorched leaves [19]. However, at present, there are no studies concerning the effect of leaf scorch on fruit quality.

In recent years, field investigation found that a kind of leaf disease, showing the scorching symptom between leaf margins and leaf veins (Figure 1), has become increasingly serious in chestnut orchards in the Yanshan Mountains, which is completely different from the previously reported leaf scorch [12,14]. However, as described by our previous study, no pathogenic bacteria were detected in the isolation and identification of microorganisms in the tested leaf-scorched orchards from five experimental locations in the Yanshan Mountains in 2019 [20]. Therefore, we analyzed and compared (1) the differences in the nutrients of leaves, roots, and soils, and (2) changes in nut phenotypes, nutrients, and antioxidants, from healthy and leaf-scorched trees, respectively, in the same chestnut orchards, to investigate the main factors leading to leaf scorch, and its effects on nut characteristics. This study would provide a basis for the prevention and control of chestnut leaf scorch disease and the scientific management of chestnut orchards in the future.

Figure 1. Healthy and scorched leaves of chestnut trees. ZF: Yanshanzaofeng; ZQ: Zhongqing No. 1.

2. Materials and Methods

2.1. Study Sites and Sample Collection

Sample collections were conducted in two adjacent chestnut orchards where no pathogenic bacteria resulting in leaf scorch were detected in 2019 [20], located on the half-sunny slope of Xiaolugou village (118°45′ E, 40°21′ N), Qinglong Manchu Autonomous County, Hebei province, in early September 2020, and in which there were healthy and scorched trees, respectively. The soil type is leaching cinnamon soil, the texture of the soil is sandy loam, and the pH is 5.9–6.1. The surrounding area is characterized by a temperate climate with an annual precipitation of 715 mm and an average temperature of 8.9 °C. Both orchards were established in 2006 and grafted with the cultivars Yanshanzaofeng (ZF) and Zhongqing No. 1 (ZQ) for 14 years, respectively. The spacing between plants and rows of chestnut trees was approximately 4 m × 5 m. Both orchards were managed using a similar measure by the same farmer, including the application of approximately 450 kg of a compound fertilizer of potassium sulfate (N:P:K = 15:15:15) and 22,500 kg of compost consisting of sheep manure for each hectare after autumn harvest, the spraying of herbicide (glufosinate ammonium) in the early of June and August, and the extra supplement of boron element. In recent years, leaf scorch symptoms were found within the chestnut trees of these two orchards, which occurred piecemeal with an increasing trend in the number of diseased trees.

The chestnut trees were divided into two treatments (groups) of healthy and diseased when collecting samples. The trees with more than 1/3 scorched leaves were identified as the symptomatic (diseased) ones, while the trees without obvious scorched leaves were identified as the asymptomatic (healthy) ones. Three sampling sites for healthy and diseased trees, respectively, were randomly selected in each orchard, with each sampling site consisting of three healthy or scorched trees. The samples were randomly collected from the east, west, south, and north sides of each tree. Leaves and fine roots (<3 mm diameter) and nuts were collected in the selected trees as well as the corresponding soils surrounding the roots. The collected samples in each sampling site were mixed and brought back to the laboratory.

2.2. Determination of Nutrient Elements

The collected leaves and roots were rinsed with tap water and deionized water, in turn, heated at 105 °C for 30 min, and then dried continuously at 75 °C in an oven until a constant weight was reached. Soil samples were air-dried naturally, crushed, and passed through an 80-mesh sieve prior to the element analysis. Total N was determined by Kjeldahl's method [21]. Available potassium (AK), available phosphorus (AP), phosphorus (P), potassium (K), sodium (Na), calcium (Ca), magnesium (Mg), copper (Cu), zinc (Zn), iron (Fe), manganese (Mn), sulfur (S), boron (B), and molybdenum (Mo) were subjected to wet mineralization by treating 0.5 g of dry sample with 6 mL of hydrochloric acid (37%), 2 mL of nitric acid (65%) and 2 mL of hydrogen peroxide (30%). After the mixed solutions were filtered, the element concentrations were determined by ICP-OES (Ametek Spectro, Arcos, Kleve, Germany) [22].

2.3. Determination of Nut Qualities

Morphological traits: The transverse diameter, vertical diameter, and lateral diameter of the fresh nuts were determined with vernier calipers. Every treatment included three replicates, each consisting of 30 nuts, and the measurement accuracy is 0.1 mm. The fruit shape index was calculated by the formula: nut shape index = vertical diameter/transverse diameter, and the nut volume was calculated by the formula: nut volume = transverse diameter × vertical diameter × lateral diameter. Single seed fresh was calculated by randomly weighing 10 fresh nuts, which were then used for the determination of the water content. The formula of water content = (fresh weight − dry weight)/fresh weight × 100%.

Nutrient compositions: Total soluble sugars and starch were analyzed according to the anthrone-sulfuric acid method [23,24], and total amino acids were analyzed according to the Ninhydrin hydrate method [25], by using commercial assay kits (Comin Biotechnology, Suzhou, China). Total soluble proteins were determined by using the Bradford Protein Assay Kit (Sangon Biotech, Shanghai, China).

Antioxidant compounds: Total polyphenols were determined by using the Folin–Ciocalteu colorimetric method [26]. Total flavonoids were determined by using the aluminum nitrate method [27,28]. Ascorbic acid (ASA), superoxide dismutase (SOD), peroxidase (POD), catalase (CAT), and polyphenol oxidase (PPO) activities were determined by using commercial assay kits (Comin Biotechnology, Suzhou, China).

2.4. Statistical Analysis

The statistical analysis was performed by one-way ANOVA and factor analysis at a 5% level of significance using the SPSS software (v25.0). GraphPad Prism (GraphPad Software Inc., San Diego, CA, USA) was used for plotting. Correlation heatmaps and redundancy analysis (RDA) were performed using the OmicStudio tools at https://www.omicstudio.cn/tool, accessed on 18 August 2022.

3. Results

3.1. Leaf Phenotypes of Healthy and Diseased Chestnut Trees

The leaves of the two cultivars, ZF (Yanshanzaofeng) and ZQ (Zhongqing No. 1), in the areas of the Yanshan Mountains, showed different degrees of yellowish brown color after scorching (Figure 1), and the main symptoms of leaf margins and interveinal tissues showed a continuous brown dry state, but this did not occur in the leaf veins and petioles. Other than brown and green, there was no obvious transition color.

3.2. Relationship between the Occurrence of Leaf Scorch and Nutrients in Chestnut Orchards

3.2.1. Comparison of Nutrient Elements of Healthy and Disease Trees

The variance analysis of nutrient elements from soils (Table S1), roots (Table S2), and leaves (Table S3), demonstrated that there were no significant differences in the contents of most elements between healthy and diseased trees of ZF and ZQ cultivars, indicating that the varieties were not the main factor of leaf scorch symptoms. Thus, the difference in elements between the two varieties was ignored in this study.

The coefficients of variation (C.V.) between the elements varied greatly in leaves (Table 1). Compared with the healthy trees, the contents of leaf N, Fe, B, and Zn in diseased trees significantly increased by 27.61%, 20.74%, 61.66%, and 31.37%, respectively, while the Mg content significantly decreased by 16.92%, but no significant changes were detected in other elements between healthy and diseased trees.

Table 1. Leaf nutrient elements in healthy and diseased chestnut trees.

Elements	Healthy Trees (g/kg)			Diseased Trees (g/kg)		
	Range of Variations	The Average ± Standard Error	C.V.%	Range of Variations	The Average ± Standard Error	C.V.%
N	20.105–24.285	22.582 ± 0.606	6.572	26.693–31.619	28.816 ± 0.769 ***	6.535
P	1.189–2.289	1.698 ± 0.166	23.910	1.564–2.44	2.061 ± 0.143	16.982
K	4.073–6.644	5.465 ± 0.436	19.543	5.643–11.475	7.793 ± 0.976	30.694
Ca	13.193–22.272	16.617 ± 1.306	19.257	11.338–16.533	14.781 ± 0.816	13.517
Mg	6.561–9.128	7.493 ± 0.37	12.091	5.098–7.705	6.225 ± 0.4 *	15.727
Fe	0.136–0.202	0.172 ± 0.011	15.698	0.172–0.261	0.217 ± 0.015 *	17.051
B	0.082–0.196	0.143 ± 0.019	32.867	0.311–0.425	0.373 ± 0.015 ***	9.920
Cu	0.006–0.007	0.007 ± 0	14.286	0.007–0.01	0.008 ± 0	12.500
Mn	0.88–2.368	1.499 ± 0.207	33.756	0.706–2.737	1.509 ± 0.332	53.943
Zn	0.027–0.046	0.035 ± 0.003	20.000	0.036–0.059	0.051 ± 0.003 **	15.686
Mo	0.001–0.001	0.001 ± 0	0.000	0.001–0.001	0.001 ± 0	0.000

Note: "***": $p < 0.001$; "**": $p < 0.01$; "*": $p < 0.05$. The same below in Tables 2 and 3.

Table 2. Root nutrient elements in healthy and diseased chestnut trees.

Elements	Healthy Trees (g/kg)			Diseased Trees (g/kg)		
	Range of Variations	The Average ± Standard Error	C.V.%	Range of Variations	The Average ± Standard Error	C.V.%
N	1.571–10.164	7.448 ± 1.247	41.018	7.733–13.87	11.439 ± 0.897 *	19.206
P	0.819–1.54	1.036 ± 0.107	25.193	0.558–1.154	0.896 ± 0.085	23.103
K	3.876–5.202	4.62 ± 0.201	10.671	4.172–6.88	5.576 ± 0.384	16.876
Ca	9.342–13.696	11.964 ± 0.676	13.850	10.537–13.369	12.305 ± 0.469	9.330
Mg	2.277–3.635	2.871 ± 0.201	17.102	2.218–2.863	2.496 ± 0.107	10.457
Fe	0.62–1.991	1.198 ± 0.236	48.247	0.334–1.595	0.88 ± 0.207	57.727
B	0.009–0.049	0.024 ± 0.006	62.500	0.051–0.114	0.081 ± 0.011 **	34.568
Cu	0.007–0.012	0.01 ± 0.001	20.000	0.007–0.009	0.008 ± 0	12.500
Mn	0.054–0.103	0.079 ± 0.008	25.316	0.035–0.061	0.051 ± 0.004 *	19.608
Zn	0.02–0.04	0.028 ± 0.003	28.571	0.018–0.024	0.022 ± 0.001	13.636
Mo	0.001–0.002	0.001 ± 0	100.000	0–0.002	0.001 ± 0	100.000

Note: "**": $p < 0.01$; "*": $p < 0.05$.

Table 3. Soil nutrient elements in healthy and diseased chestnut trees.

Elements	Healthy Trees (g/kg)			Diseased Trees (g/kg)		
	Range of Variations	The Average ± Standard Error	C.V.%	Range of Variations	The Average ± Standard Error	C.V.%
N	10.198–11.346	10.765 ± 0.183	4.171	10.283–12.068	11.14 ± 0.316	6.957
P	0.367–3.299	1.847 ± 0.538	71.305	0.613–1.427	0.99 ± 0.121	29.798
K	16.968–23.042	20.643 ± 1.051	12.464	20.207–40.026	29.372 ± 3.035 *	25.306
AK	0.041–0.109	0.07 ± 0.01	35.714	0.099–0.187	0.136 ± 0.017 *	30.882
AP	0.008–0.087	0.042 ± 0.014	83.333	0.015–0.06	0.045 ± 0.007	37.778
Na	11.104–16.145	12.452 ± 0.759	14.937	6.827–15.256	11.156 ± 1.49	32.718
Ca	6.81–22.967	14.939 ± 2.389	39.173	5.807–12.167	9.236 ± 0.932	24.708
Mg	12.736–21.32	17.506 ± 1.171	16.383	12.344–16.328	14.092 ± 0.607 *	10.545
Cu	0.025–0.056	0.041 ± 0.004	24.390	0.013–0.02	0.017 ± 0.001 ***	17.647
Zn	0.086–0.156	0.124 ± 0.011	21.774	0.088–0.127	0.104 ± 0.006	13.462
Fe	45.268–64.584	54.392 ± 2.728	12.283	37.756–48.599	44.177 ± 1.529 *	8.480
Mn	0.456–1.075	0.754 ± 0.087	28.117	0.511–0.759	0.63 ± 0.034	13.016
B	0.048–0.072	0.060 ± 0.004	15.000	0.04–0.053	0.048 ± 0.002 *	10.417
S	0.085–0.32	0.174 ± 0.035	50.000	0.058–0.741	0.26 ± 0.103	96.923

Note: "***": $p < 0.001$; "*": $p < 0.05$.

Similarly, the C.V. values between the elements varied greatly in roots (Table 2). Compared with the healthy trees, the contents of root N and B in diseased trees significantly increased by 34.89% and 70.37%, respectively, while the Mn content significantly decreased by 35.44%, but no significant changes were found in other elements between healthy and diseased trees.

The C.V. values between the elements also varied greatly in soils (Table 3). Compared with the healthy trees, the soil K and AK content in diseased trees significantly increased by 29.72% and 48.53%, respectively, while the Mg, Cu, Fe, and B content significantly decreased by 19.50%, 58.54%, 18.78%, and 20.00%, but no significant changes were shown in other elements between healthy and diseased trees.

3.2.2. Correlation Analysis between Elements in Various Parts

There was a significant positive correlation between soil Mg and root Mn as well as soil AK and root B, while soil Cu, B, and Fe were significantly negatively correlated with root B. (Figure 2a). The root N was significantly positively correlated with the leaf Zn, Fe, and N, and root B was significantly positively correlated with the leaf Zn, Fe, N, and B, but root Mn was significantly negatively correlated with the leaf Zn, N, and B (Figure 2b). The soil AK was significantly positively correlated with leaf B, and soil K was significantly positively correlated with leaf N and B, but negatively correlated with leaf Mg. The soil Cu was significantly negatively correlated with leaf Zn, Cu, N, and B, but positively correlated with leaf Mg. The soil B and Fe were significantly negatively correlated with leaf B (Figure 2c).

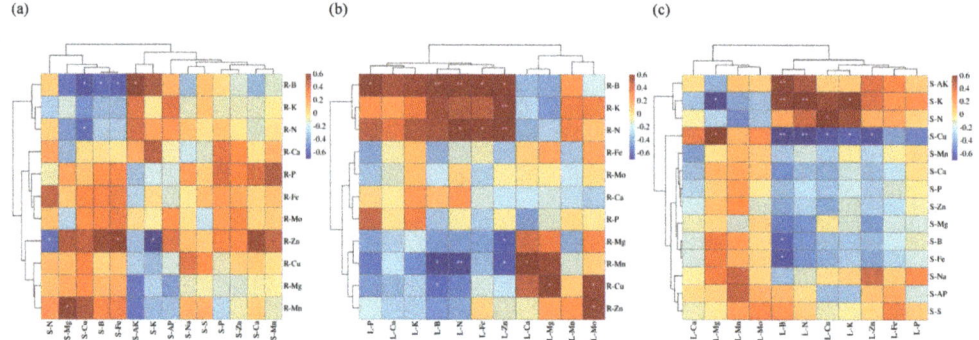

Figure 2. Clustering correlation heatmap with signs in various parts: (**a**) Soil (S) and root (R); (**b**) Root and leaf (L); (**c**) Soils and leaf. "**": $p < 0.01$; "*": $p < 0.05$.

3.2.3. Chestnut Leaf Factor Analysis

Factor analysis was performed based on the results of 11 determined elements in healthy leaves (Table 4) and diseased leaves (Table 5). In healthy leaves, it can be seen that the cumulative variance contribution rate of the first four main factors accounted for 96.78% of the total variance of the original variables, which largely retains the characteristics, differences, and interrelationships of the original 11 variables (Table 4), so the complex relationships between the 11 elements in the healthy leaves can be converted into 4 unrelated comprehensive indicators. From the loading point of view of the factors, the first main factor is mainly determined by N, K, and B, the second main factor is mainly determined by Fe, Mn, and Zn, the third main factor is mainly determined by P and Cu, and the fourth main factor is mainly determined by Mo.

Table 4. Analysis of nutrient factors in healthy leaves.

Characters	Principal Component 1	Principal Component 2	Principal Component 3	Principal Component 4
N	0.991	0.035	0.037	−0.116
P	0.469	0.035	0.674	−0.397
K	0.847	−0.479	0.208	0.043
Ca	−0.963	0.024	−0.222	−0.121
Mg	−0.918	0.083	−0.348	0.132
Fe	−0.019	0.814	0.131	0.525
B	0.841	0.014	−0.184	−0.500
Cu	0.120	0.199	0.940	−0.128
Mn	−0.182	0.940	0.058	−0.258
Zn	0.065	0.752	0.594	0.277
Mo	−0.109	0.065	−0.261	0.932
Eigenvalues	5.129	2.915	1.597	1.005
Variance contribution rate/%	46.631	26.500	14.515	9.138
Cumulative contribution rate/%	46.631	73.130	87.645	96.784

In diseased leaves, the cumulative contribution rate of the 11 variances of the first 4 main factors accounted for 90.63% of the total variance of the original variables (Table 5). Similarly, the relationship between the 11 elements in the scorched leaf can be converted into 4 unrelated comprehensive indicators. The first main factor is mainly determined by N, K, and Zn, the second main factor is mainly determined by Mg and Mn, and the third main factor is mainly determined by P, Fe, and Cu. In this way, the contribution of 11 elements to the scorched leaves of chestnut leaves can be divided into three categories: {N, K, Zn}, {Mg, Mn}, {P, Fe, Cu}.

Table 5. Analysis of nutrient element factors of diseased leaves.

Characters	Principal Component 1	Principal Component 2	Principal Component 3
N	0.718	−0.596	0.306
P	0.408	0.584	0.668
K	0.741	−0.405	0.469
Ca	−0.861	0.359	−0.169
Mg	−0.303	0.865	0.093
Fe	−0.207	0.500	0.810
B	−0.983	0.023	0.136
Cu	0.131	−0.301	0.913
Mn	0.033	0.990	0.022
Zn	0.815	0.431	0.056
Mo	−0.109	−0.055	−0.802
Eigenvalues	4.624	3.376	1.969
Variance contribution rate/%	42.037	30.693	17.903
Cumulative contribution rate/%	42.037	72.730	90.633

3.3. Effects of Leaf Scorch Occurrence on Nut Qualities of Chestnuts

3.3.1. Changes in Nut Phenotypic Index

Leaf scorch showed significant effects on nut phenotypes (Figure 3), including a significant reduction in the transverse diameter in both cultivars, the fruit shape index in ZQ, and the vertical diameter, nut volume, and single seed fresh in ZF. Compared with those in the healthy ones, the nut volume and single seed fresh considerably decreased by 59.80% and 36.78%, respectively, indicating that the occurrence of leaf scorch severely affected the nut production of the ZF cultivar.

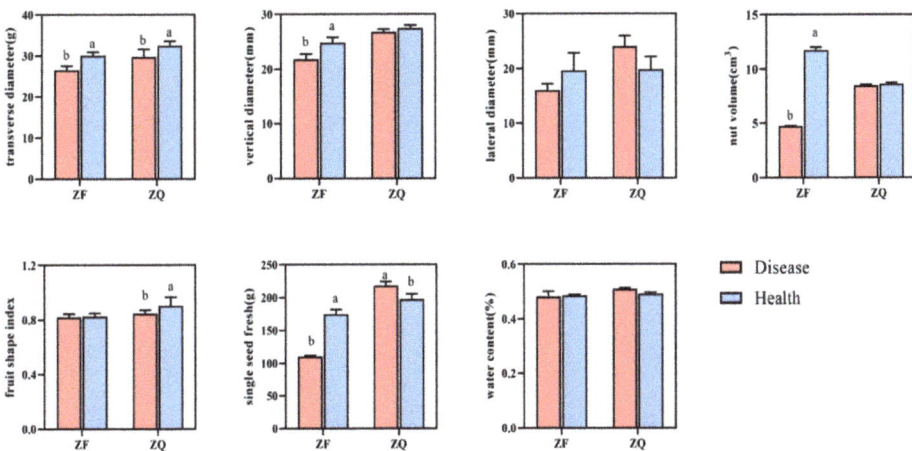

Figure 3. Changes in nut phenotypic traits in healthy and diseased chestnut trees. The significance of difference was labeled as 'a' and 'b' between healthy and diseased trees.

3.3.2. Changes in Nut Nutrients and Antioxidants

At the nutrient level, leaf scorch had almost no significant effect on starch, soluble sugars, and total amino acids, as well as total soluble proteins, except for its significant decrease in ZQ (Figure 4). Similarly, leaf scorch had no significant influences on the total polyphenols, total flavonoids, SOD, and POD in the two cultivars, but significantly increased the CAT activity in ZF and the ASA content in ZQ, compared with the healthy trees, indicating that the different cultivars may resist leaf scorch disease through different antioxidants.

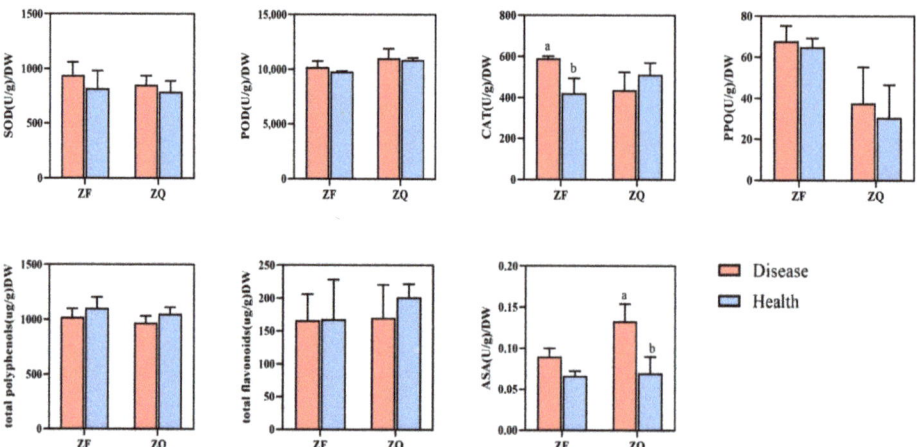

Figure 4. Changes in nut nutrients and antioxidants in healthy and diseased chestnut trees. The significance of difference was labeled as 'a' and 'b' between healthy and diseased trees.

3.3.3. Correlation Analysis of Soil Elements and Nut Qualities

Redundancy analysis (RDA) showed that RDA1 and RDA2 explained 33.88% and 18.39% of the total variations, respectively (Figure 5a). In healthy trees, Mg, Fe, B, Cu, Zn, Ca, P, total amino acids, total polyphenols, total flavonoids, and CAT were closely related to each other, revealing their importance in the health of chestnuts, while N, K, AK, AP, PPO, SOD, ASA, POD, and soluble sugars were closely related to each other in diseased trees, indicating that these parameters played important roles in the response to the occurrence of leaf scorch. The Pearson correlation analysis (Figure 5b) demonstrated that N was significantly positively correlated with total soluble proteins, but AP was contrary; N was negatively correlated with POD, but AK and K were significantly positively correlated with ASA and CAT, respectively, indicating the application of N may increase growth but decrease the resistance, but K can increase the resistance by enhancing the activities of antioxidants. Mg was positively correlated with starch but negatively correlated with ASA, as were Cu, Fe, and B.

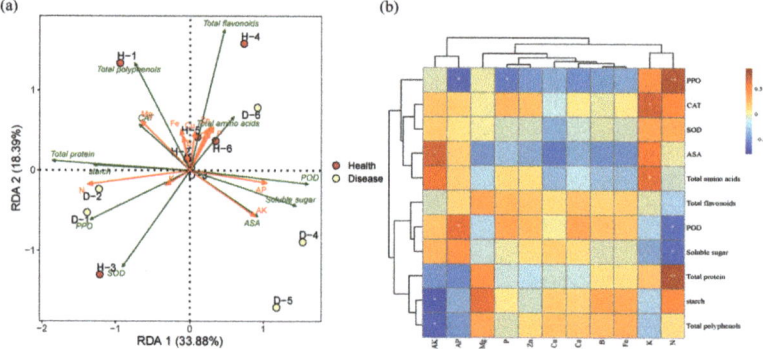

Figure 5. Correlation of nut characteristics with soil elements. (**a**) Redundancy analysis (RDA). The color ranges of red and green lines (**a**) represented positive and negative correlation, respectively; (**b**) Pearson correlation heatmap. * and ** represented the significance at 0.05 and 0.01 levels, respectively.

3.3.4. Correlation Analysis of Leaf Elements and Nut Qualities

The analysis of RDA showed that RDA1 and RDA2 explained the total change of 33.88% and 18.39% respectively (Figure 6a). The nut characteristics of healthy trees were significantly different from those of leaf-scorched trees, which were initially separated by axis 2 and negatively correlated with B, Mn, Fe, and Zn. In addition, B and Mg had a significant positive correlation with ASA and soluble sugars, but a negative correlation with CAT and total soluble proteins. Zn was significantly positively correlated with SOD and ASA, but significantly negatively correlated with CAT, total polyphenols, and total flavonoids. The Pearson correlation analysis showed that Zn was significantly positively correlated with SOD, and Fe was significantly negatively correlated with total polyphenols, furthermore, B was significantly positively correlated with ASA, and Mg was negatively correlated with CAT (Figure 6b).

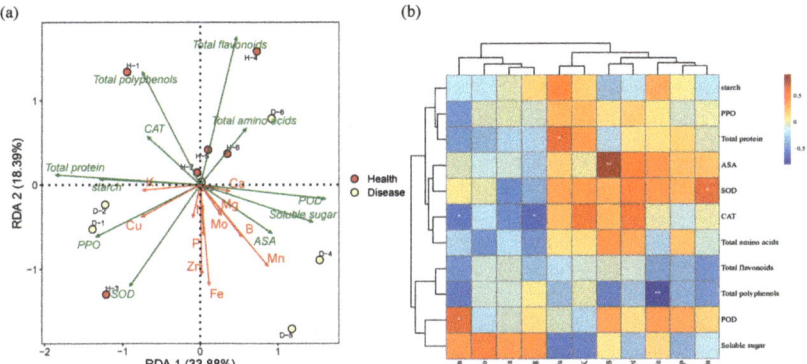

Figure 6. Correlation of nut characteristics with leaf elements. (**a**) Redundancy analysis (RDA). The color ranges of red and green lines (**a**) represented positive and negative correlation, respectively; (**b**) Pearson correlation heatmap. * and ** represented the significance at 0.05 and 0.01 levels, respectively.

4. Discussion

During the management of chestnut orchards, B is the most important trace element because of its role in pollination and fertilization; thus, farmers often apply extra B fertilizer to reduce the empty percentage of chestnut fruits. However, it is very easy to cause B toxicity due to the excessive application in practice because farmers have not mastered scientific fertilization methods. It was also found in our field investigation that excessive application of B fertilizer by chestnut farmers caused the whole leaves to wither. B toxicity is regarded as one of the limiting factors affecting plant growth [29]. After an overdose, B is first concentrated in the old leaves, starting from the leaf tip and leaf margin of the plants. A common symptom of B overdose is the appearance of yellow borders along the leaf margins, and symptoms of loss of green color, scorching, and necrosis occur at sites with high B concentrations [30–32]. In this study, leaf B was higher by over 160.8% in scorched trees compared with healthy trees. Excess B can arouse leaf scorch on old leaves and delay development in spinach [33], significantly inhibit chlorophyll content and cause the tips of leaves to undergo chlorosis and necrosis in rice seedlings [34,35], and lead to yellow areas at the top of the leaves and their margins in citrus plants [36]. The symptoms described in these studies are similar to the symptoms of leaf scorch in the leaf-scorched ZF and ZQ cultivars. In addition, B toxicity can induce membrane damage, altering the plant's antioxidant defense system [37] as, for example, in the increase in CAT activity of sunflowers leaves [38], *Cucurbita pepo* and *Cucumis sativus* [39] and pepper plants [40], and the increase in ASA of citrus plants [41], which are consistent with results in this study. It is also interesting to note that leaf B in the healthy trees reached 0.143 g/kg, far higher than that in roots (0.024 g/kg) and soils (0.060%), while B contents in leaf, root, and soils are all lower than 0.100 g/kg in other chestnut orchards without leaf scorch occurrence (our unpublished data), further indicating the excessive B application in the selected chestnut orchards. The nut phenotypes of scorched chestnut trees were also inhibited to a certain extent in this study, which was similar to the previous results in wheat [42,43], tomatoes, and cucumbers [44].

Magnesium is one of the key elements that make up chlorophyll and has an important role in photosynthetic organs [45,46]. At the beginning of leaf Mg deficiency, the inter-vein color of the leaf tip and leaf margin fades, while the veins remain green, forming clear veins on the leaves; the leaves even dry and fall off in severe cases [47]. In this study, a lower Mg content of soils (Table 3) and roots (Table 2) may result in a significant decrease in leaf Mg content (Table 1), which may be one of the key reasons for the occurrence of leaf scorch in the diseased chestnut trees.

Zinc and Fe are essential elements required for chestnut growth, but excess Zn and Fe can inhibit growth and affect the colors of leaves [42,43,45,46]. In this study, however, the ranges of Zn and Fe contents in both leaves and roots were within the normal level of Zn of between 0.030 and 0.200 mg/kg DW [48,49] and Fe of between 64 and 250 mg/kg (DW) [50], which were also around or lower than the values of Zn and Fe in other chestnut orchards without leaf scorch occurrence (our unpublished data), although both were higher in scorched leaves compared to healthy ones. This indicated that the symptom of chestnut leaf scorch in the present work may be not caused by the levels of zinc or Fe.

Overall, the difference in B and Mg contents in the scorched leaves, based on this study, suggested that synergistic and/or antagonistic effects on micro- and macro-nutrients existed in soils and chestnut trees (Figure 7). Interactions among mineral elements affect plant nutrient status and plant health and yield [51]. Previous studies showed that applying K fertilizer in soils can lead to an increase in B concentration in soybean leaves [52,53], accompanied by a reduction in soil Mg concentration, and thus, decrease the accumulation and concentration of Mg in leaves [45,52]. In this work, the correlation analysis showed that the increase in K and AK in soils had a promotive role in the accumulation of B in leaves and roots by sacrificing B in soils, but had an inhibiting role in leaf Mg content (Figure 7; Tables 1–3), which was consistent with a previous report [52].

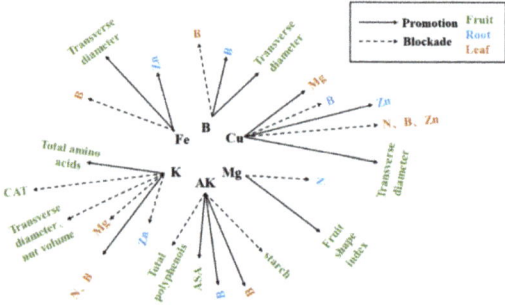

Figure 7. An action model of the altered soil elements on root elements, leaf elements, and nut characteristics.

An increasing N content prompts the demand for B in the plant [51], and the N content is significantly increased in roots (Table 2), while the Mn content in the roots, which is not conducive to boron increase [51], is significantly reduced (Table 2), coupled with the accumulation of B in roots, resulting in an extremely significant difference in the B content in leaves. The Zn content can be affected by fertilization and the balance of nutrient elements in plants [51]. The application of N fertilizer increases Zn concentration and bioavailability in wheat [54]. In this study, an increase in leaf Zn may be caused by the increase in root N of diseased trees, which may result from the excessive application of N fertilizer in order to improve the yield of chestnut orchards.

Soil is the carrier of tree growth, and the nutritional status of soil directly affects the growth, yield, and quality [55]. The available B content in soils was significantly positively correlated with citrus fruit diameter [56], similar to the results in this study. The AK content in soils is the determinant of ASA content in peppers [57], and excess K can significantly increase the ASA content in apples and *Prunus persica* [58,59]. In this study, the AK content was also positively correlated with the ASA content in the fruit. In addition, excess K leads to a decrease in the CAT activity in rapeseed but increases the free amino acid content in wheat [60,61], consistent with the results of this work. In general, leaf scorch increased the ASA content and CAT activity in chestnut nuts, which can enhance fruit quality to a certain extent, although these changes were the consequence of tree adaptation to the occurrence of leaf scorch.

5. Conclusions

This study showed that chestnut leaf scorch may be caused by excessive boron content and Mg deficiency in the leaves, which may be caused by the alteration of the AK, B, Mg, Cu, and Fe balances in soils, of which the reduction of Mg is most likely caused by AK. In addition, leaf scorch had a significant impact on the phenotypes and characteristics of the chestnut tree nuts. Thus, we would hope that the present work could provide a theoretical basis and practical guidance for the prevention of chestnut leaf scorch disease. In addition, it is also urgent to conduct a deep investigation into the specific factors causing leaf scorch in the future.

Supplementary Materials: The following supporting information can be downloaded at https://www.mdpi.com/article/10.3390/f14010071/s1, Table S1: Analysis of leaf element variance of ZF and ZQ; Table S2: Analysis of root element variance of ZF and ZQ; Table S3: Analysis of soil element variance of ZF and ZQ.

Author Contributions: Manuscript draft: R.C., J.Z. (Jiabing Zhao) and S.S.; experiment: J.Z. (Jiabing Zhao), R.C., J.Z. (Jingle Zhu), Y.Z., Y.J., H.L. and Y.W.; analysis: R.C., J.Z. (Jiabing Zhao), J.Z. (Jingle Zhu), C.M., X.S., Y.Z., J.Y., L.P., W.S., Z.J; conception and supervision: S.S., Z.J., C.M. and W.S. All authors have read and agreed to the published version of the manuscript.

Funding: This work was supported by grants from the National Key R & D Program of China (2020YFD1000702), Project of Science and Technology Development Center of State Forestry and grassland administration (KJZXSA2019045); and the Science & Technology Research and Development Program of Hebei Province (16226312D).

Data Availability Statement: Not applicable.

Acknowledgments: We thank Ruzhen Jiao's group for their kind help during element analysis, and we also thank Zhiwei Wang, the chestnut orchardist from Xiaolugou village, Qinglong Manchu Autonomous County, Hebei province, for providing us with access to the experimental sites.

Conflicts of Interest: The authors declare no conflict of interest.

References

1. Li, R.; Sharma, A.K.; Zhu, J.C.; Zheng, B.; Xiao, G.S.; Chen, L. Nutritional biology of chestnuts: A perspective review. *Food Chem.* **2022**, *395*, 133575. [CrossRef] [PubMed]
2. Fernandes, Â.; Barreira, J.C.M.; Antonio, A.L.; Bento, A.; Botelho, M.L.; Ferreira, I.C.F.R. Assessing the effects of gamma irradiation and storage time in energetic value and in major individual nutrients of chestnuts. *Food Chem. Toxicol.* **2011**, *49*, 2429–2432. [CrossRef]
3. Li, M.; Yang, S.Z.; Peng, L.T.; Zeng, K.F.; Feng, B.R.; Yang, J.J. Compositional shifts in fungal community of chestnuts during storage and their correlation with fruit quality. *Postharvest Biol. Technol.* **2022**, *191*, 111983. [CrossRef]
4. Hepting, G.H. Death of the American Chestnut. *J. For. Hist.* **1974**, *18*, 60–67. [CrossRef]
5. Anagnostakis, S.L.; Hillman, B. Evolution of the chestnut tree and its blight. *Chestnut Blight* **1992**, *52*, 2–10.
6. Sicard, A.; Zeilinger, A.R.; Vanhove, M.; Schartel, T.E.; Beal, D.J.; Daugherty, M.P.; Almeida, R.P.P. Xylella fastidiosa: Insights into an emerging plant pathogen. *Annu. Rev. Phyto. Pathol.* **2018**, *56*, 181–202. [CrossRef]
7. Gong, S.; Zhang, X.T.; Jiang, S.X.; Chen, C.; Ma, H.B.; Ni, Y. A new species of *Ophiognomonia* from Northern China inhabiting the lesions of chestnut leaves infected with *Diaporthe eres*. *Mycol. Prog.* **2017**, *16*, 83–91. [CrossRef]

8. Güney, I.G.; Özer, G.; Türkolmez, S.; Dervis, S. Canker and leaf scorch on olive (*Olea europaea* L.) Caused by Neoscytalidium Dimidiatum Turkey. *Crop Prot.* **2022**, *157*, 105985. [CrossRef]
9. Bucci, E.M. *Xylella fastidiosa*, a new plant pathogen that threatens global farming: Ecology, molecular biology, search for remedies. *Biochem. Biophys. Res. Commun.* **2018**, *502*, 173–182. [CrossRef]
10. Anguita-Maeso, M.; Ares-Yebra, A.; Haro, C.; Román-Écija, M.; Olivares-García, C.; Costa, J.; Marco-Noales, E.; Ferrer, A.; Navas-Cortés, J.A.; Landa, B.B. *Xylella fastidiosa* infection reshapes microbial composition and network associations in the xylem of almond trees. *Front. Microbiol.* **2022**, *13*, 866085. [CrossRef]
11. Kumar, B.; Choudhary, M.; Kumar, K.; Kumar, P.; Kumar, S.; Bagaria, P.K.; Sharma, M.; Lahkar, C.; Singh, B.K.; Pradhan, H.; et al. Maydis leaf blight of maize: Update on status, sustainable management and genetic architecture of its resistance. *Physiol. Mol. Plant Pathol.* **2022**, *121*, 101889. [CrossRef]
12. Jiang, S.; Liu, C.; Wang, Q.; Jia, N.; Li, C.; Ma, H. A new chestnut disease—brown margin leaf blight and the pathogen identification. *Sci. Silvae Sin.* **2011**, *47*, 177–180, (Abstract in English).
13. Ma, H.; Liu, G.; Gong, S.; Zhang, X.; Wang, Q.; Jiang, S. Pathogenic effects of *Ophiognomonia castaneae* coordinated with *Phomopsis castaneae-mollissimae*. *Shandong Agric. Sci.* **2016**, *48*, 109–112, 115, (Abstract in English).
14. Kazempour, M.N.; Khodaparast, S.A.; Salehi, M.; Amanzadeh, B.; Nejat-Salary, A.; Shiraz, B.K. First record of chestnut blight in Iran. *J. Plant Pathol.* **2006**, *88*, 121–125.
15. Hacisalihoglu, G.; Kochian, L.V. How do some plants tolerate low levels of soil zinc? Mechanisms of zinc efficiency in crop plants. *New Phytol.* **2003**, *159*, 341–350. [CrossRef]
16. Srivastava, K.; Singh, S. Zinc nutrition, a global concern for sustainable citrus production. *J. Sustain. Agric.* **2013**, *25*, 37–41. [CrossRef]
17. Liang, C. A Preliminary Study on Leaf Scorch Phenomenon of Roadside Ginkgo Trees in Beijing Urban Area. Master's Thesis, Beijing Forestry University, Beijing, China, 2016. (Abstract in English).
18. Guo, Q.E.; Guo, T.W.; Nan, L.L.; Wang, Y.Q. Preliminary survey on the cause of withered and die off of apple leaf in Qin'an county of Gansu. *Acta Agric. Boreali-Occident. Sin.* **2006**, *15*, 68–70, (Abstract in English).
19. Li, Y.; Pu, S.H.; Ma, X.P.; Zhang, J.F.; Li, Q.J. Study on characteristics and causation of walnut withered leaf symptom. *Xinjiang Agric. Sci.* **2022**, *59*, 1475–1481, (Abstract in English).
20. Ren, F.; Dong, W.; Shi, S.Q.; Wang, H.H.; Dou, G.M.; Yan, D.H. Primary study on causes and associated pathogens for chestnut leaf scorch. *For. Res.* **2021**, *34*, 185–192, (Abstract in English).
21. Zeng, S.C.; Chen, B.G.; Jiang, C.A.; Wu, Q.T. Impact of fertilization on chestnut growth, N and P concentrations in runoff water on degraded slope land in South China. *J. Environ. Sci.* **2007**, *19*, 827–833. [CrossRef]
22. Baldi, E.; Cavani, L.; Mazzon, M.; Marzadori, C.; Quartieri, M.; Toselli, M. Fourteen years of compost application in a commercial nectarine orchard: Effect on microelements and potential harmful elements in soil and plants. *Sci. Total Environ.* **2021**, *752*, 141894. [CrossRef] [PubMed]
23. Yemm, E.W.; Willis, A.J. The estimation of carbohydrates in plant extracts by anthrone. *Biochem. J.* **1954**, *57*, 508–514. [CrossRef] [PubMed]
24. López-Delgado, H.; Zavaleta-Mancera, H.A.; Mora-Herrera, M.E.; Vázquez-Rivera, M.; Flores-Gutiérrez, F.X.; Scott, I.M. Hydrogen peroxide increases potato tuber and stem starch content, stem diameter, and stem lignin content. *Am. J. Potato Res.* **2005**, *82*, 279–285. [CrossRef]
25. Chen, Y.Y.; Fu, X.M.; Mei, X.; Zhou, Y.; Cheng, S.H.; Zeng, L.T.; Dong, F.; Yang, Z.Y. Proteolysis of chloroplast proteins is responsible for accumulation of free amino acids in dark-treated tea (*Camellia sinensis*) leaves. *J. Proteom.* **2017**, *157*, 10–17. [CrossRef] [PubMed]
26. Chan, Y.T.; Huang, J.T.; Wong, H.C.; Li, J.; Zhao, D.Y. Metabolic fate of black raspberry polyphenols in association with gut microbiota of different origins in vitro. *Food Chem.* **2022**, *404*, 134644. [CrossRef]
27. Ma, H.; Xu, X.M.; Wang, S.M.; Wang, J.Z. Effects of microwave irradiation on the expression of key flavonoid biosynthetic enzyme genes and the accumulation of flavonoid products in *Fagopyrum tataricum* sprouts. *J. Cereal Sci.* **2021**, *101*, 103275. [CrossRef]
28. Wang, S.M.; Wang, S.M.; Wang, J.Z.; Peng, W.P. Label-free quantitative proteomics reveals the mechanism of microwave-induced Tartary buckwheat germination and flavonoids enrichment. *Food Res. Int.* **2022**, *160*, 111758. [CrossRef]
29. Riaz, M.; Kamran, M.; Rizwan, M.; Ali, S.; Wang, X.R. Foliar application of silica sol alleviates boron toxicity in rice (*Oryza sativa*) seedlings. *J. Hazard. Mater.* **2022**, *423*, 127175. [CrossRef]
30. Brdar-Jokanovi'c, M. Boron toxicity and deficiency in agricultural plants. *Int. J. Mol. Sci.* **2020**, *21*, 1424. [CrossRef]
31. Riaz, M.; Lei, Y.N.; Wu, X.W.; Hussain, S.; Aziz, O.; El-Desouki, Z.; Jiang, C.C. Excess boron inhibited the trifoliate orange growth by inducing oxidative stress, alterations in cell wall structure, and accumulation of free boron. *Plant Physiol. Biochem.* **2019**, *141*, 105–113. [CrossRef]
32. Mamani-Huarcaya, B.M.; Gonzalez-Fontes, A.; Navarro-Gochicoa, M.T.; Camacho-Cristobal, J.J.; Ceacero, C.J.; Herrera-Rodriguez, M.B.; Cutire, O.F.; Rexach, J. Characterization of two Peruvian maize landraces differing in boron toxicity tolerance. *Plant Physiol. Biochem.* **2022**, *18*, 167–177. [CrossRef]
33. Gunes, A.; Inal, A.; Bagci, E.G.; Coban, S.; Pilbeam, D.J. Silicon mediates changes to some physiological and enzymatic parameters symptomatic for oxidative stress in spinach (*Spinacia oleracea* L.) grown under B toxicity. *Sci. Hortic.* **2007**, *113*, 113–119. [CrossRef]

34. Kayaa, C.; Ashraf, M. Exogenous application of nitric oxide promotes growth and oxidative defense system in highly boron stressed tomato plants bearing fruit. *Sci. Hortic.* **2015**, *185*, 43–47. [CrossRef]
35. Riaz, M.; Kamran, M.; El-Esawi, M.A.; Hussain, S.; Wang, X.R. Boron-toxicity induced changes in cell wall components, boron forms, and antioxidant defense system in rice seedlings. *Ecotoxicol. Environ. Saf.* **2021**, *216*, 112192. [CrossRef]
36. Martínez-Cuenca, M.R.; Martínez-Alcántara, B.; Quiñones, A.; Ruiz, M.; Iglesias, D.J.; Primo-Millo, E.; Forner-Giner, M.A. Physiological and molecular responses to excess boron in *Citrus macrophylla* W. *PLoS ONE* **2015**, *10*, E0134372. [CrossRef]
37. Karabal, E.; Yücel, M.; Öktem, H.A. Antioxidant responses of tolerant and sensitive barley cultivars to boron toxicity. *Plant Sci.* **2003**, *164*, 925–933. [CrossRef]
38. Bozca, F.D.; Leblebici, S. Interactive effect of boric acid and temperature stress on phenological characteristics and antioxidant system in *Helianthus annuus* L. *S. Afr. J. Bot.* **2022**, *147*, 391–399. [CrossRef]
39. Landi, M.; Remorini, D.; Pardossi, A.; Guidi, L. Boron excess affects photosynthesis and antioxidant apparatus of greenhouse *Cucurbita pepo* and *Cucumis sativus*. *J. Plant Res.* **2013**, *126*, 775–786. [CrossRef]
40. Çatav, S.S.; KÖskeroglu, S.; Tuna, A.L. Selenium supplementation mitigates boron toxicity induced growth inhibition and oxidative damage in pepper plants. *South Afr. J. Bot.* **2022**, *146*, 375–382. [CrossRef]
41. Kayıhan, D.S.; Kayıhan, C.; Çiftçi, Y.O. Excess boron responsive regulations of antioxidative mechanism at physio-biochemical and molecular levels in *Arabidopsis thaliana*. *Plant Physiol. Biochem.* **2016**, *109*, 337–345. [CrossRef]
42. Grieve, C.M.; Poss, J.A. Wheat response to interactive effects of boron and salinity. *J. Plant Nutr.* **2000**, *23*, 1217–1226. [CrossRef]
43. Masood, S.; Wimmer, M.A.; Witzel, K.; Zorb, C.; Muhling, K.H. Interactive effects of high boron and NaCl stresses on subcellular localization of chloride and boron in wheat leaves. *J. Agron. Crop Sci.* **2012**, *198*, 227–235. [CrossRef]
44. Alpaslan, M.; Gunes, A. Interactive effects of boron and salinity stress on the growth, membrane permeability and mineral composition of tomato and cucumber plants. *Plant Soil* **2001**, *236*, 123–128. [CrossRef]
45. Billard, V.; Maillard, A.; Coquet, L.; Jouenne, T.; Cruz, F.; Garcia-Mina, J.M.; Yvin, J.C.; Ourry, A.; Etienne, P. Mg deficiency affects leaf Mg remobilization and the proteome in *Brassica napus*. *Plant Physiol. Biochem.* **2016**, *107*, 337–343. [CrossRef] [PubMed]
46. Liu, X.M.; Wang, L.; Ma, F.Y.; Guo, J.Y.; Zhu, H.; Meng, S.Y.; Bi, S.S.; Wang, H.T. Magnetic treatment improves the seedling growth, nitrogen metabolism, and mineral nutrient contents in *Populus* × *euramericana* 'Neva' under cadmium stress. *Forests* **2022**, *13*, 947. [CrossRef]
47. Siefermann-Harms, D.; Boxler-Baldoma, C.; Wilpert, K.V.; Heumann, H.G. The rapid yellowing of spruce at a mountain site in the Central Black Forest (Germany). *J. Plant Physiol.* **2004**, *161*, 423–437. [CrossRef]
48. Kaur, H.; Garg, N. Zinc toxicity in plants: A review. *Planta* **2021**, *253*, 129. [CrossRef]
49. Schwalbert, R.; Milanesi, G.D.; Stefanello, L.; Moura-Bueno, J.M.; Drescher, G.L.; Marques, A.C.R.; Kulmann, M.S.D.S.; Berghetti, A.P.; Tarouco, C.P.; Machado, M.C.; et al. How do native grasses from South America handle zinc excess in the soil? A physiological approach. *Environ. Exp. Bot.* **2022**, *195*, 104779. [CrossRef]
50. Hajara, E.W.I.; Sulaimanb, A.Z.B.; Sakinah, A.M.M. Assessment of heavy metals tolerance in leaves, stems and flowers of *Stevia rebaudiana* plant. *Procedia Environ. Sci.* **2014**, *20*, 386–393. [CrossRef]
51. Souza, G.A.D.; Carvalho, J.G.D.; Rutzke, M.; Albrecht, J.C.; Guilherme, L.R.G.; Li, L. Evaluation of germplasm effect on Fe, Zn and Se content in wheat seedlings. *Plant Sci.* **2013**, *210*, 206–213. [CrossRef]
52. Shaibur, M.R.; Shamim, A.H.M.; Kawai, S. Growth response of hydroponic rice seedlings at elevated concentrations of potassium chloride. *J. Agric. Rural. Dev.* **2008**, *6*, 43–53. [CrossRef]
53. Godoi, N.M.I.; Gazola, R.D.N.; Buzetti, S.; Jalal, A.; Celestrino, T.D.S.; Oliveira, C.E.D.S.; Nogueira, T.A.R.; Panosso, A.R.; Filho, M.C.M.T. Residual influence of nitrogen, phosphorus and potassium doses on soil and *Eucalyptus* nutrition in coppice. *Forests* **2021**, *12*, 1425. [CrossRef]
54. Wang, X.S.; Guo, Z.K.; Hui, X.L.; Wang, R.Z.; Wang, S.; Kopittke, P.M.; Wang, Z.H.; Shi, M. Improved Zn bioavailability by its enhanced colocalization and speciation with S in wheat grain tissues after N addition. *Food Chem.* **2023**, *404*, 134582. [CrossRef]
55. Chen, M.Y.; Zhao, T.T.; Peng, J.; Zhang, P.; Han, F.; Liu, X.L.; Zhong, C.H. Multivariate analysis of relationship between soil nutrients and fruit quality in 'Dong Hong' kiwifruit. *Plant Sci. J.* **2021**, *39*, 192–200, (Abstract in English).
56. Wang, C.; Song, F.; Wang, Z.J.; He, L.G.; Jiang, Y.C.; Wu, L.M. The influence of soil nutrient status on Ehima No. 28 quality in Hubei province. *Hubei Agric. Sci.* **2021**, *60*, 81–90, (Abstract in English).
57. Liu, Z.B.; Huang, Y.; Tan, F.J.; Chen, W.C.; Ou, L.J. Effects of soil type on trace element absorption and fruit quality of pepper. *Front. Plant Sci.* **2021**, *30*, 698796. [CrossRef]
58. Qiao, X.Y. Research on Effects of Potassium on Tree Nutrient and Quality of Prunuspersica (L) batsch. Master's Thesis, Northwest Agricultural and Forestry University, Xianyang, China, 2016. (Abstract in English).
59. Zhang, W. Effects of Potassium on Fruit Quality and Its Association with Metabolic Pathway of Trehalose 6-Phosphate in Apple. Ph.D. Dissertation, Northwest Agricultural and Forestry University, Xianyang, China, 2017. (Abstract in English).

60. Lu, J.W.; Chen, F.; Liu, D.B.; Wan, Y.F.; Cao, Y.P. Effect of potash application on some enzyme content in rapeseed leaf. *Chin. J. Crop Sci.* **2002**, *24*, 61–62, (Abstract in English).
61. Li, L.J. Studies on Effect of Potassium on Phenolic Metabolism of Wheat and Its Relationship with Dynamic of Aphid Population. Master's Thesis, Henan Agricultural University, Zhengzhou, China, 2002. (Abstract in English).

Disclaimer/Publisher's Note: The statements, opinions and data contained in all publications are solely those of the individual author(s) and contributor(s) and not of MDPI and/or the editor(s). MDPI and/or the editor(s) disclaim responsibility for any injury to people or property resulting from any ideas, methods, instructions or products referred to in the content.

Promotion Effects of *Taxus chinensis* var. *mairei* on *Camptotheca acuminata* Seedling Growth in Interplanting Mode

Chunjian Zhao [1,2,3,4], Sen Shi [1,2,3,4], Naveed Ahmad [5], Yinxiang Gao [6], Chunguo Xu [7], Jiajing Guan [1,2,3,4], Xiaodong Fu [1,2,3,4] and Chunying Li [1,2,3,4,*]

1. College of Chemistry, Chemical Engineering and Resource Utilization, Northeast Forestry University, Harbin 150040, China
2. Key Laboratory of Forest Plant Ecology, Ministry of Education, Northeast Forestry University, Harbin 150040, China
3. Engineering Research Center of Forest Bio-Preparation, Ministry of Education, Northeast Forestry University, Harbin 150040, China
4. Heilongjiang Provincial Key Laboratory of Ecological Utilization of Forestry-Based Active Substances, Northeast Forestry University, Harbin 150040, China
5. Department of Chemistry, Division of Science and Technology, University of Education, Lahore 54660, Pakistan
6. Institute of Jiangxi Oil-Tea Camellia, Jiujiang University, Jiujiang 332000, China
7. Dasuhe Forest Farm, Qingyuan, Liaoning 113312, China
* Correspondence: lcy@nefu.edu.cn; Tel./Fax: +86-451-8219-1387

Abstract: Wild *Camptotheca acuminata* Decne (*C. acuminata*) resources are becoming endangered and face poor growth. Preliminary investigation results found that the growth of *C. acuminata* in an artificial mixed forest of *Taxus chinensis* var. *mairei* (Lemee et Levl.), Cheng et L. K. Fu (*T. chinensis* var. *mairei*) and *C. acuminata* was significantly higher than that in pure forests. Understanding the reasons for the above differences can help create a mixed forest of *T. chinensis* var. *mairei* and *C. acuminata* to solve the problem of depleting *C. acuminata* resources. In this study, the growth and soil indexes under two different modes (*C. acuminata*/*T. chinensis* var. *mairei* interplanted and monocultured *C. acuminata* seedlings) were compared. The results showed that plant height, basal diameter, photosynthesis rate and chlorophyll content of *C. acuminata* under the interplanting mode were higher than those under monoculture. The growth rates of plant height and basal diameter that were calculated from interplanted specimens increased by 25% and 40%, respectively, compared with those from specimens that were monocultured. Photosynthetic rates from different light intensities under interplanting were higher than those in seedlings under monoculture. The contents of chlorophylls a and b and total chlorophyll under interplanting were 1.50, 1.59, and 1.47 times higher than those under monoculture, respectively. The numbers of bacteria and fungi in the interplanted culture were higher than those in the monoculture. Furthermore, the differences in microbial diversity under different planting modes were analyzed via the amplicon sequencing method. Soil enzyme activities increased under interplanting compared with that in the monoculture. Taxane allelochemicals were detected in the range of 0.01–0.67 μg/g in the interplanting mode from April to September. *T. chinensis* var. *mairei* may increase the establishment and productivity of *C. acuminata* seedlings under interplanting mode through improvements in enzyme activity, changes in microorganism population structure, and release of allelochemicals.

Keywords: *Camptotheca acuminata* Decne; *Taxus chinensis* var. *mairei* (Lemee et Levl.); Cheng et L. K. Fu; interplanting; allelochemicals; enzyme activity; microorganism

1. Introduction

Plantation management usually involves single cultivation, leading to reduced productivity and reforestation problems. It is highly desirable to develop specific techniques to minimize and overcome the problems of managed plantations. One of the most developed techniques is using a mixed plantation rather than monocultures. Successful interplanting modes, based on carefully designed species mixtures, reveal many potential advantages in long-term practices [1]. The interspecific relationship among trees is one of the important factors that affect the survival of artificial mixed forests. Plant interactions are one of the important research topics in interspecific relationships, where plants can sense and recognize coexisting conspecifics or heterospecifics and thus adjust their growth, reproduction, and defense strategies [2]. Throughout its life cycle, a plant can interact simultaneously and sequentially, directly or indirectly, with many plant neighbors, whether in forests or in more natural environments. The nature and intensity of plant interactions are important factors that affect t plant populations and community structure, and the nature and intensity of plant interactions are influenced by abiotic/biotic environmental factors. This includes competition with heterospecifics (different species), reciprocal helping (i.e., mutually beneficial interactions), commensalism (i.e., facilitation) and asymmetric interactions such as parasitism and allelopathy [3]. The mechanisms of plant interactions have been investigated mainly from above-ground plants and below-ground soils [4]. Given the myriad of interactions between above-ground and below-ground communities and their well-known impacts on ecosystem function [5], it is often assumed that the composition of below-ground bacterial communities, allelochemicals, soil nutrients, enzyme activity, and above-ground plant communities will reflect one another [6–8]. Recently, studies have shown that some tree species have synergistic promotional effects. The intercropping with garlic promoted cucumber plant growth and attenuated damage caused by soil sickness [9]. As well as their promotion potential, Brassica species can be utilized to achieve higher productivity by using them as cover crops, companion crops, and intercrops for mulching and residue incorporation or simply by including them in crop rotations [10]. Therefore, it is useful to address poor forest productivity from the perspective of interspecific promotion.

Camptotheca acuminata Decne (*C. acuminata*) is a second-class key protected tree species in China. Camptothecin (CPT) is an important secondary metabolite in *C. acuminata*, and it has been used in the treatment of cancer [11]. The global demand for CPT antitumor drugs is increasing year by year, and its economic value is high. The process of isolating CPT from *C. acuminata* has become particularly important [12,13]. *C. acuminata* is used as a raw material for anti-cancer drugs; however, the unreasonable exploitation of wild-type *C. acuminata* has significantly reduced the distribution of wild resources [14,15]. Currently, researchers are focusing on the cultivation of *C. acuminata* plantations to alleviate the resource crisis of *C. acuminata* [16]. However, the poor growth of pure *C. acuminata* forests and the small amount of fruit per plant makes it necessary to choose the appropriate mixed tree species for cultivating *C. acuminata* mixed forests [15]. Researchers have found that *C. acuminata* growth can be improved by interplanting it with *Taxus chinensis* var. *mairei* (Lemee et Levl.) Cheng et L. K. Fu (*T. chinensis* var. *mairei*). The plant height and basal diameter of *C. acuminata* under interplanting are always higher than that of pure planting, and the average growth rate of the basal diameter in pure species of *C. acuminata* plantations is only 29.4% of that in mixed-species plantations [17]. Wang Jikun and other researchers improved the soil conditions of the mixed-planting mode among three species, *C. acuminata*, *T. chinensis* var. *mairei*, and Rosmarinus officinalis, and compared the average maximum photosynthetic rate of *C. acuminata* with that in pure planting mode; it was 16.95µmol·m^2 s^{-1}, higher than that the planting mode of *C. acuminata* pure species [18]. Therefore, *T. chinensis* var. *mairei* has been advocated for interplanting in mixed *C. acuminata* mode, and it has become a successful interplanting strategy [17]. However, the potential mechanism involved in the interspecific relationship among the trees needs to be further studied.

Taxus chinensis var. *mairei*, which is listed as a first-class protected species, is an evergreen arbor that is ubiquitous in the southern region of China [19]. Previous studies have shown that the interplanting of *T. chinensis* var. *mairei* and *C. acuminata* can promote *C. acuminata* growth, which was explained from the perspective of improving the microclimate of mixed forest land [20]. Our previous studies have found that cephalomannine, 10-deacetylbacatin III, paclitaxel, and 7-Epi-10-deacetyl-paclitaxel (7-Epi-10-DAT) in *T. chinensis* var. *mairei* were identified as allelochemicals that play a promotion role on *C. acuminata* seedlings [21]. This study measured the content of organic carbon, total nitrogen, the number of microbial populations, and enzymatic activities in soil under interplanting and monoculture in order to determine the effect of *T. chinensis* var. *mairei* on *C. acuminata* seedling growth from the perspectives of microbial population structure and soil chemical properties. The objective was to provide further theoretical support for interspecific relationships in the interplanting of *T. chinensis* var. *mairei* and *C. acuminata*.

2. Materials and Methods

2.1. Pot Experiment

This experiment was conducted in a greenhouse (45°43′9″ N, 126°38′5″ E) from April to September 2018 at the Key Laboratory of Forest Plant Ecology, Ministry of Education, Northeast Forestry University, China. The soil was collected from Xialian village, Xukou town, Hangzhou City, Zhejiang Province, China, and the soil type was yellow soil. *T. chinensis* var. *mairei* (5-year) and *C. acuminata* seedlings (1-year) were cross-interplanted at a 1:1 ratio in pots (40 × 40 × 23 cm, 50 kg of soil) in April. The plant spacing was kept to 20 ± 1 cm. *C. acuminata* seedlings under monoculture mode were used as control (Figure 1). Tap water was irrigated to each pot at the start, and the pots were kept moistened throughout the experimental period. All pots were randomly placed in the greenhouse, and plants were grown under day/night temperatures of 23/18 °C, with a relative humidity of 70%–75%, and a 12-h photoperiod. The growth index of the *C. acuminata* seedlings was measured within 1–3 days of the beginning of each month. Each treatment had 30 biologically effective replications.

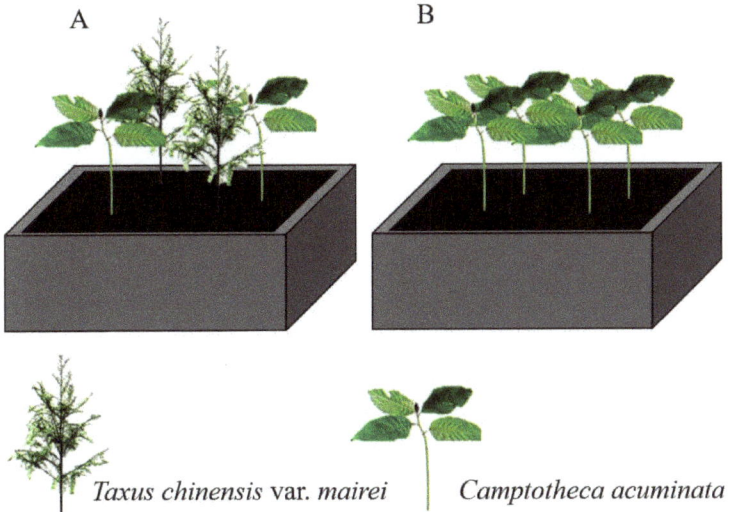

Figure 1. Diagram of the experimental planting modes. (**A**). Interplanted *C. acuminata* and *T. chinensis* var. *mairei*; (**B**). Monocultured *C. acuminata*.

2.2. *C. acuminata* Seedlings Bioassay

2.2.1. Determination of Plant Height and Basal Diameter

The plant height was measured with a tape measure (units: cm; precision: 0.1 cm), and the basal diameter was measured with a vernier caliper (units: mm; precision: 0.02 mm) (Pulisen Measuring Tools Co. Ltd., Harbin, China).

2.2.2. Determination of Main Photosynthetic Indicators and Chlorophyll Content

The LI-6400 portable photosynthetic system (LI-COR Inc., Lincoln, NE, USA) was used to test the photosynthetic characteristics of the plants in different cultivation modes (interplanted *T. chinensis* var. *mairei*/*C. acuminata*; monoculture *C. acuminata*). The leaves were collected from the third leaf at the top of a *C. acuminata* plant. This plant material was cleaned, shade-dried, and stored in a freezer at $-20\ °C$ for further analysis. Chlorophylls of tissue samples were extracted with 80% acetone and determined using the Sosnowski method [22]. The time each day to measure the photosynthetic parameters of plant leaves was from 9 to 11 a.m.

2.2.3. Determination of Total Nitrogen and Organic Carbon in Soil

Five soil samples were randomly collected from a depth of 5–15 cm in the pots. Organic carbon content in rhizosphere soil was determined according to the $K_2Cr_2O_7$–H_2SO_4 wet oxidation method [23]. Total nitrogen concentration was analyzed via a colorimetric method using a continuous-flow autoanalyzer (AutoAnalyzer III, Bran, Luebbe GmbH, Germany) after the samples were digested with the Kjeldahl method [24].

2.2.4. Dynamic Changes in Soil Microbial Population and Enzyme Activities

Five soil cores (5–15 cm depth) were taken from each plot and thoroughly mixed. The samples were sieved (20-mesh) to remove plant residues and then homogenized. A sample of soil was stored at $-80\ °C$ immediately after soils were brought to the laboratory for determination of the soil parameters. The composition of the soil microbial population was evaluated using plate colony-counting methods [25,26]. All microorganisms were cultured at $28\ °C$.

Soils were sampled in August 2018. Soil DNA was extracted from 0.4 g of frozen soil using the FastDNA™ SPIN Kit (MP Biomedicals TMInc, Eschwege, Germany), following modified according to manufacturers' recommendations [27]. Aliquots (50 µL) of DNA extracts were purified with the OneStep™ PCR Inhibitor Removal Kit (Zymo Research Biotech, Orange County, California, USA) and subsequently quantified using a microtiter plate assay with Quant-iT™ PicoGreen® dsDNA Reagent (Thermo Fisher Scientific, Bremen, Germany). DNA templates of each sample were prepared by diluting the purified DNA to 10 ng $µL^{-1}$ with nuclease-free water (Carl Roth, Roth-sur-Auhl, Rhineland, Germany). The 515 F 5′-barcode- (GTGCCAGCMGCCGCGG)-3′ and 5′-CTTGGTCATTTAGAGGAAGTAA-3′ primers were used to amplify 16 S rRNA genes at the hypervariable V4 regions for the bacterial and internal transcribed spacer 1 (ITS1) region. Then, they were amplified via polymerase chain reaction (PCR) via multiplexed barcoded amplicon sequencing by BMKCloud (www.biocloud.net) (accessed on Dec 10, 2022) on the Illumina MiSeq platform (Illumina, San Diego, CA, USA), as described by Castano et al. (2020) [28].

All five enzymes' (urease, sucrase, protease, phosphatase, and dehydrogenase) activities were determined via the colorimetric method [29]. The controls used distilled water instead of substrates. Each enzyme measurement had five replicates.

2.3. Quantitation of Allelochemicals in Rhizosphere and Non-Rhizosphere Soils

The quantitation of allelochemicals in interplanted *C. acuminata* and *T. chinensis* var. *mairei* rhizosphere soils and non-rhizosphere soils was completed via the UPLC-MS/MS method developed by our team [30].

2.4. Statistical Analysis

The experiments were carried out via a completely randomized design. Data were reported as the mean, standard error (S.E) of five replicates. For pot trials and the *C. acuminata* seedlings bioassay, SPSS software version 20.00 was used for statistical analysis. A student's *t*-test was used to evaluate differences among means. Differences were considered statistically significant at $p < 0.05$.

3. Results and Discussion

3.1. Effect of Interplanting on Plant Height and Basal Diameter of C. acuminata Seedlings

The heights and basal diameters of *C. acuminata* seedlings through interplanting (*T. chinensis* var. *mairei*/*C. acuminata*) in pot experiments on different days were measured. The results showed that *T. chinensis* var. *mairei* significantly increased *C. acuminata* seedling height and the basal diameter under interplanting mode; peaking appeared in September. From June to September, the heights and basal diameters of interplanted *C. acuminata* seedlings were all significantly higher than those under the *C. acuminata* monoculture (Tables 1 and 2). In September, the plant heights of the monoculture *C. acuminata* seedlings and interplanted seedlings increased by 95% and 120%, respectively, compared with April. The growth rate of plant height calculated from the interplanted mode increased by 25% compared with that from the monoculture mode. In September, the basal diameters measured from the monoculture and interplanted *C. acuminata* increased by 47% and 66%, respectively, compared with April. The basal diameter from the interplanted specimens increased by 19% compared with those from the monoculture. Our findings are consistent with the trends of previous studies. Kong's studies indicated that interplanting could effectively improve the growth of some plants. It has been observed that *J. mandshurica* establishment and management can be improved by interplanting with larch (*Larix gmelini*). The height, diameter at breast height, individual tree volume, and the stand stocking per hectare of *J. mandshurica* in the interplanting mode were 1.33, 1.87, 4.71, and 1.69 times higher than those in the monoculture mode, respectively, indicating that *L. gmelini* accelerates the growth of *J. mandshurica* in the interplanting mode [31]. *Pinus sylvestris* var. and *Broussonetia papyrifera* Linn. could be used to promote the growth of *A. pedunculata* seedlings, as well as enhance the construction of mixed plantations in coal mine degradation areas [32]. In our experiment, the interplanting of *C. acuminata* seedlings and *T. chinensis* var. *mairei* can significantly increase the growth rate of *C. acuminata* seedlings, and the positive effects of plant allelopathy depend to a large extent on the selection of mixed species.

Table 1. Effect of *C. acuminata* seedling height in different planting modes at different months.

Planting Mode	Plant Height (cm)					
	April	May	June	July	August	September
MC	50.1 ± 1.2	55.9 ± 2.3	73.9 ± 2.6	92.1 ± 2.4	95.7 ± 2.4	97.9 ± 2.5
IC	49.8 ± 1.1	56.7 ± 2.1	80.9 ± 2.3 *	105.8 ± 2.6 *	108.3 ± 2.6 *	109.8 ± 2.7 *

Each value is the mean value ± SE (n = 5); * Significant differences in different planting modes at these months (*t*-test, $p < 0.05$); MC, monoculture *C. acuminata*; IC, interplanted *C. acuminata*.

Table 2. Effect of *C. acuminata* seedling basal diameter in different planting modes at different months.

Planting Mode	Basal Diameter (mm)					
	April	May	June	July	August	September
MC	7.04 ± 0.04	7.23 ± 0.32	8.73 ± 0.44	9.04 ± 0.36	10.12 ± 0.32	10.33 ± 0.44
IC	7.02 ± 0.06	8.09 ± 0.34 *	9.23 ± 0.42 *	10.84 ± 0.42 *	11.36 ± 0.38 *	11.65 ± 0.56 *

Each value is the mean value ± SE (n = 5); * Significant differences in different planting modes at these months (*t*-test, $p < 0.05$); MC, monoculture *C. acuminata*; IC, interplanted *C. acuminata*.

3.2. Effect of Interplanting on Photosynthesis of C. acuminata Seedlings

Under different light intensities, the net photosynthetic rate of *C. acuminata* seedlings interplanted with *T. chinensis* var. *mairei* was higher than that of *C. acuminata* seedlings in monoculture (Figure 2A); the net photosynthetic rate increased by 28%, 33%, 27%, 38% and 29% from May to September, respectively. In August, the net photosynthetic rate of *C. acuminata* seedlings interplanted with *T. chinensis* var. *mairei* was the highest. Compared with the monoculture, interplanting increased the stomatal conductance of *C. acuminata* seedlings by 24%, 25%, 27%, 27%, and 28% from May to September (Figure 2B), respectively, while the stomatal conductance of *C. acuminata* interplanted and monoculture seedlings under 100% light intensities (1600 µmol m^2·s^{-1}) was lower than that under other light intensities conditions (800 µmol m$^2 \cdot$s^{-1} and 480 µmol m$^2 \cdot$s^{-1}, respectively). Under different light intensities, interplanting with *C. acuminata* seedlings all enhanced intercellular CO$_2$ concentrations compared with seedlings in monoculture (Figure 2C); however, the growth rates decreased with an increase in processing time, which were 115%, 98%, 83%, 93% and 36% from May to September, respectively; meanwhile, the intercellular CO$_2$ concentrations of *C. acuminata* interplanted and monoculture seedlings in September were the lowest. The transpiration rate of *C. acuminata* interplanted with *T. chinensis* var. *mairei* was lower than that of *C. acuminata* seedlings in monoculture (Figure 2D), which was similar to the trend in intercellular CO$_2$ concentrations, which increased by 75%, 59%, 44%, 36% and 38% from May to September, respectively.

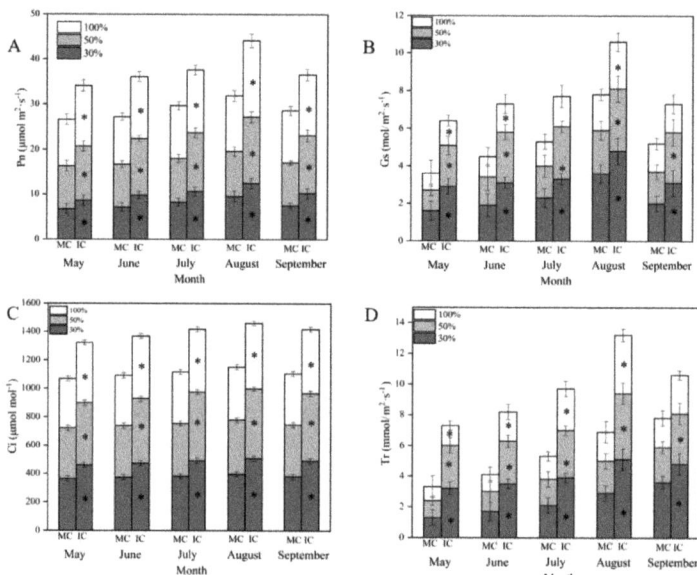

Figure 2. Effects of *C. acuminata* seedling-related indicators on the net photosynthetic rate (**A**), the stomatal conductance (**B**), the intercellular CO$_2$ concentrations (**C**) and the transpiration rate (**D**) in plant leaves under different planting modes at different months and light intensities. Each value is the mean value ± SE (n = 5). * Significant differences in different planting modes during these months (t-test, $p < 0.05$). MC, monoculture *C. acuminata*; IC, interplanted *C. acuminata*.

Light energy is the driving force of photosynthesis in green plants. In plant cultivation, rational utilization of light energy can help green plants fully photosynthesize [33]. Net photosynthetic rate, stomatal conductance, intercellular CO$_2$ concentration, and transpiration rate are important indicators of plant photosynthesis; they are closely related to plant growth and photosynthetic function. Interplanting (maize/soybean) decreased the net photosynthetic rate of soybean, but the rate recovered to varying degrees after the maize harvest [34]. In an interplanting system, two planting modes may affect the photosynthesis of each plant type. Our test showed that *T. chinensis* var. *mairei*/*C. acuminata* interplanting increased the net photosynthetic rate, stomatal conductance, intercellular CO$_2$ concentration, and transpiration rate of *C. acuminata* seedlings (Figure 2).

3.3. Effect of Interplanting on Chlorophyll Content of C. acuminata Seedlings

The chlorophyll content is an important indicator for evaluating plant growth. The contents of chlorophyll a, chlorophyll b, and total chlorophyll under different planting modes are shown in Figure 3. The chlorophyll content in the *C. acuminata* seedlings leaves under the different planting modes (interplanted *T. chinensis* var. *mairei*/*C. acumi-*

nata; monoculture *C. acuminata*) exhibited similar trends. The contents of chlorophyll a (Figure 3A), chlorophyll b (Figure 3B), and total chlorophyll (Figure 3C) were higher in the interplanted *C. acuminata* seedlings than those in the monoculture *C. acuminata* seedlings during the growth period. The content of chlorophyll was increased first and then decreased under different planting moods (interplanted *T. chinensis* var. *mairei*/*C. acuminata*; monoculture *C. acuminata*). Chlorophyll a, b, and total chlorophyll contents were 1.50, 1.59, and 1.47 times than those of the monoculture from May to September. The maximum content was obtained in July and August. The reason is possible that the soil in mixed planting contains more ingredients beneficial to plant growth, which promotes the growth of leaves and increases the chlorophyll content [35]. Chlorophyll in photosynthesis directly influences photosynthetic ability. The contents of chlorophyll in *C. acuminata* seedlings increased by interplanting with *T. chinensis* var. *mairei*, indicating enhanced photosynthetic rate leading to a higher biomass of *C. acuminata* seedlings. Thus, *T. chinensis* var. *mairei* exerts positive allelopathic effects on *C. acuminata* seedlings through allelochemicals in interplanting.

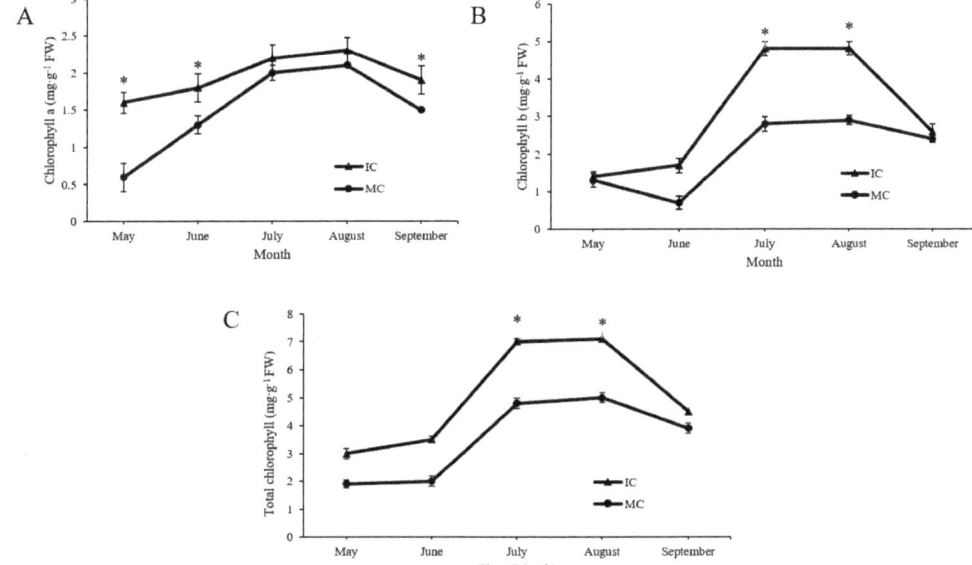

Figure 3. Effects on chlorophyll an (**A**) and b (**B**) content, and total chlorophyll content (**C**) *C. acuminata* seedlings under different planting modes at different months. Each value is the mean value ± SE (n = 5). * Significant differences in different planting modes during these months (t-test, $p < 0.05$). MC, monoculture *C. acuminata*; IC, interplanted *C. acuminata*.

3.4. Effect of Interplanting on Content of Organic Carbon and Total Nitrogen in Soil of C. acuminata Seedlings

The content of organic carbon and total nitrogen in both the monoculture and interplanted modes initially increased and then decreased during the growth period of the *C. acuminata* seedlings (Figure 4A,B). In the monoculture, the organic carbon level ranged from 17 to 24 mg·g^{-1}, with the lowest level occurring at the early stage of *C. acuminata* seedlings. In the interplanted mode, the organic carbon level was in the range from 19 mg·g^{-1} to 32 mg·g^{-1}, with the lowest level occurring at the early stage of *C. acuminata* seedlings, and the organic carbon content in interplanted seedlings was higher than that in monoculture seedlings (Figure 4A). Similar to the results with organic carbon, variation in total nitrogen showed the same trends between the monoculture and interplanted treatments (Figure 4B). For the monoculture, the content of total nitrogen initially increased from May to September, peaking at 8.0 mg·g^{-1} in August. For the interplanting mode, the total nitrogen content increased gradually from May to August, peaked at 9.3 mg·g^{-1}, and then fell slowly after August.

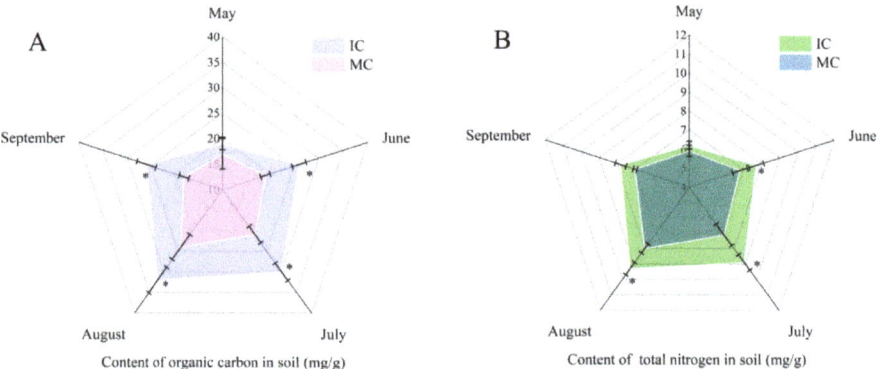

Figure 4. Effects on soil organic carbon (**A**) and total nitrogen contents (**B**) under different planting modes in different months. Each value is the mean value ± SE (n = 5). * Significant differences in different planting modes during these months (t-test, $p < 0.05$); MC, monoculture *C. acuminata*; IC, interplanted *C. acuminata*.

Sufficient but not excessive nutrient element levels in the soil are vital for *C. acuminata* seedling growth. The results showed that the contents of total nitrogen and organic carbon in the interplanted soil were higher than those in the monoculture soil. These nutrient levels were all-sufficient for *C. acuminata* seedling growth, and interplanting *T. chinensis* var. *mairei* with *C. acuminata* was helpful to avoid salinization of the soil under mixed cultivation and improve fertilizer use efficiency. Higher total nitrogen and organic carbon levels were observed in interplanted *C. acuminata* seedlings with *T. chinensis* var. *mairei*. These results are in almost agreement with those of Wei et al. [36] and indicate that interplanted *T. chinensis* var. *mairei* and *C. acuminata* seedlings play an important role in activating and balancing nutrients in the soil. We think that this is due to the fact that plants and soils have close interactions, such as the exchange of substances and nutrient cycling.

3.5. Effect of Interplanting on Enzyme Activity in the Soil of C. acuminata Seedlings

C. acuminata seedlings urease, sucrase, dehydrogenase, protease, and phosphatase activities between interplanted and monoculture modes from May to September were determined, and the results are shown in Figure 5. It can be seen from Figure 5 that the enzyme activity of the interplanted soil was significantly higher than that of the monoculture soil. In August, enzyme activity from the interplanted mode was significantly higher than that monoculture mode ($p < 0.05$).

Soil enzyme activity is an important indicator of soil quality. Generally, extracellular enzymes are secreted by soil microorganisms in order to decompose large, polymeric compounds and are closely related to the cycling of carbon, nitrogen, and phosphorus [37,38]. These results are in agreement with those of Zeng et al.'s research. These results showed that intercropping turmeric and ginger with patchouli could improve soil enzymes. They also modify the soil's physical and chemical properties through changes in enzyme activity [39]. Zhao et al. reported that higher activities of catalase, urease, invertase, and phosphatase were detected in maize/hot pepper interplanted soil [40]. Significantly higher activities of invertase, urease, and phosphatase in the soil of a garlic/cotton interplanting system throughout the course of growth in comparison to a cotton monoculture were observed. Interplanting garlic enhanced its phosphatase activity and cellulase activity [41].

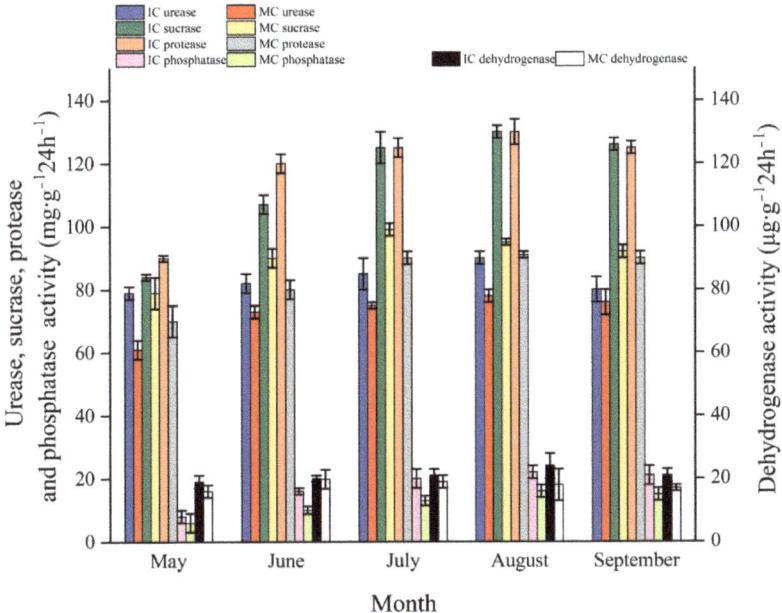

Figure 5. Effects on urease, sucrase, protease, and phosphatase (mg·g^{-1} 24 h^{-1}), dehydrogenase (μg·g^{-1} 24 h^{-1}) activities in soil under different planting modes at different months. Each value is the mean value ± SE (n = 5). MC, monoculture C. acuminata; IC, interplanted C. acuminata.

3.6. Effect of Interplanting on Microbial Populations in the Soil of C. acuminata Seedlings

Bacteria populations in soil appeared to change during the different months between the monoculture and interplanted C. acuminata seedlings. However, the bacterial population reached a sharp peak for both modes in August, and the bacterial population in the interplanted soil was significantly higher than that in the monoculture soil (27.6 × 10^6 > 19.6 × 10^6 cfu·g^{-1}) (Figure 6A). Fungal populations in the soil were significantly different between the monoculture and interplanted modes from July to September (p < 0.05), and the trend was similar between the monoculture and interplanted modes (Figure 6B). In the interplanted culture, the fungal population increased significantly to 11.9 × 10^5 cfu·g^{-1} in September. In the monoculture, the number of fungi showed a similar trend to that in the interplanted mode each month and peaked in August at 7.51 × 10^5 cfu·g^{-1}.

The compositions of bacterial and fungal communities were tested for their taxonomy in interplanted and monoculture C. acuminata modes. The relative abundances of bacterial phyla Acidobacteriota (0.320), Proteobacteria (0.271), Methylomirabilota (0.104), Myxococcota (0.063), Actinobacteriota (0.044) and Gemmatimonadota (0.028) in monoculture C. acuminata, as well as the archaeal phylum Acidobacteriota (0.260) and Proteobacteria (0.270) in interplanted C. acuminata, encompassed the largest proportion of sequences of bacterial communities (Figure 6C). In interplanted C. acuminata, members of phyla Ascomycota (0.748), Basidiomycota (0.128), Chytridiomycota (0.029), and Glomeromycota (0.016) were prevalent fungal groups across the investigated soils. Unlike the interplanted C. acuminata, the monoculture C. acuminata showed a fungal composition of mainly Ascomycota (0.664), Basidiomycota (0.150), Chytridiomycota (0.051), Mortierellomycota (0.045), Glomeromycota (0.025) and Calcarisporiellomycota (0.012) (Figure 6D).

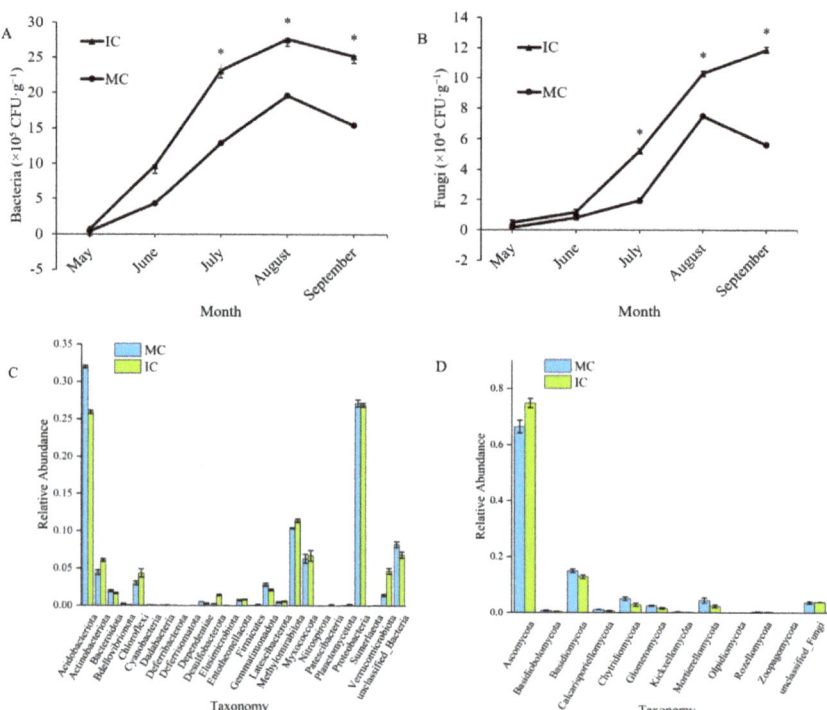

Figure 6. Effects on bacteria (**A**) and fungi (**B**) in soil under different planting modes at different months. * Significant ($p < 0.05$) differences in planting modes during these months (t-test). MC, monoculture C. acuminata; IC, interplanted C. acuminata. Relative abundance of bacterial (**C**) and fungal (**D**) communities at the phylum level, based on Illumina sequencing of the 16S rRNA gene and the ITS1 region from two plant modes (MC, monoculture C. acuminata; IC, interplanted C. acuminata). Each value is the mean value ± SE (n = 3).

Microorganisms are one of the important components of the soil ecosystem. Soil microbial populations play an important role in many soil functions, including organism decomposition and nutrient cycling [42,43]. A previous study showed that interplanting might affect the number of soil microbial populations under continuously interplanted modes. Intercropping between sugarcane and soybean is widely used to increase crop yield and promote the sustainable development of the sugarcane industry. Intercropping improved the bacterial population in the rhizosphere in comparison with a monoculture [44]. Elshahawy et al. showed that interplanting cucumber with onion or garlic increased the population of bacteria [45]. Tian and Boparai et al. stated that wheat or garlic interplanting with cotton significantly enhanced the population of bacteria and actinomycetes in the rhizosphere compared to monoculture planting; garlic interplanting had even more remarkable effects [46,47]. Ali et al. reported that the soil population of bacteria and actinomycetes increased under a cucumber/garlic interplanting system compared to a cucumber monoculture [48]. Interplanting increased the number of rhizosphere fungi in aromatic plants [49]. In our study, similar results were also obtained, where interplanting T. chinensis var. mairei with C. acuminata seedlings in soil enhanced the populations of bacteria and fungi in comparison with monoculture C. acuminata seedlings. A comparison of the diversity of microbial composition in monoculture C. acuminata and interplanted C. acuminata showed that there were differences between them. It is very clear that plant–microbe interactions in the rhizosphere are influenced by several factors. As an indispensable energy and nutrient source for microorganisms, soil organic matter content was reported to play an important role in shaping microbial communities [50]. Secondary metabolites in plant root secretions also affect the growth of particular microbes [51]. As several studies have established, the rhizosphere microbiome composition greatly affects plant growth, and plants employ several mechanisms to recruit specific microflora [52]. Our results fully demonstrated that the balance of microorganisms in the soil would not be destroyed under interplanting modes; moreover, the reasonable microbial population distribution played a role in promoting the growth of *C. acuminata*. It has been shown that methylomirabilota can enhance soil carbon and nitrogen cycling to a certain extent, and ascomycota is significant in promoting plant growth [53]. In this study, both methylomirabilota and ascomycota were found in higher percentages in the soil of interplanted *C. acuminata* than in the soil of monoculture *C. acuminata*; therefore, we infer that the better growth of interplanted *C. acuminata* comes from these two microbials, which may be promoting nutrient utilization and uptake, further affecting the growth of *C. acuminata*.

3.7. Release of Taxane Allelochemicals from T. chinensis var. Mairei in Interplanting Mode

All allelochemicals in the rhizosphere and non-rhizosphere soil were determined, and the results are shown in Table 3. The content of 10-DAB III peaked at 0.67 μg/g in rhizosphere soil and 0.21 μg/g in non-rhizosphere soil in August. In the same month, the content of 7-Epi-10-DAT was the lowest; only 0.32 and 0.12 μg/g were found in the rhizosphere and non-rhizosphere soil, respectively. The contents of cephalomannine and paclitaxel were less than 0.01 μg/g in both rhizosphere and non-rhizosphere soils. It is possible that soil microorganisms degrade the allelochemicals, or it may be related to the adsorption of the soil to them [54]. In addition, the physicochemical properties, release concentrations, environmental factors, and retention and transformation in the soil have an impact on whether allelochemicals are detected in the soil or not [55]. Therefore, there were two allelochemicals (10-DAB III and 7-Epi-10-DAT) in the rhizosphere and non-rhizosphere soil, which suggested that they may be present in the interplanting mode, which consequently affects the growth of *C. acuminata* seedlings.

The soil ecosystem is a complex composition in which soil enzymes and soil microorganisms play an important role in promoting the soil nutrient cycle [56]. The physical and chemical properties of soil have a great impact on soil enzyme activity and soil microbial community structure [57]. In contrast to competition for resources from soil, allelopathy involves the release of allelochemicals from plants into the environment [58]. Accordingly, the detection of allelochemicals from rhizosphere soil and non-rhizosphere soil is a key to understand plant–plant interspecific interactions [59]. A series of interactions between allelochemicals and soil abiotic and biotic factors may occur when allelochemicals are released through the soil [55]. In particular, soil microbial interactions radically alter the environment and provide a much better indication of real effects [60]. The presence of taxane medium allelochemicals may be one of the important factors for *T. chinensis* var. *mairei* to promote *C. acuminata* seedling growth under the interplanting mode. However, the interactions among allelochemicals, soil enzymes and soil microorganisms require further clarification.

Table 3. Contents of cephalomannine, 10-DAB III, paclitaxel and 7-Epi-10-DAT in rhizosphere soils and non-rhizosphere soil.

Compounds (μg/g)	April		May		June		July		August		September	
	Rhizosphere Soil	Non-Rhizosphere Soil	Rhizosphere Soil	Non-Rhizosphere Soil	Rhizosphere Soil	Non-Rhizosphere Soil	Rhizosphere Soil	Non-Rhizosphere Soil	Rhizosphere Soil	Non-Rhizosphere Soil	Rhizosphere Soil	Non-Rhizosphere Soil
Cephalomannine	<0.01 ± 0.01	<0.01 ± 0.01	<0.01 ± 0.01	<0.01 ± 0.01	<0.01 ± 0.01	<0.01 ± 0.01	<0.01 ± 0.01	<0.01 ± 0.01	<0.01 ± 0.01	<0.01 ± 0.01	<0.01 ± 0.01	<0.01 ± 0.01
10-DAB III	0.23 ± 0.01	0.09 ± 0.01	0.34 ± 0.01	0.12 ± 0.01	0.46 ± 0.01	0.15 ± 0.01	0.65 ± 0.01	0.20 ± 0.01	0.67 ± 0.01	0.21 ± 0.01	0.63 ± 0.01	0.19 ± 0.01
Paclitaxel	<0.01 ± 0.01	<0.01 ± 0.01	<0.01 ± 0.01	<0.01 ± 0.01	<0.01 ± 0.01	<0.01 ± 0.01	<0.01 ± 0.01	<0.01 ± 0.01	<0.01 ± 0.01	<0.01 ± 0.01	<0.01 ± 0.01	<0.01 ± 0.01
7-Epi-10-DAT	0.13 ± 0.01	0.04 ± 0.01	0.17 ± 0.01	0.06 ± 0.01	0.22 ± 0.01	0.09 ± 0.01	0.30 ± 0.01	0.11 ± 0.01	0.32 ± 0.01	0.12 ± 0.01	0.31 ± 0.01	0.11 ± 0.01

4. Conclusions

The *T. chinensis* var. *mairei* and *C. acuminata* seedling interplanted cultivation resulted in significant positive effects and improvements in the establishment and productivity of *C. acuminata* seedlings and involved multiple factors. In this study, biomass, net photosynthetic rate, chlorophyll, organic carbon, total nitrogen content, microbial populations, and enzyme activities in the soil of interplanted *C. acuminata* seedlings increased. The change in soil microbial diversity under the interplanting of *C. acuminata* was found to be beneficial to the growth of *C. acuminata*. Such effects should be correlated to the joint action of microbial and enzymatic activities in the soil after interplanting with *T. chinensis* var. *maire*. Meanwhile, taxane allelochemicals released from the root exudate of *T. chinensis* var. *maire* into the soil in the interplanted culture may be one of the reasons for enhanced growth in *C. acuminata* seedlings. Thus, our study provides theoretical support for clarification of the promotional effects of *T. chinensis* var. *mairei* on *C. acuminata* seedling growth in interplanted mode. It also offers many potential implications and applications in managed tree ecosystems. However, how microorganisms enhance enzymatic activity or transfer nutrients and interact with allelochemicals to promote the growth of *C. acuminata* seedlings in interplanted soils are mechanisms that require further exploration.

Author Contributions: Conceptualization, C.Z. and S.S.; methodology, S.S.; software, S.S.; validation, C.L.; formal analysis, Y.G.; investigation, J.G.; resources, C.X.; data curation, S.S. and J.G.; writing—original draft preparation, S.S.; writing—review and editing, C.Z., C.L. and Naveed Ahmad; visualization, S.S.; super-vision, X.F.; project administration, C.L.; funding acquisition, C.Z. and C.L. All authors have read and agreed to the published version of the manuscript.

Funding: This work was financially supported by the Fundamental Research Funds for the Central Universities (2572022AW25), the National Natural Science Foundation (31870609), Science and Technology Program of Jiangxi Administration of Traditional Chinese Medicine (2020A0379), and the 111 Project, China (B20088).

Data Availability Statement: Data are available on request from the corresponding author.

Conflicts of Interest: The authors declare no conflict of interest.

References

1. Chmura, D.J.; Guzicka, M.; Rokowski, R. Accumulation of standing aboveground biomass carbon in *Scots pine* and *Norway spruce* stands affected by genetic variation. *For. Ecol. Manag.* **2021**, *496*, 119476. [CrossRef]
2. Xu, Y.; Cheng, H.F.; Kong, C.H.; Meiners, S.J. Intra-specific kin recognition contributes to inter-specific allelopathy: A case study of allelopathic rice interference with paddy weeds. *Plant Cell Environ.* **2021**, *44*, 3479–3491. [CrossRef] [PubMed]
3. Subrahmaniam, H.J.; Libourel, C.; Journet, E.P.; Morel, J.B.; Munos, S.; Niebel, A.; Raffaele, S.; Roux, F. The genetics underlying natural variation of plant-plant interactions, a beloved but forgotten member of the family of biotic interactions. *Plant J.* **2018**, *93*, 747–770. [CrossRef] [PubMed]
4. Rasmann, S. As above so below: Recent and future advances in plant-mediated above- and belowground interactions. *Am. J. Bot.* **2022**, *109*, 672–675. [CrossRef]
5. Barberán, A.; Mcguire, K.L.; Wolf, J.A.; Jones, F.A.; Wright, S.J.; Turner, B.L.; Essene, A.; Hubbell, S.P.; Faircloth, B.C.; Fierer, N. Relating belowground microbial composition to the taxonomic, phylogenetic, and functional trait distributions of trees in a tropical forest. *Ecol. Lett.* **2015**, *18*, 1397–1405. [CrossRef]
6. Zhang, X.; Liu, S.R.; Huang, Y.T.; Fu, S.L.; Wang, J.X.; Ming, A.G.; Li, X.Z.; Yao, M.J.; Li, H. Tree species mixture inhibits soil organic carbon mineralization accompanied by decreased r-selected bacteria. *Plant Soil* **2018**, *431*, 203–216. [CrossRef]
7. Shi, S.; Cheng, J.B.; Ahmad, N.; Zhao, W.Y.; Tian, M.F.; Yuan, Z.Y.; Li, C.Y.; Zhao, C.J. Effects of potential allelochemicals in a water extract of Abutilon theophrasti Medik. on germination and growth of *Glycine max* L., *Triticum aestivum* L., and *Zea mays* L. *J. Sci. Food Agric.* **2022**. [CrossRef]
8. Reynolds, H.L.; Packer, A.; Bever, J.D.; Clay, K. Grassroots ecology: Plant-microbe-soil interactions as drivers of plant community structure and dynamics. *Ecology* **2003**, *84*, 2281–2291. [CrossRef]
9. Du, L.; Huang, B.; Du, N.; Guo, S.; Shu, S.; Sun, J. Effects of garlic/cucumber relay intercropping on soil enzyme activities and the microbial environment in continuous cropping. *Hortscience* **2017**, *52*, 78–84. [CrossRef]
10. Rehman, S.; Shahzad, B.; Bajwa, A.A.; Hussain, S.; Rehman, A.; Cheema, S.A.; Abbas, T.; Ali, A.; Shah, L.; Adkins, S.; et al. Utilizing the allelopathic potential of brassica species for sustainable crop production: A review. *J. Plant Growth Regul.* **2019**, *38*, 343–356. [CrossRef]

11. Li, S.; He, H.; Xi, Y.; Li, L. Chemical constituents and pharmacological effects of the fruits of *Camptotheca acuminata*: A review of its phytochemistry. *Asian J. Tradit. Med.* **2018**, *13*, 40–48.
12. Krishnan, J.J.; Gangaprasad, A.; Satheeshkumar, K. In vitro mass multiplication and estimation of camptothecin (CPT) in *Ophiorrhiza mungos* L. var. *angustifolia (Thw.) Hook. f. Ind. Crops Prod.* **2018**, *119*, 64–72. [CrossRef]
13. Zhao, C.; Li, C.; Wang, L.; Zu, Y.; Yang, L. Determination of camptothecin and 10-hydroxycamptothecin in *Camptotheca acuminata* by LC-ESI-MS/MS. *Anal. Lett.* **2010**, *43*, 2681–2693. [CrossRef]
14. Sadre, R.; Magallanes-Lundback, M.; Pradhan, S.; Salim, V.; Mesberg, A.; Jones, A.D.; Dellapenna, D. Metabolite diversity in alkaloid biosynthesis: A multilane (diastereomer) highway for camptothecin synthesis in *Camptotheca acuminata*. *Plant Cell* **2016**, *28*, 1926–1944. [CrossRef] [PubMed]
15. Wen, B.; Yang, P. Implications of seed germination ecology for conservation of *Camptotheca acuminata*, a rare, endemic, and endangered species in China. *Plant Ecol.* **2021**, *222*, 209–219. [CrossRef]
16. Trueman, S.J.; Richardson, D.M. Propagation and chlorophyll fluorescence of *Camptotheca acuminata* cuttings. *J. Med. Plants Res.* **2011**, *5*, 1–6.
17. Li, L.; Yang, F.; Pang, H.; Gao, Y.; Sun, J. Relationship between dynamic change of growth and meteorological factors in artificial composite community of *Camptotheca acuminata*. *Chin. J. Plant Res.* **2008**, *28*, 486–490.
18. Wang, J. *Studies on Directional Cultivation of Active Substances for Artificial Complex Community of Camptotheca chinensis and Rosemary*; Northeast Forestry University: Harbin, China, 2006.
19. Li, L.; Chen, Y.; Ma, Y.; Wang, Z.; Wang, T.; Xie, Y. Optimization of Taxol extraction process using response surface methodology and investigation of temporal and spatial distribution of taxol in *Taxus mairei*. *Molecules* **2021**, *26*, 5485. [CrossRef]
20. Yang, F.; Liu, W.; Wang, W.; Zu, Y. Characteristics of life history type spectrum of mixed communities of *Camptotheca acuminata* and *Taxus chinensis*. *Bull. Bot. Res.* **2011**, *31*, 711–715.
21. Shi, S.; Gao, Y.X.; Li, C.Y.; Zhang, J.J.; Guan, J.J.; Zhao, C.J.; Fu, Y.J. Allelopathy of *Taxus chinensis* var. mairei on *Camptotheca acuminata* seedling growth and identification of the active principles. *J. Plant Interact.* **2022**, *17*, 33–42. [CrossRef]
22. Lee, T.-C.; Shih, T.-H.; Huang, M.-Y.; Lin, K.-H.; Huang, W.-D.; Yang, C.-M. Eliminating interference by anthocyanins when determining the porphyrin ratio of red plant leaves. *J. Photochem. Photobiol. B-Biol.* **2018**, *187*, 106–112. [CrossRef] [PubMed]
23. Martz, M.; Heil, J.; Marschner, B.; Stumpe, B. Effects of soil organic carbon (SOC) content and accessibility in subsoils on the sorption processes of the model pollutants nonylphenol (4-n-NP) and perfluorooctanoic acid (PFOA). *Sci. Total Environ.* **2019**, *672*, 162–173. [CrossRef] [PubMed]
24. Srivastava, R.K.D.; Panda, R.K.P.; Chakraborty, A.P. Quantification of nitrogen transformation and leaching response to agronomic management for maize crop under rainfed and irrigated condition. *Environ. Pollut.* **2020**, *265*, 114866. [CrossRef] [PubMed]
25. Borowik, A.; Wyszkowska, J.; Kucharski, J.; Baćmaga, M.; Tomkiel, M. Response of microorganisms and enzymes to soil contamination with a mixture of terbuthylazine, mesotrione, and S-metolachlor. *Environ. Sci. Pollut. Res.* **2017**, *24*, 1910–1925. [CrossRef] [PubMed]
26. Shi, J.; Zhang, F.; Wu, S.; Guo, Z.; Huang, X.; Hu, X.; Holmes, M.; Zou, X. Noise-free microbial colony counting method based on hyperspectral features of agar plates. *Food Chem.* **2019**, *274*, 925–932. [CrossRef]
27. Spohn, M.; Klaus, K.; Wanek, W.; Richter, A. Microbial carbon use efficiency and biomass turnover times depending on soil depth—Implications for carbon cycling. *Soil Biol. Biochem.* **2016**, *96*, 74–81. [CrossRef]
28. Castano, C.; Berlin, A.; Durling, M.B.; Ihrmark, K.; Lindahl, B.D.; Stenlid, J.; Clemmensen, K.E.; Olson, A. Optimized metabarcoding with Pacific biosciences enables semi-quantitative analysis of fungal communities. *New Phytol.* **2020**, *228*, 1149–1158. [CrossRef]
29. Gromes, R.; Mueller, K.; Nagel, E.; Uhlmann, J. The use of photometric test kits for investigations of soil enzyme activities. *J. Plant Nutr. Soil Sci.* **2015**, *164*, 431–433. [CrossRef]
30. Bin, Q.; Siming, N.; Qianqian, L.; Zahid, M.; Jiabo, C.; Zhanyu, Y.; Chunying, L.; Chunjian, Z. Quick and in-situ detection of different polar allelochemicals in Taxus soil by microdialysis combined with UPLC-MS/MS. *J. Agric. Food Chem.* **2022**. [CrossRef]
31. Kong, C. Allelopathic potential of root exudates of larch (*Larix gmelini*) on Manchurian walnut (*Juglans mandshurica*). *Allelopath. J.* **2007**, *20*, 127–134.
32. Wang, X.; Zhang, R.; Wang, J.; Di, L.; Sikdar, A. The effects of leaf extracts of four tree species on *Amygdalus pedunculata* seedlings growth. *Front. Plant Sci.* **2021**, *11*, 587579. [CrossRef]
33. Walter, J.; Kromdijk, J. Here comes the sun: How optimization of photosynthetic light reactions can boost crop yields. *J. Integr. Plant Biol.* **2022**, *64*, 564–591. [CrossRef] [PubMed]
34. Yu, L.; Tang, Y.; Wang, Z.; Gou, Y.; Wang, J. Nitrogen-cycling genes and rhizosphere microbial community with reduced nitrogen application in maize/soybean strip intercropping. *Nutr. Cycl. Agroecosystems* **2019**, *113*, 35–49. [CrossRef]
35. Beneragama, C.; Goto, K. Chlorophyll a:b ratio increases under low-light in 'Shade-tolerant' *Euglena gracilis*. *Trop. Agric. Res.* **2010**, *22*, 12–25. [CrossRef]
36. Wei, B.; Zhang, J.; Wen, R.; Chen, T.; Xia, N.; Liu, Y.; Wang, Z. Genetically modified sugarcane intercropping soybean impact on rhizosphere bacterial communities and co-occurrence patterns. *Front. Microbiol.* **2021**, *12*, 742391. [CrossRef] [PubMed]
37. Wang, M.; Han, Y.; Xu, Z.; Wang, S.; Jiang, M.; Wang, G. Hummock-hollow microtopography affects soil enzyme activity by creating environmental heterogeneity in the sedge-dominated peatlands of the Changbai Mountains, China. *Ecol. Indic.* **2021**, *121*, 107187. [CrossRef]

38. Yu, P.; Tang, X.; Zhang, A.; Fan, G.; Liu, S. Responses of soil specific enzyme activities to short-term land use conversions in a salt-affected region, northeastern China. *Sci. Total Environ.* **2019**, *687*, 939–945. [CrossRef]
39. Zeng, J.; Liu, J.; Lu, C.; Ou, X.; Luo, K.; Li, C.; He, M.; Zhang, H.; Yan, H. Intercropping with turmeric or ginger reduce the continuous cropping obstacles that affect *Pogostemon cablin* (Patchouli). *Front. Microbiol.* **2020**, *11*, 579719. [CrossRef]
40. Zhao, Q.L.; Song, X.L.; Sun, X.Z. Studies on soil microorganism quantities and soil enzyme activities in the garlic-cotton and wheat-cotton intercropping systems. *Plant Nutr. Fertil. Sci.* **2011**, *17*, 1474–1480.
41. Liu, T.; Cheng, Z.; Meng, H.; Ahmad, I.; Zhao, H. Growth, yield and quality of spring tomato and physicochemical properties of medium in a tomato/garlic intercropping system under plastic tunnel organic medium cultivation. *Sci. Hortic.* **2014**, *170*, 159–168. [CrossRef]
42. Gong, H.; Du, Q.; Xie, S.; Hu, W.; Deng, J. Soil microbial DNA concentration is a powerful indicator for estimating soil microbial biomass C and N across arid and semi-arid regions in northern China. *Appl. Soil Ecol.* **2021**, *160*, 1–8. [CrossRef]
43. Ghani, M.I.; Ali, A.; Atif, M.J.; Ali, M.; Amin, B.; Anees, M.; Khurshid, H.; Cheng, Z. Changes in the soil microbiome in eggplant monoculture revealed by high-throughput Illumina MiSeq Sequencing as influenced by raw garlic stalk amendment. *Int. J. Mol. Sci.* **2019**, *20*, 2125–2151. [CrossRef] [PubMed]
44. Liu, Y.; Ma, W.; He, H.; Wang, Z.; Cao, Y. Effects of Sugarcane and soybean intercropping on the nitrogen-fixing bacterial community in the Rhizosphere. *Front. Microbiol.* **2021**, *12*, 3349–3353. [CrossRef] [PubMed]
45. Elshahawy, I.E.; Osman, S.A.; Abd-El-Kareem, F. Protective effects of silicon and silicate salts against white rot disease of onion and garlic, caused by *Stromatinia cepivora*. *J. Plant Pathol.* **2020**, *103*, 27–43. [CrossRef]
46. Boparai, A.K.; Manchanda, J.S. Response of cotton and wheat cultivars to soil-applied boron in a boron-deficient, noncalcareous typic ustochrept. *Commun. Soil Sci. Plant Anal.* **2019**, *50*, 108–118. [CrossRef]
47. Tian, X.; Li, C.; Zhang, M.; Li, T.; Lu, Y.; Liu, L. Controlled release urea improved crop yields and mitigated nitrate leaching under cotton-garlic intercropping system in a 4-year field trial. *Soil Tillage Res.* **2018**, *175*, 158–167. [CrossRef]
48. Ali, A.; Ghani, M.I.; Ding, H.; Fan, Y.; Cheng, Z.; Iqbal, M. Co-amended synergistic interactions between arbuscular mycorrhizal fungi and the organic substrate-induced cucumber yield and fruit quality associated with the regulation of the AM-fungal community structure under anthropogenic cultivated soil. *Int. J. Mol. Sci.* **2019**, *20*, 1539. [CrossRef]
49. Sun, Y.; Chen, L.; Zhang, S.; Miao, Y.; Zhang, Y.; Li, Z.; Zhao, J.; Yu, L.; Zhang, J.; Qin, X.; et al. Plant interaction patterns shape the soil microbial community and nutrient cycling in different intercropping scenarios of aromatic plant species. *Front. Microbiol.* **2022**, *13*, 888789. [CrossRef]
50. Burns, K.N.; Bokulich, N.A.; Cantu, D.; Greenhut, R.F.; Kluepfel, D.A.; O'Geen, A.T.; Strauss, S.L.; Steenwerth, K.L. Vineyard soil bacterial diversity and composition revealed by 16S rRNA genes: Differentiation by vineyard management. *Soil Biol. Biochem.* **2016**, *103*, 337–348. [CrossRef]
51. Jiang, Y.; Li, S.; Li, R.; Zhang, J.; Liu, Y.; Lv, L.; Zhu, H.; Wu, W.; Li, W. Plant cultivars imprint the rhizosphere bacterial community composition and association networks. *Soil Biol. Biochem.* **2017**, *109*, 145–155. [CrossRef]
52. Gupta, A.; Singh, U.B.; Sahu, P.K.; Paul, S.; Kumar, A.; Malviya, D.; Singh, S.; Kuppusamy, P.; Singh, P.; Paul, D.; et al. Linking soil microbial diversity to modern agriculture practices: A review. *Int. J. Environ. Res. Public Health* **2022**, *19*, 3141. [CrossRef] [PubMed]
53. Song, Y.; Liu, C.; Wang, X.; Ma, X.; Jiang, L.; Zhu, J.; Gao, J.; Song, C. Microbial abundance as an indicator of soil carbon and nitrogen nutrient in permafrost peatlands. *Ecol. Indic.* **2020**, *115*, 106362. [CrossRef]
54. Chen, K.J.; Zheng, Y.Q.; Kong, C.H.; Zhang, S.Z.; Li, J.; Liu, X.G. 2,4-Dihydroxy-7-methoxy-1,4-benzoxazin-3-one (DIMBOA) and 6-Methoxy-benzoxazolin-2-one (MBOA) levels in the wheat rhizosphere and their effect on the soil microbial community structure. *J. Agric. Food Chem.* **2010**, *58*, 12710–12716. [CrossRef] [PubMed]
55. Scavo, A.; Abbate, C.; Mauromicale, G. Plant allelochemicals: Agronomic, nutritional and ecological relevance in the soil system. *Plant Soil* **2019**, *442*, 23–48. [CrossRef]
56. Thompson, A.R.; Roth-Monzón, A.J.; Aanderud, Z.T.; Adams, B.J. Phagotrophic protists and their associates: Evidence for preferential grazing in an abiotically driven soil ecosystem. *Microorganisms* **2021**, *9*, 1555. [CrossRef]
57. Qi, L.; Zhou, P.; Yang, L.; Gao, M. Effects of land reclamation on the physical, chemical, and microbial quantity and enzyme activity properties of degraded agricultural soils. *J. Soil Sediments* **2020**, *20*, 973–981. [CrossRef]
58. Wu, C.; Chen, Y.; GrenierHéon, D. Effects of allelopathy and competition for nutrients and water on the survival and growth of plant species in a natural *Dacrydium* forest. *Can. J. For. Res.* **2021**, *52*, 100–108. [CrossRef]
59. Kong, C.H.; Xuan, T.D.; Khanh, T.D.; Tran, H.D.; Trung, N.T. Allelochemicals and signaling chemicals in plants. *Molecules* **2019**, *24*, 2737–2746. [CrossRef]
60. Vicua, R.; González, B. The microbial world in a changing environment. *Rev. Chil. de Hist. Nat.* **2021**, *94*, 2–5. [CrossRef]

Article

Effects of Intercropping *Pandanus amaryllifolius* on Soil Properties and Microbial Community Composition in *Areca Catechu* Plantations

Yiming Zhong [1,2,3], Ang Zhang [1,2,3,*], Xiaowei Qin [1,2,3], Huan Yu [1,2,3], Xunzhi Ji [1,2,3], Shuzhen He [1,2,3], Ying Zong [1,2,3], Jue Wang [1,2,3] and Jinxuan Tang [1,2,3]

[1] Spice and Beverage Research Institute, Chinese Academy of Tropical Agricultural Sciences, Wanning 571533, China
[2] Hainan Provincial Key Laboratory of Genetic Improvement and Quality Regulation for Tropical Spice and Beverage Crops, Spice and Beverage Research Institute, Chinese Academy of Tropical Agricultural Sciences, Wanning 571533, China
[3] Key Laboratory of Genetic Resource Utilization of Spice and Beverage Crops of Ministry of Agriculture and Rural Affairs, Spice and Beverage Research Institute, Chinese Academy of Tropical Agricultural Sciences, Wanning 571533, China
* Correspondence: angzhang_henu@163.com

Citation: Zhong, Y.; Zhang, A.; Qin, X.; Yu, H.; Ji, X.; He, S.; Zong, Y.; Wang, J.; Tang, J. Effects of Intercropping *Pandanus amaryllifolius* on Soil Properties and Microbial Community Composition in *Areca Catechu* Plantations. *Forests* 2022, 13, 1814. https://doi.org/10.3390/f13111814

Academic Editors: Chunjian Zhao, Zhi-Chao Xia, Chunying Li and Jingle Zhu

Received: 4 October 2022
Accepted: 27 October 2022
Published: 31 October 2022

Publisher's Note: MDPI stays neutral with regard to jurisdictional claims in published maps and institutional affiliations.

Copyright: © 2022 by the authors. Licensee MDPI, Basel, Switzerland. This article is an open access article distributed under the terms and conditions of the Creative Commons Attribution (CC BY) license (https://creativecommons.org/licenses/by/4.0/).

Abstract: The areca nut (*Areca catechu* L.) and pandan (*Pandanus amaryllifolius* Roxb.) intercropping cultivation system has been widely practiced to improve economic benefits and achieve the development of sustainable agriculture in Hainan Province, China. However, there is a lack of research on the relationships among soil properties, soil enzyme activities, and microbes in this cultivation system. Therefore, a random block field experiment of pandan intercropped with areca nut was established to investigate the effects of environmental factors on the diversity and functions of soil microbial communities in Lingshui county, Hainan Province. The diversity and composition of soil microbial communities under different cropping modes were compared using Illumina sequencing of 16S rRNA (bacteria) and ITS-1 rRNA (fungi) genes, and FAPROTAX and FUNGuild were used to analyze and predict the bacteria and fungi community functions, respectively. Correlation analysis and redundancy analysis were used to explore the responses of soil microbial communities to soil environmental factors. The results showed that the bacterial community was more sensitive to the areca nut and pandan intercropping system than the fungal community. The functional predictions of fungal microbial communities by FAPROTAX and FUNGuild indicated that chemoheterotrophy, aerobic chemoheterotrophy, and soil saprotroph were the most dominant functional communities. The intercropping of pandan in the areca nut plantation significantly enhanced the soil bacterial Ace and Chao indices by reducing the soil organic carbon (SOC) and total phosphorus (TP) content. In the intercropping system, urease (UE) and acid phosphatase were the key factors regulating the soil microbial community abundance. The dominant bacterial and fungal phyla, such as Firmicutes, Methylomirabilota, Proteobacteria, Actinobacteria, Chloroflexi, Verrucomicrobia, and Ascomycota significantly responded to the change in planting modes. Soil properties, such as UE, total nitrogen, and SOC had a significant stimulating effect on Proteobacteria, Chloroflexi, and Ascomycota. In summary, soil bacteria responded more significantly to the change in cropping modes than soil fungi and better reflected the changes in soil environmental factors, suggesting that intercropping with pandan positively affects soil microbial homeostasis in the long-term areca nut plantation.

Keywords: cultivation mode; soil physicochemical properties soil enzyme activity; soil microbial diversity; microbial community structure

1. Introduction

With the development of large-scale and intensive agricultural production modes, the degradation of farmland soil microbial communities has become a prominent problem that restricts the sustainable development of agriculture. As a farming mode based on the principle of promoting and complementing ecology, intercropping allows two or more crops to be planted on the same field, thereby, not only optimizing the utilization of resources such as sunlight, water, nutrients, and shared space [1], improving the net effect in the tradeoff between interspecies competition and the facilitation of crop growth [2,3], and considerably increasing yield [4], but also stimulating the interactions among soil nutrients, enzymes, microbes, and several coexisting crops [5], thus, maintaining the relative balance of soil microbial community [6,7]. Areca nut is an important cash crop in the tropical regions of South and Southeast Asia [8,9], which is often intercropped with vegetables, cocoa, banana, black pepper, and cardamom [10]. Among them, pandan is a tropical spice crop with high economic value; it is shade-tolerant and suitable for intercropping in areca nut forests [11–13]. Therefore, exploring the relationship among soil properties, enzyme activities, and microbes, as well as the mechanism of the intercropping mode to maintain soil health, is conducive to maintaining the efficient production of areca nut and pandan. Soil enzyme activity is a vital indicator of soil quality and is essential in evaluating soil health [14]. Common soil enzymes, such as catalase (CAT), acid phosphatase (ACP), urease (UE), and invertase, play a catalytic role in the decomposition of plant and animal residues, accelerating their biochemical reactions [15]. UE participates in the ammoniation of organic nitrogen in the nitrogen cycle in the farmland ecosystem to produce plant-available nitrogen [16]. UE activity determines the transfer rate of soil nutrients [17]. Peroxidase (POD) degrades lignin and coupled polysaccharides and is related to the degradation of polyphenols produced by soil fungi [18,19].

Microbial community composition is related to soil function and ecosystem sustainability because it is involved in soil organic matter dynamics and nutrient cycling processes, as well as in the metabolism of the soil system [20–22]. Soil microbes are diverse and functionally valuable, containing various species of bacteria, archaea, fungi, microfauna, and viruses [23]. As the two major categories of the farmland soil microbial system, bacteria and fungi usually represent the soil microbial community and are used to analyze and compare the soil microbial diversity indices and community structure [24]. Soil management measures can directly affect soil properties by changing the relationship between soil microbial community and soil properties [21]. For example, planting modes affect the development and vitality of soil microbes, mainly by changing soil properties. Lower pH affects bacterial and fungal densities. A moderate improvement of soil organic carbon (SOC) and soil nitrogen content stimulate soil microbial abundance and diversity [25], whereas excessive nutrient addition inhibits microbial diversity. There is a close relationship between soil enzyme activities and soil microbial characteristics because soil microbes are capable of secreting a range of enzymes, and changes in soil enzyme activities reflect changes in the nutrient requirements and metabolic activity of soil microbes [26]. Intercropping affects the relationship between soil enzyme activity and the microbial community by changing soil properties and microenvironments and then regulates the structure and function of the soil microbial community [27]. It is noteworthy that the intercropping of different crops has various effects on soil microbial content: intercropping of Kura clover with prairie cordgrass increases the abundance of arbuscular mycorrhizal fungi [28], intercropping wolfberry with Gramineae plants increases bacterial alpha diversity [29], and a melon/cowpea intercropping system enhanced the content of beneficial microbes [2]. Legumes and nitrogen fixation may increase the nitrogen content of the soil, but in other intercropping systems, different effects may occur [30].

Soil microbial diversity is inextricably linked to microbial function, and increased microbial diversity implies the improved soil biochemical response and sustainability of soil function [31]. Previous studies considered that the major functions of soil microbes are regulating soil functional diversity [32], decomposing plant residues [22], maintaining soil fertility and productivity, participating in carbon and nitrogen cycling [33], and inhibiting pathogens [34]. The effect of intercropping on soil microbial diversity also can significantly alter soil function [35]. Complex interactions exist among soil resources, soil enzymes, and soil microbes in the intercropping system [5]. Intercropping causes changes in soil properties and nutrients, significantly affecting the metabolic activities of soil microbes [6], including the production of cellulase that decomposes polysaccharides, UE, ACP [36], and neutral and alkaline phosphatases that participate in nitrogen and phosphorus cycles [37]. Moreover, the interactions among microbes, soil nutrients, and various enzymes in the intercropping system cause changes in the abundance of microbes and enzyme activities, which can improve the soil micro-ecological environment and functions [38].

However, the effects of intercropping of cash crops on soil microbial communities are still unclear in tropical farmland. Therefore, the areca nut and pandan intercropping field experiment was established to: (1) clarify the effects of intercropping on soil properties, enzyme activities, and microbial community diversity and structure; (2) explore the key mechanism of how the intercropping system alters soil microbial community diversity and structure; and (3) investigate the effect of soil microbial community functional change in a tropical intercropping system.

2. Materials and Methods

2.1. Study Site

The experiment was performed in Sanjiaowei Village, Lingshui County (109°56′ E, 18°31′ N, a.s.l. 36 m) in southeastern Hainan Province, China, from 2015. The mean annual temperature was 25 °C and the mean annual precipitation was 1700 mm at the experimental site. The soil was tidal sand–mud (US Soil Taxonomy classification) with a pH of 6.00, organic matter of 20.04 g·kg^{-1}, electrical conductivity (EC) of 96.68 S·m^{-1}, and soil-available nitrogen (SAN), soil-available phosphorus (SAP), and soil-available potassium (SAK) concentrations of 77.78 mg·kg^{-1}, 17.22 mg·kg^{-1}, and 51.46 mg·kg^{-1}, respectively. Total nitrogen (TN), total phosphorus (TP), and total potassium (TK) were 1.33 g·kg^{-1}, 0.82 g·kg^{-1}, and 11.93 g·kg^{-1}, respectively.

2.2. Experimental Design and Management

The experiment adopted a randomized block design. Each block had three plots and each block was replicated 6 times, and one plot was set for each planting mode: areca nut monocropping (AM), pandan monocropping (PM), and areca nut and pandan intercropping (I), and the block was repeated 6 times. The cultivation period of areca nut is about 6 years, and the cultivation period of fragrant pandan is about 3 years. The planting density was 2.5 m × 2.5 m for areca nut and 50 cm × 50 cm for pandan. During the experiment, water and fertilizer management, pest control, and other field management practices remained the same in the three planting modes.

2.3. Soil Sampling

Soil samples were collected in June 2021. Five soil samples (0–20 cm) were randomly collected from each plot with a diameter of 5 cm and then mixed as one soil sample. After sieving (<2 mm, <0.20 mm) to remove plant roots and other visible foreign bodies, all soil samples were immediately brought back to the laboratory. Soil samples were divided into two parts: one was used to analyze soil physicochemical properties after air drying, and the other was stored in a −80 °C freezer for soil microbial community analysis.

2.4. Analysis of Soil Physicochemical Properties and Soil Enzyme Activities

Soil pH was measured using a pH/conductivity meter (FE28, China; soil: water ratio was 1:2.5). After weighing the fresh weight, the soil samples were oven-dried at 105 °C for 24 h and weighed again to calculate the soil water content (SWC) [39]. Electrical conductivity (EC) was measured using the pH/conductivity meter (DDS-307A conductivity meter, China) [40]. Soil organic matter was determined by a total organic carbon analyzer (Multi N/C 3100, Jena, Germany) [41], and bulk density (BD) was measured (BD, g/cm^3 = soil dry weight/soil volume). Alkali-hydrolyzed nitrogen (SAN) was determined using the alkaline hydrolysis diffusion method. SAP was assessed using Bray's method (UV2310 II, Shanghai,

China) [42]. SAK was determined using flame photometry (6400A, Changsha, China) [43]. TN was determined by Kelvin distillation, TP using the molybdenum blue colorimetric method, and TK by flame photometry [44]. Soil catalase (CAT), soil polyphenol oxidase (PPO), soil peroxidase (POD), soil acid phosphatase (ACP), and soil urease (UE) were determined by ultraviolet–visible spectrophotometry using kits (Suzhou Comin Biotechnology Co., Ltd., Suzhou, China). PPO can catalyze pyrogallol to produce colored species with characteristic light absorption at 430 nm. H_2O_2 has a characteristic absorption peak at 240 nm. By measuring the change in the absorbance of the solution at this wavelength after reacting with the soil, the activity level of CAT can be reflected. POD catalyzes the oxidation of organic substances to quinones, which have characteristic light absorption at 430 nm. Using the indophenol blue colorimetric method, the NH_3-N generated by the urease hydrolysis of urea was identified (www.cominbio.com).

2.5. Soil DNA Extraction and Sequencing

Total soil DNA was extracted and purified using the EZNA® Soil DNA Extraction Kit (Omega, Norwalk, CT, USA). Using barcode-tagged primer sequences for bacteria: 338F (5′-ACTCCTACGGGAGGCAGCAG-3′) and 806R (5′-GGACTACHVGGGTWTCTAAT-3′), and fungi: internal transcribed spacer (ITS) 1F (5′-CTTGGTCATTTAGAGGAAGTAA-3′) and ITS2R (5′-GCTGCGTTCTTCATCGATGC-3′), the corresponding soil bacterial 16S rRNA V3-V4 region and fungal ITS-1 region sequences were amplified, and 2% agarose gel electrophoresis was used to detect the length of the amplified products. According to the quantitative detection results, the amplified products were mixed into one sample, and a clone library was constructed. The loading amount for each library was calculated based on the library search results, and the paired-end sequencing method was used on the Illumina MiSeq high-throughput platform for sequencing. The data were analyzed using the Majorbio cloud platform (www.Majorbio.com (accessed on 22 September 2022)).

2.6. Bioinformatics Analysis

Paired-end reads of raw DNA fragments were merged using FLASH 1.2.11 [45] software and quality filtered using QIIME 1.9.1 software [46]. Valid sequences were obtained, and reads that could not be assembled were discarded. Unique sequences with 97% or greater similarity were clustered into operational taxonomic units (OTUs) using UPARSE 7.0.1090 software. MOTHUR 1.30.2 [47] annotated each OTU using the small subunit rRNA SILVA database (v 138) [48] and UNITE 8.0 fungi database [49]. The sample with the least data was used as the standard for normalization (normalization using the normalization method: the sequences of all samples are randomly selected to that amount of data according to the minimum number of sample sequences). Soil microbial community diversity and richness were calculated using QIIME.

2.7. Statistical Analysis

Taxonomic alpha diversity was calculated as the estimated community diversity by the Shannon index using the Mothur software package (v.1.30.1), and nonmetric multidimensional scaling (NMDS) was selected to reflect the changes in the microbial structure under intercropping modes, these changes were referred to as microbial beta diversity. Network interaction analysis of microbial composition was analyzed and painted by SPSS 23.0 and Cytoscape V3.8.2, respectively. FAPROTAX (v1.2.1) [50] and FUNGuild (v1.0) [51] were used to analyze and predict the microbial community functions, respectively. The same community with different guild annotations was selected for all annotations in FUNGuild using three classification levels: highly probable, probable, and possible. The microbial community was divided into bacteria and fungi in this study (after confirmation, no archaeal taxa were found in the 16 s dataset). The experimental indicator (soil physical and chemical properties, soil enzyme activity, soil fungal–bacterial diversity and community structure, and prediction of soil fungal–bacterial functional communities) was analyzed by one-way ANOVA to determine differences between intercropping and monocropping modes. NMDSs were statistically assessed using a permutational analysis of variance (PERMANOVA). The statistical significance ($p < 0.05$) was calculated using Duncan's test. Correlations between soil properties, soil enzyme activities, and soil microbial community diversity were calculated and analyzed using a Spearman correlation matrix. Redundancy analysis was performed and mapped using the analysis of soil microbial community composition about environmental factors; the model was assessed for 999 iterations based on Monte Carlo permutations. Data analyses were performed using SAS V8 and Canaco 5.0. The graphs were plotted using Origin 2021b and R.4.0.5.

3. Results

3.1. Changes in Soil Physicochemical Properties and Enzyme Activities

One-way ANOVA revealed the significant effects of planting modes on soil properties and enzyme activities. Compared to the AM mode, intercropping significantly increased pH, BD, and TK by 0.47 (absolute difference, $p < 0.001$), 16.71% (absolute difference, $p < 0.001$), and 18.44% (relative difference, $p < 0.05$), respectively, whereas EC, SOC, SAK, SAN, SOP, TN, and TP were significantly decreased by 41.52% ($p < 0.001$), 22.28% ($p < 0.001$), 47.80% ($p < 0.001$), 36.98% ($p < 0.001$), 23.79% ($p < 0.01$), 22.33% ($p < 0.001$), and 47.24% ($p < 0.01$) under the intercropping mode, respectively (Figure 1). Most of the soil physical and chemical properties were significantly lower under the intercropping mode when compared with PM monoculture ($p < 0.01$), except the soil TK content was significantly increased by 9.94% ($p < 0.001$), and SWC, SAK, and SAN were not affected. Compared to the AM mode, intercropping significantly increased ACP and UE activity by 36.64% and 8.27% (relative difference, $p < 0.01$). However, the activity of CAT, PPO, and UE were significantly decreased by 14.61%, 37.07%, and 20.04% when compared with PM, while POD and ACP in the intercropping mode were significantly higher than in the PM mode by 70.64% and 33.48%, respectively (Figure 2). There is a gigantic difference between the AM and PM modes in the physicochemical characteristics and enzyme activity. The soil properties and enzyme activities of the areca nut forest were altered dramatically after intercropping with pandan in this study.

3.2. Changes in Soil Microbial Community Diversity

The number of soil bacterial community sequences per sample ranged from 29,519 to 62,515 (mean = 41,633), whereas the number of fungal community sequences ranged from 54,987 to 87,289 (mean = 66,394). The intercropping mode significantly increased the bacterial Ace and Chao indices by 28.30% and 27.26% (relative difference), whereas other soil microbial diversity indices did not change significantly (Figure 3). One-way ANOVA revealed that intercropping did not affect the Shannon or Simpson index when compared with AM or PM, whereas it significantly increased Ace and Chao indices by 28.24% and 28.67%, respectively, in the bacterial community when compared with PM. When compared with AM, intercropping increased the bacterial Shannon index by 5.08%. Ace and Chao indices were increased by 72.17% and 69.95%, respectively, in the fungal community after intercropping, when compared with PM ($p < 0.05$, Figure 3).

Nonmetric multidimensional scaling analysis (NMDS) was conducted to reflect microbial beta diversity (Appendix A, Figure A1). The soil bacterial characteristics for the AM and intercropping treatments were nearly the same, whereas the soil bacterial characteristics under PM treatment were quite different from the intercropping AM mode. The soil fungal characteristics were not greatly affected by the intercropping modes in this study (Appendix A, Figure A1).

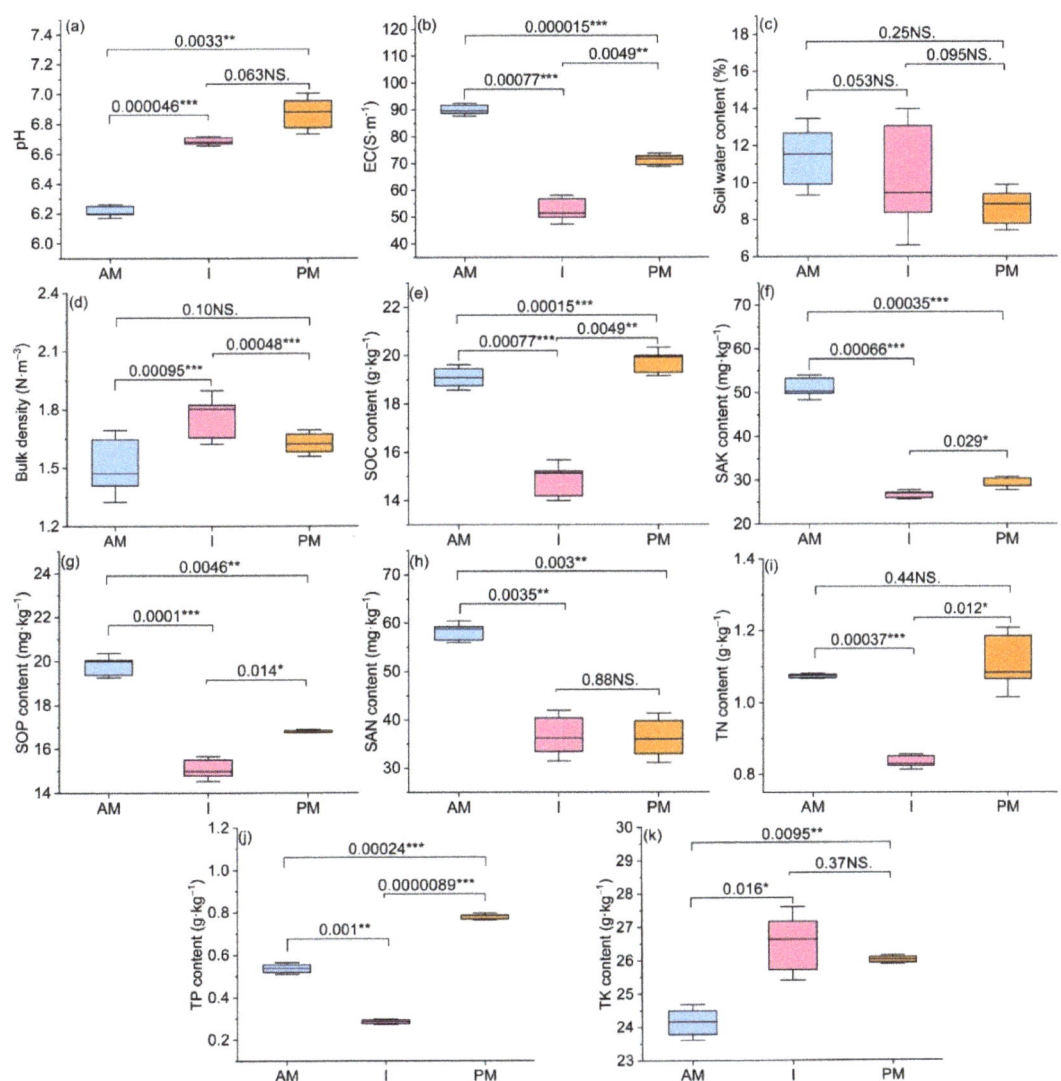

Figure 1. Soil properties under different cropping patterns (n = 9). * is significant at the 0.05 level; ** is significant at the 0.01 level; and *** is significant at the 0.001 level. AM represents areca nut monocropping; I represents areca nut intercropping with pandan; and PM represents pandan monocropping. Note: (**a**)-pH, (**b**)-EC, (**c**)-SWC, (**d**)-BD, (**e**)-SOC, (**f**)-SAK, (**g**)-SOP, (**h**)-SAN, (**i**)-TN, (**j**)-TP, (**k**)-TK.

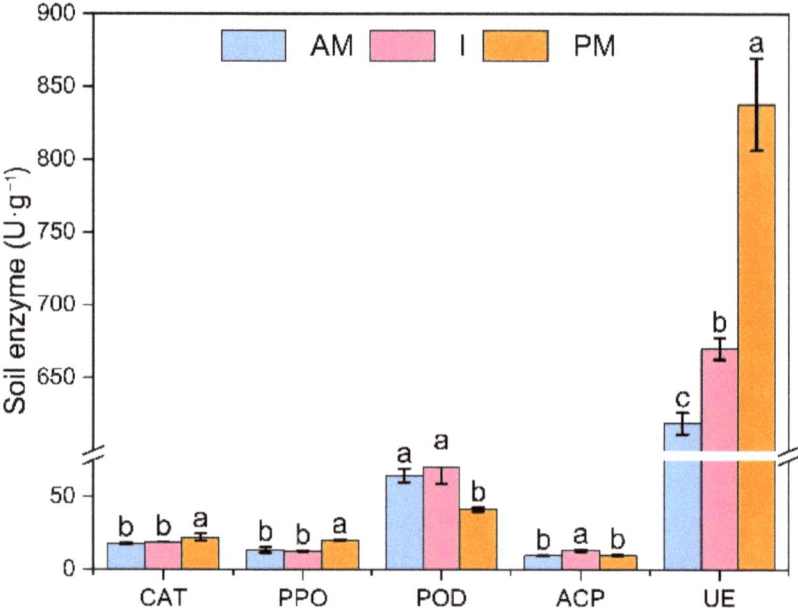

Figure 2. Soil enzyme activity under different cropping patterns. Different letters indicate significant differences (ANOVA, $p < 0.05$, and Tukey's HSD post hoc analysis) among different planting modes. AM represents areca nut monocropping; I represents areca nut intercropping with pandan; and PM represents pandan monocropping. Different lowercase letters indicate significant differences between treatments under the same index ($p < 0.05$).

3.3. Changes in the Composition and Structure of Soil Microbial Community

The phyla with relative abundance greater than 1% in soil bacterial and fungal communities are usually considered the dominant phyla. The 12 dominant phyla in the bacterial community were Proteobacteria (25.44%), Actinobacteria (20.94%), Acidobacteria (15.64%), Firmicutes (10.40%), Chloroflexi (7.89%), Bacteroides (4.24%), Myxococcota (3.24%), Methylomirabilota (1.96%), Verrucomicrobia (1.59%), Gemmatimonadota (1.25%), Planctomycetota (1.13%), and Bdellovibrionota (1.07%). The four dominant phyla in the fungal community were Ascomycota (76.77%), Basidiomycota (11.33%), unclassified fungi (7.62%), and Rozellomycota (2.75%) (Figure 4). Compared with AM, Firmicutes in intercropping significantly decreased by 12.61%, whereas Methylomirabilota and unclassified bacteria were significantly increased by 2.88% and 0.68%, respectively, and Acidobacteria abundance increased by 5.86%. Compared with PM, Proteobacteria, Ascomycota, and Chloroflexi were significantly reduced by 1.62%, 16.45%, and 1.89%. Methylomirabilota and Verrucomicrobia were significantly increased by 1.35% and 1.93% (absolute difference, all $p < 0.05$), respectively, after intercropping (Figure 4, Appendix A, Table A1). There was a strong positive correlation among the dominant bacteria groups: Acidobacteriota, Actinobacteriota, Bacteroidota, Chloroflexi, Firmicutes, Methylomirabilota, Myxococcota, and Proteobacteria. However, the dominant fungal community Ascomycota showed a strong negative correlation with other fungal groups except Zoopsgomycota (Figure 5).

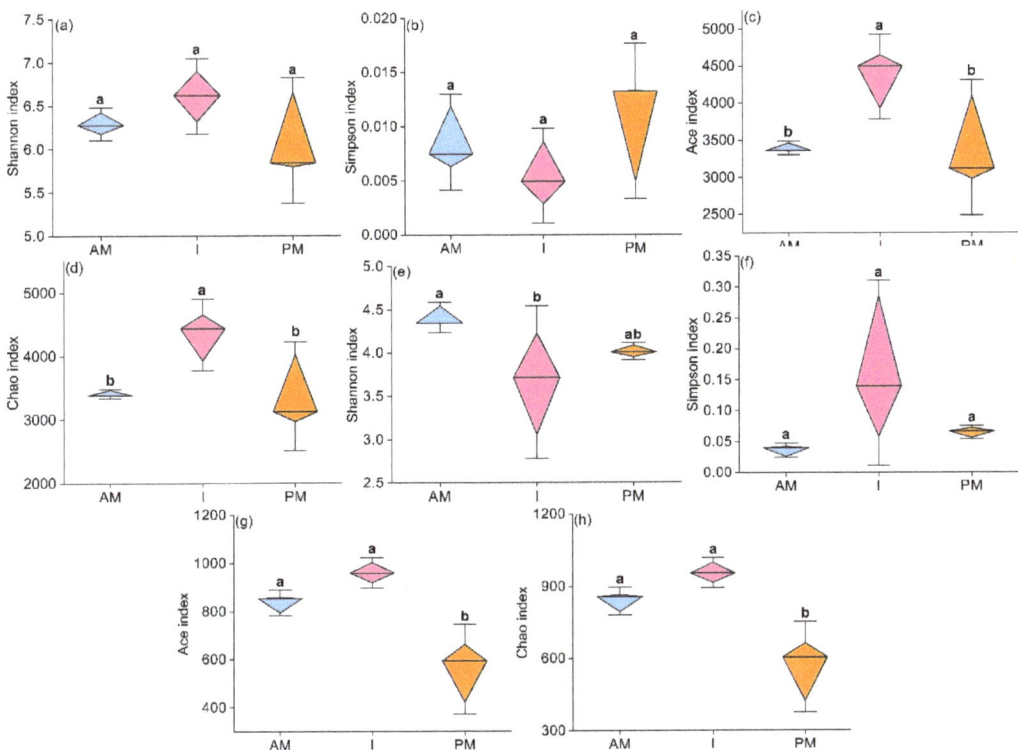

Figure 3. Changes in soil microbial alpha diversity index under different planting patterns ((**a–d**): bacteria; (**e–h**): fungi). Different letters indicate significant differences (ANOVA, $p < 0.05$, and Tukey's HSD post hoc analysis) among different planting modes. AM represents areca nut monocropping; I represents areca nut intercropping with pandan; and PM represents pandan monocropping. Different lowercase letters indicate significant differences between treatments under the same index ($p < 0.05$).

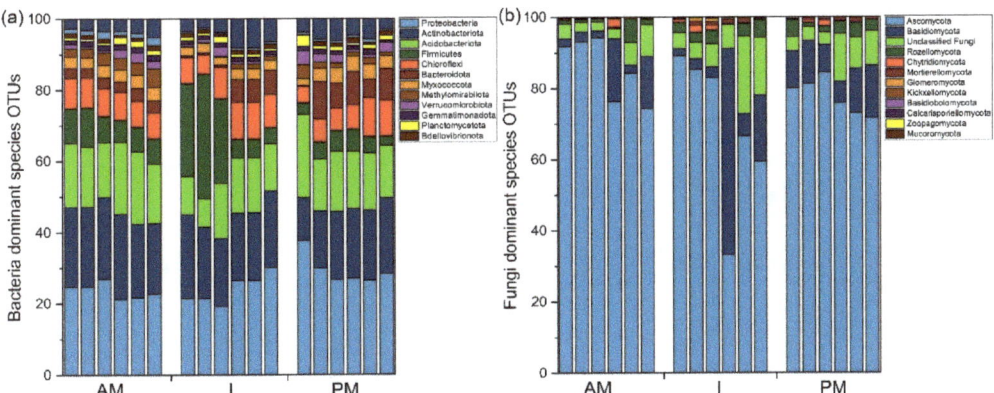

Figure 4. Soil microbial community composition under different planting patterns ((**a**)-bacteria, (**b**)-fungi). The abundance of each taxon was calculated as the percentage of sequences per gradient for a given microbial group. AM represents areca nut monocropping; I represents areca nut intercropping with pandan; and PM represents pandan monocropping.

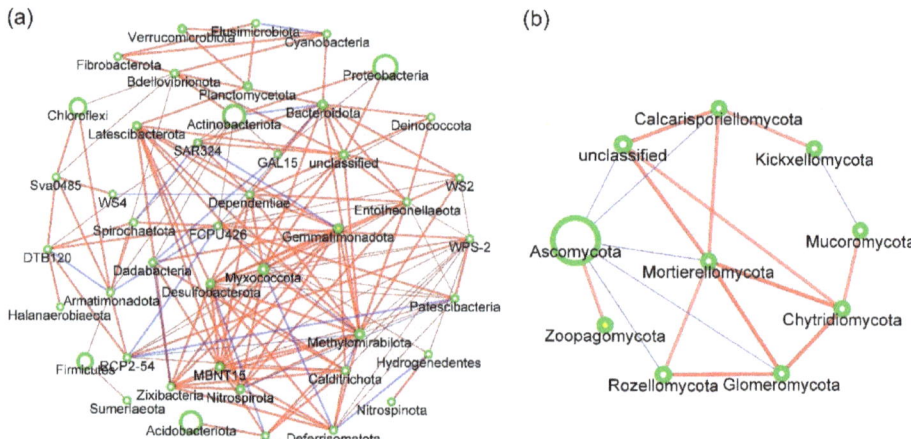

Figure 5. Network interaction diagram of dominant bacterial (**a**) and fungal (**b**) phyla. The line between the circles indicates that there is a correlation, red lines indicate a positive correlation, while the blue lines show a negative correlation. The size of the points represents the magnitude of phyla abundance, while the thickness of the line represents the correlation size. Each circle represents a microbial phylum. Red indicates a positive correlation, and blue indicates a negative correlation ($p < 0.05$).

3.4. Changes in Soil Microbial Functional Profiles

The functional prediction of the soil bacterial community showed that the main functional groups in each plot were "chemoheterotrophy" and "aerobic chemoheterotrophy". The majority of bacteria in nature are chemoheterotrophic bacteria, and their energy comes from the oxidation and decomposition of soil organic matter. The relative content of chemoisomeric bacteria in the AM and I modes was significantly lower than in the PM (Figure 6a, Appendix A, Table A2). The main functional prediction of the soil fungal community was "soil saprotroph". Soil saprophytic fungi absorb nutrients from dead plant residues or other organic substances, which are also chemoautotrophic microbes in nature. The relative abundance of soil saprotroph fungi in the PM treatment was slightly lower than that in the AM and I treatments in this study. It was worth noting that the relative abundance of "Symbiotroph" in fungi under the intercropping treatment was significantly lower than that in the AM or PM treatment (Figure 6b, Appendix A, Table A3).

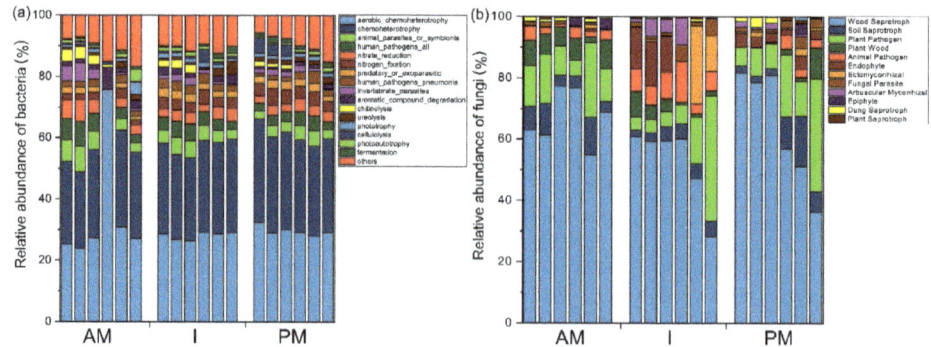

Figure 6. Predicted functional profiles of the soil bacteria (**a**) and fungi (**b**) with different planting mode. AM represents areca nut monocropping; I represents areca nut intercropping with pandan; and PM represents pandan monocropping.

3.5. Relationship between Soil Properties and Soil Enzymes

A close relationship was observed between soil properties and enzyme activities. There was a positive and negative correlation between TP ($R = 0.87$), SWC ($R = -0.66$), and PPO, respectively. POD was significantly negatively correlated with TP ($R = -0.84$). ACP was highly significantly negatively correlated with SOC ($R = -0.96$), TN ($R = -0.91$), EC ($R = -0.84$), and TP ($R = -0.82$), SOP ($R = -0.78$), whereas it was positively correlated with BD. UE was significantly positively correlated with TP ($R = 0.71$) and pH ($R = 0.85$). However, CAT was not correlated with soil physicochemical properties (all $p < 0.05$, Figure 7).

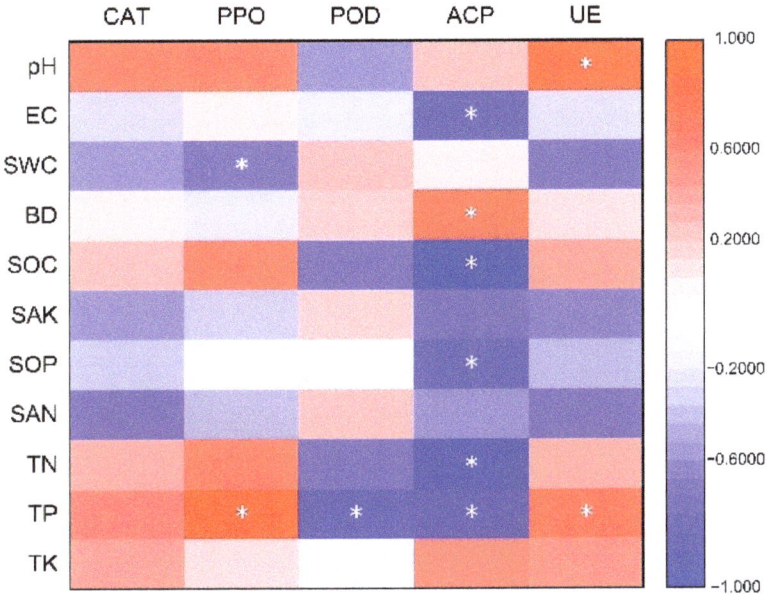

Figure 7. Relationship between soil environmental factors and soil enzyme activities. * Correlation is significant at the 0.05 level. Red indicates a positive correlation and blue indicates a negative correlation, and the darker the color, the stronger the correlation. Electrical conductivity (EC), soil water content (SWC), soil bulk density (BD), soil organic carbon (SOC), soil-available potassium (SAK), soil-available phosphorus (SAP), alkali-hydrolyzed nitrogen (SAN), total nitrogen (TN), total phosphorus (TP), total potassium (TK), soil catalase (CAT), soil polyphenol oxidase (PPO), soil peroxidase (POD), soil acid phosphatase (ACP), and soil urease (UE). Same below.

3.6. Influence of Soil Biological and Abiotic Factors on Soil Microbial Community Diversity

The Ace and Chao indices of the bacterial community were negatively correlated with SOC ($R = -0.79, -0.80$; $p < 0.05, 0.01$) and TP ($R = -0.71, -0.73$; $p < 0.05$), respectively, but positively correlated with ACP ($R = 0.83, 0.84, p < 0.01$). Fungal Ace and Chao indices were negatively correlated with SOC ($R = -0.75, -0.74$; $p < 0.05$), TN ($R = -0.68, -0.67$; $p < 0.05$), TP ($R = -0.91, -0.91$; $p < 0.001$), CAT ($R = -0.69, -0.70$; $p < 0.05$), PPO ($R = -0.91, -0.92$; $p < 0.001$), and UE ($R = -0.84, -0.84$; $p < 0.01$), respectively, but positively correlated with POD ($R = 0.88, 0.88$; $p < 0.01$). The fungal Shannon index was positively correlated with EC ($R = 0.67$; $p < 0.05$), SAK ($R = 0.68$; $p < 0.05$), and SOP ($R = 0.73$; $p < 0.05$), respectively, but negatively correlated with ACP ($R = -0.76$; $p < 0.05$). The fungal Simpson index was positively correlated with ACP ($R = 0.82$; $p < 0.01$, Table 1).

Table 1. Relationship between soil microbial alpha diversity and environmental factors.

Soil Properties	Bacteria				Fungi			
	Shannon	Simpson	Ace	Chao	Shannon	Simpson	Ace	Chao
pH	−0.19	0.17	0.12	0.10	−0.53	0.30	−0.45	−0.46
EC	−0.31	0.28	−0.64	−0.63	0.67 *	−0.60	−0.26	−0.24
SWC	0.05	−0.03	−0.02	0.00	0.16	−0.02	0.48	0.49
BD	0.28	−0.16	0.60	0.59	−0.58	0.59	0.26	0.25
SOC	−0.61	0.58	−0.79 *	−0.80 **	0.56	−0.65	−0.75 *	−0.74 *
SAK	−0.16	0.14	−0.48	−0.46	0.68 *	−0.53	0.11	0.12
SOP	−0.31	0.27	−0.64	−0.63	0.73 *	−0.65	−0.15	−0.14
SAN	−0.09	0.06	−0.39	−0.37	0.54	−0.37	0.21	0.22
TN	−0.42	0.38	−0.64	−0.66	0.59	−0.66	−0.68 *	−0.67 *
TP	−0.63	0.57	−0.71 *	−0.73 *	0.35	−0.53	−0.91 ***	−0.91 ***
TK	0.27	−0.25	0.62	0.61	−0.40	0.35	0.36	0.34
CAT	−0.11	0.09	−0.05	−0.09	−0.16	−0.02	−0.69 *	−0.70 *
PPO	−0.50	0.45	−0.47	−0.50	0.04	−0.27	−0.91 ***	−0.91 ***
POD	0.50	−0.50	0.52	0.54	0.07	0.13	0.88 **	0.88 **
ACP	0.64	−0.59	0.83 **	0.84 **	−0.76 *	0.82 **	0.64	0.63
UE	−0.48	0.45	−0.30	−0.33	−0.22	−0.03	−0.84 **	−0.84 **

Note: * Correlation is significant at the 0.05 level; ** Correlation is significant at the 0.01 level; and *** Correlation is significant at the 0.001 level. Electrical conductivity (EC), soil water content (SWC), soil bulk density (BD), soil organic carbon (SOC), soil-available potassium (SAK), soil-available phosphorus (SAP), alkali-hydrolyzed nitrogen (SAN), total nitrogen (TN), total phosphorus (TP), total potassium (TK), soil catalase (CAT), soil polyphenol oxidase (PPO), soil peroxidase (POD), soil acid phosphatase (ACP), and soil urease (UE).

3.7. Responses of Soil Microbial Community Structure to Three Planting Modes

The soil TP ($F = 6.6$, $p = 0.004$) and pH ($F = 5.5$, $p = 0.012$) significantly affected the soil bacterial community structure in this study (Figure 8a). Soil enzyme activities such as POD ($F = 3.6$, $p = 0.022$) had significant effects on soil bacteria (Appendix A, Table A5).

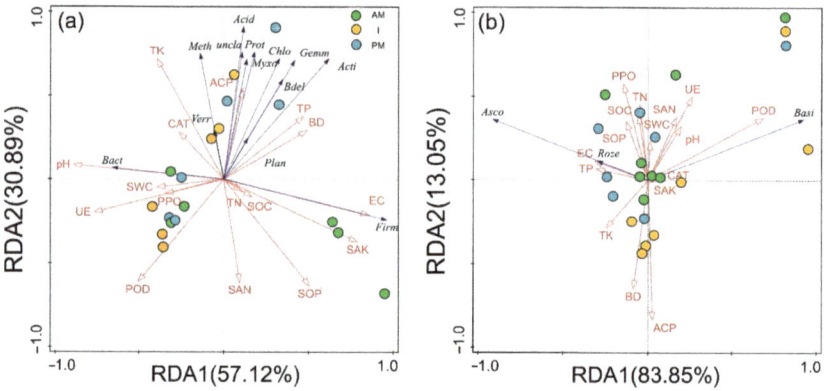

Figure 8. Ordination plots of the results from the redundancy analysis (RDA) to identify the relationships among the microbial (bacterial and fungal) taxa (blue arrows) and the soil properties and enzyme activities (red arrows) at the phylum level. (**a**) Relationships between soil bacterial communities and environmental variables and (**b**) relationships between soil fungal communities and environmental variables. Bacterial taxa: Proteobacteria (Prot), Actinobacteriota (Acti), Acidobacteriota (Acid), Firmicutes (Firm), Chloroflexi (Chlo), Bacteroidota (Bact), Myxococcota (Myxo), Methylomirabilota (Meth), Verrucomicrobiota (Verr), Gemmatimonadota (Gemm), Planctomycetota (Plan), and Bdellovibrionota (Bdel). Fungal taxa: Ascomycota (Asco), Rozellomycota (Roze), and Basidiomycota (Basi). Soil properties: electrical conductivity (EC), soil water content (SWC), soil bulk density (BD), soil organic carbon (SOC), soil-available potassium (SAK), soil-available phosphorus (SAP), alkali-hydrolyzed nitrogen (SAN), total nitrogen (TN), total phosphorus (TP), total potassium (TK), soil catalase (CAT), soil polyphenol oxidase (PPO), soil peroxidase (POD), soil acid phosphatase (ACP), and soil urease (UE). Same below.

Most bacterial phyla, such as Proteobacteria and Actinobacteria, were negatively correlated with TN, SOC, PPO, and POD, respectively. Ascomycota was negatively correlated with SOC, SAK, and TN, respectively (Figure 8b). The results of the multiple stepwise regression analysis of the soil microbial phyla indicated that TK, TP, TN, SOC, SAK, EC, UE, and pH were the main factors affecting the abundance of soil bacterial communities such as Actinobacteria, Acidobacteria, and Ascomycota. Among these, the most significant factors affecting soil microbes were TK and SOC (Appendix A, Table A4). Soil fungi community structure was not affected by soil physicochemical properties and soil enzyme activities (Appendix A, Table A5).

4. Discussion

4.1. Effects of Planting Modes on Soil Enzyme Activities

Soil enzymes are biologically active substances and catalysts involved in soil biochemical processes such as organic matter decomposition and nutrient cycling [52]. The enhancement of soil enzyme activities can accelerate the transformation of organic nutrients and improve the utilization efficiency of nutrients. However, the activities of different soil enzymes differ based on the different crops and planting modes, and the response of soil enzyme activities to soil management varies greatly [29]. In general, researchers consider that soil enzymes were increased with intercropping in the chestnut – tea or cereal–legume intercropping systems [15,53]. However, a meta-analysis indicated that intercropping had an increase, decrease, or neutral effect on soil enzyme activities in most intercropping systems [54].

Specifically, soil ACP catalyzes the mineralization of SOP compounds into inorganic phosphorus, and its activity directly affects SOP decomposition, transformation, and bioavailability [55]. A significant negative correlation was observed between ACP and SOP and TP, indicating that the demand for phosphorus significantly increases in crops under the intercropping mode, thus, stimulating the activity of ACP in the present study (Figure 7). The increase in ACP activity is conducive to the turnover of phosphorus between plants and soil. The soil UE is usually related to the soil nitrogen cycle, and it hydrolyzes urea into ammonia for plant utilization [56,57]. Compared with the AM mode, the soil UE activity of PM was significantly increased, but the soil alkaline-hydrolyzed nitrogen content was significantly reduced, indicating that the demand for nitrogen might be significantly higher in pandan than in areca nut. The soil CAT is mainly related to the degradation of hydrocarbons and heavy metals in soil, and PPO decomposes organic matter and accelerates soil humification [58]. Compared with the PM mode, the decreased CAT and PPO activities under intercropping indicated that the intensity of SOC metabolism (mineralization) significantly decreased when pandan was planted between the areca nut forest in this study (Figures 2 and 7).

4.2. Regulatory Mechanisms of Planting Modes on Soil Microbial Diversity

Plant cultivation methods are the key factors affecting soil microbial communities and biological health [59]. The increase in crop varieties and the rational allocation of time and space between crops improve the soil rhizosphere microenvironment and nutrients, regulate the nutrient metabolism balance of microbes, and promote the functional potential as well as the relative stability of soil microbial communities [60]. The interactions among crop roots, rhizosphere soil, and soil microbes in the intercropping mode promote the accumulation of root exudates (i.e., organic acids) and the activity of soil catalytic substances (i.e., soil enzymes), increase the stability and anti-interference ability of the soil ecosystem, and improve microbial diversity [61]. Soil bacterial diversity was significantly increased after intercropping, which might be attributed to the regulation of soil properties and enzyme activities by soil bacteria (Figure 8). At the same time, more diverse plant litter and root secretions may have a positive impact on bacterial diversity [10,62].

SOC are a nutrient source for plants and soil microbes. The reduced SOC content reflects the decline in soil bacteria utilization of the carbon source and metabolic rate [63].

The negative correlation between SOC and the Ace and Chao indices indicated that the reduced content of SOC after intercropping was one of the main reasons for the addition in bacterial diversity. Phosphorus plays an important role in root development, stem growth, production of root secretions, and ATP synthesis [64]. The reduction in elemental phosphorus increased the complexity of the soil bacterial symbiotic network and affected the original metabolic level of soil bacteria [65]. Therefore, the soil bacterial diversity index was significantly and negatively correlated with both soil TP and SOP, thereby suggesting that soil bacteria were sensitive to the phosphorus content at the experimental site. Intercropping could significantly improve bacterial diversity by further reducing the phosphorus content in this study (Figures 1 and 5).

4.3. Key Regulatory Factors of Different Planting Modes on Soil Microbial Community Composition and Structure

The composition, structure, and function of the soil microbial community in the farmland ecosystem are closely related in the current study [66]. Reasonable intercropping is mainly performed indirectly by changing nutrient content and soil enzyme activities [67], which is beneficial to keep the soil microbial community structure stable, thereby improving the metabolic activity and functional diversity of beneficial microbes and, thus, inhibiting the growth of anaerobic bacteria, denitrifying bacteria, and other harmful microbes that occur in monoculture cultivation [68].

The decrease in Proteobacteria and Actinobacteria, the two abundant bacterial phyla, may be related to the biological properties of soil bacteria and different optimal growth environments. Proteobacteria perform the function of nitrogen fixation in the soil bacterial community and, using UE catalysis, convert soil's organic nitrogen to ammonia for plant uptake [69]. The decrease in Proteobacteria abundance after intercropping may be attributed to the significant reduction of soil TN content, because Proteobacteria, soil UE activity, and plants maintained the balance of soil nitrogen content in this study (Figure 8a). Actinobacteria genera such as *Actinomyces*, *Micromonospora*, and *Streptomyces* produce enzymes that dissolve phosphorus and accelerate the effective degradation of organic matter [70]. Thus, the decrease in soil TP content in the intercropping mode may have been one of the main reasons for the decrease in Actinobacteria-relative abundance in this study (Figures 1i and 6a) [71,72]. Acidobacteria are slow-growing oligotrophic bacteria with a K-selected life strategy, and Acidobacteria abundance is higher when the soil organic matter content is low [73]. The above conclusion was confirmed by the fact that the content of soil organic matter decreased, but Acidobacteria abundance increased in areca nut soil after the intercropping with pandan in this study. Species of Firmicutes are often found in nutrient-rich soil environments and can produce antimicrobial substances that promote plant growth and reduce the growth of pathogenic bacteria, while the acid soil environment may have a negative impact on Firmicutes abundance and activity [74]. The reduction of soil pH after intercropping may be the main reason for the decrease in Firmicutes in this study (Figures 4 and 5).

Soil bacteria and fungi responded differently to the modes of pandan intercropped with areca nut. In terms of fungal community composition, Ascomycota, Basidiomycota, and Rozellomycota were the three dominant fungal phyla, which were consistent with previous studies [75]. Compared with the significant changes in the bacterial community, the fungal community, except Ascomycota, was insensitive to changes in soil physicochemical properties, nutrients, and enzyme activities caused by intercropping. Ascomycota comprises decomposing fungi that decompose lignin and other organic substances that are not easily decomposed in soil, and it was also closely related to soil organic matter [76]. Thus, the decrease in the soil organic matter content might be the main reason for the decrease in Ascomycota abundance under the intercropping mode in this study (Figures 1 and 5).

4.4. Effects of Different Planting Modes on Soil Microbial Functional Groups

The FAPROTAX database was created to generate functional profiles by connecting individual organisms to ecologically relevant metabolic activities and applies to the functional annotation of bacteria associated with environmental samples [77]. The FAPROTAX prediction, which has been utilized frequently by other researchers, is arguably the best method for predicting probable microbial roles in samples [78]. Soil bacterial community function is highly correlated with the type of plants in the land, and changes in apoplectic inputs and root secretions in the intercropping system bring changes in the environment for soil bacteria to survive, which may lead to changes in soil bacterial community function [79]. Chemoheterotrophy and aerobic chemoheterotrophy have been found to be the most significant functions of the soil bacterial population in this study. Aerobic chemoheterotrophy can speed up the biodegradation of organic materials, and both are involved in the C cycle process. Chemoheterotrophic bacteria are decomposers in nature and are responsible for in situ restoration in all ecosystems [77]. The leaf litter of areca nut affects the growth of pandan, which needs to be cleared regularly. Therefore, there was a lack of carbon input from the areca nut litter in the intercropping model. During the experiment, the leaves of pandan were also harvested, so almost no litter material was produced, and the organic matter from litter material in the intercropping system was reduced, and the organic matter content decreased. In this study, the SOC, TN, and SAN contents in the soil were lower in the I mode than in the AM mode, and the closely related aerobic chemoheterotrophy and chemoheterotrophy functional communities were also significantly reduced, further demonstrating the close relationship between the bacterial functional communities and environmental factors (Figures 1 and 6).

For the functional determination of fungi, FUNGuild is an effective tool because it can identify the functional group roles of fungi from the perspective of trophic guilds, rather than from individual OTUs [80]. The results obtained from the FUNGuild procedure showed that soil saprotrophs dominate the functions exercised by the fungi, which may play a central role in organic decomposition [51]. The proportion of functional groups of wood saprotroph fungi and lichenized fungi was increased under the intercropping mode, which related to the increased crop root biomass, but the specific functions still need to be further investigated in this study.

5. Conclusions

Intercropping pandan with areca nut had a positive impact on soil microbial diversity and dynamic balance, despite the fact that the bacterial community was more sensitive to the intercropping mode than the fungal community in the tropical plantations. We suggest that the decrease in soil nutrient content under the intercropping mode was the main reason for the increase in soil microbial diversity. Moreover, the change in soil enzyme activity may have changed the competitive relationships between the different kinds of microbes and nutrients, and then significantly changed the microbial community structure and functional groups. Complex interactions among soil properties, enzyme activity, and microbial communities not only resist the impact of intercropping management on soil functions but are also conducive to improving biodiversity in the tropical plantation.

Author Contributions: Writing—original draft preparation, Y.Z. (Yiming Zhong); conceptualization, X.Q.; methodology, A.Z.; software, Y.Z. (Yiming Zhong) and A.Z.; validation, J.T. and H.Y.; formal analysis, X.J.; investigation, J.T and Y.Z. (Yiming Zhong); resources, X.Q.; data curation, A.Z.; writing—review and editing, X.Q. and A.Z.; visualization, J.W. and A.Z.; supervision, S.H. and Y.Z. (Ying Zong); project administration, X.Q.; funding acquisition, H.Y. and X.J. All authors have read and agreed to the published version of the manuscript.

Funding: This research was supported by the Hainan Natural Science Foundation, China (No. 2019RC323). National Tropical Plants Germplasm Resource Center.

Data Availability Statement: Not applicable.

Acknowledgments: We thank Lihua Li, Shuangyan Qi, Jinshuang Li, Shaoguan Zhao and Jiang Zhong for their contributions to the preliminary work.

Conflicts of Interest: The authors declare no conflict of interest.

Appendix A

Table A1. Soil dominant microbial composition under different planting patterns (phylum level).

	PM	I	AM
Bacteria			
Proteobacteria	19,202.33 ± 332.93 b	20,230.33 ± 1974.88 b	24,117.33 ± 866.37 a
Actinobacteriota	16,801.00 ± 1021.31 b	15,365.33 ± 712.89 b	20,129.67 ± 298.15 a
Acidobacteriota	10,725.33 ± 1697.53 a	14,804.67 ± 2381.47 a	13,546.33 ± 2433.41 a
Firmicutes	14,715.33 ± 3308.49 a	4676.67 ± 519.08 b	6590.33 ± 1202.16 b
Chloroflexi	6552.67 ± 992.70 ab	5415.00 ± 93.15 b	7745.33 ± 1054.78 a
Bacteroidota	3069.67 ± 894.62 a	3479.67 ± 1728.98 a	4053.67 ± 167.99 a
Myxococcota	2568.67 ± 160.48 a	2770.00 ± 427.36 a	2750.33 ± 315.78 a
Methylomirabilota	458.33 ± 176.98 c	2624.33 ± 531.17 a	1812.00 ± 317.02 b
Verrucomicrobiota	1187.67 ± 936.47 ab	2081.67 ± 328.36 a	706.33 ± 296.01 c
Gemmatimonadota	920.33 ± 126.75 a	1126.67 ± 29.30 a	1074.33 ± 122.40 a
Planctomycetota	915.67 ± 307.42 a	1308.33 ± 496.08 a	602.33 ± 101.11 a
Bdellovibrionota	811.00 ± 149.08 a	751.67 ± 255.58 a	1102.67 ± 31.13 a
unclassified Bacteria	420.00 ± 51.68 b	909.67 ± 280.23 a	994.33 ± 206.93 a
Fungi			
Ascomycota	107,293.00 ± 6816.14 a	85,814.67 ± 10,179.97 b	112,730.67 ± 5197.99 a
Basidiomycota	14,402.00 ± 2356.95 a	20,973.67 ± 18,341.30 a	9764.33 ± 5609.52 a
Unclassified Fungi	10,493.67 ± 3031.47 a	13,498.33 ± 5525.81 a	6377.67 ± 2092.01 a
Rozellomycota	4314.00 ± 1480.56 a	3867.67 ± 2074.86 a	2791.00 ± 2524.00 a

Note: Different letters indicate significant differences between treatments under the same soil microbes ($p < 0.05$).

Table A2. Relative abundance of bacterial functional groups based on intercropping under the FAPROTAX tool.

Bacteria	PM	I	AM
Aerobic_chemoheterotrophy	8183.83 ± 1148.68 a	5707.67 ± 1096.84 b	6216.83 ± 1358.33 b
Chemoheterotrophy	7349.50 ± 3765.68 a	5907.00 ± 1153.43 a	6550.17 ± 1440.54 a
Animal_parasites_or_symbionts	1519.00 ± 1102.83 a	913.33 ± 291.12 a	760.67 ± 135.74 a
Human_pathogens_all	1490.5 ± 1089.56 a	892.83 ± 301.78 a	721.50 ± 146.22 a
Nitrate_reduction	1326.00 ± 944.36 a	735.67 ± 270.20 a	677.83 ± 207.45 a
Nitrogen_fixation	590.17 ± 316.09 b	726.67 ± 91.09 ab	897.17 ± 201.86 a
Predatory_or_exoparasitic	477.67 ± 291.51 a	504.67 ± 170.19 a	500.17 ± 73.51 a
Human_pathogens_pneumonia	538.50 ± 276.01 a	538.33 ± 119.02 a	398.67 ± 162.01 a
Invertebrate_parasites	882.33 ± 887.74 a	292.83 ± 211.82 a	281.83 ± 157.54 a
Aromatic_compound_degradation	561.50 ± 166.59 b	401.33 ± 208.33 ab	227.17 ± 104.81 a
Chitinolysis	636.83 ± 657.41 a	289.83 ± 315.06 a	155.50 ± 148.64 a
Ureolysis	316.00 ± 204.44 a	337.00 ± 329.12 a	326.83 ± 137.34 a
Phototrophy	403.33 ± 428.91 a	277.17 ± 69.48 a	244.33 ± 51.35 a
Cellulolysis	127.83 ± 71.19 b	208.50 ± 147.37 ab	503.67 ± 469.15 a
Photoautotrophy	380.50 ± 428.29 a	233.17 ± 66.01 a	172.00 ± 85.81 a
Fermentation	192.83 ± 104.82 a	195.67 ± 165.37 a	302.50 ± 152.66 a
Others	2946.83 ± 1281.30 a	2191.50 ± 614.30 a	2000.17 ± 1076.32 a

Note: Different letters indicate significant differences between treatments under the same bacterial functional groups ($p < 0.05$).

Table A3. Relative abundance of fungi functional groups based on intercropping under the FUN-Guild tool.

Fungi	AM	I	PM
Wood_Saprotroph	11,270.17 ± 7010.83 a	4806.83 ± 1352.41 b	7946.67 ± 1797.94 ab
Soil_Saprotroph	858.17 ± 164.50 a	417.33 ± 245.91 a	894.17 ± 827.92 a
Plant_Pathogen	1706.00 ± 565.32 a	1601.50 ± 2539.11 a	2047.33 ± 1870.07 a
Plant_Pathogen_Wood_Saprotroph	1186.83 ± 891.84 a	333.83 ± 267.63 b	344.83 ± 608.14 b
Animal_Pathogen	453.83 ± 361.55 ab	689.67 ± 280.65 a	221.83 ± 242.93 b
Endophyte	16.17 ± 24.31 b	645.67 ± 699.13 a	433.17 ± 150.1 ab
Ectomycorrhizal	28.17 ± 20.95 a	696.00 ± 1028.55 a	128.17 ± 100.88 a
Fungal_Parasite	93.50 ± 54.34 a	256.17 ± 315.36 a	263.17 ± 198.15 a
Arbuscular_Mycorrhizal	8.33 ± 5.47 b	258.67 ± 239.75 a	115.33 ± 112.58 ab
Epiphyte	274.67 ± 423.17 a	0.33 ± 0.82 a	79.50 ± 193.27 a
Dung_Saprotroph	101.00 ± 47.92 a	27.00 ± 24.76 a	134.83 ± 142.33 a
Plant_Saprotroph	55.00 ± 27.40 a	27.00 ± 21.72 a	58.00 ± 58.36 a
Others	4.17 ± 1.72 a	21.67 ± 23.75 a	36.67 ± 63.59 a

Note: Different letters indicate significant differences between treatments under the same fungi functional groups ($p < 0.05$).

Table A4. Stepwise regression analysis model of soil dominant microorganisms and environmental factors.

Soil Microorganisms	Regression Model	R^2	F Value	p Value
Actinobacteriota	*Actinobacteriota* = 2831.888 + 11,364.551 × TP + 325.103 × TK	0.951	57.843	0.000
Ascomycota	*Ascomycota* = 10,079.723 + 5134.928 × SOC	0.734	19.297	0.003
Chloroflexi	*Chloroflexi* = −906.557 + 7427.543 × TN	0.643	12.632	0.009
Firmicutes	*Firmicutes* = −5205.866 + 388.736 × SAK	0.846	38.479	0.000
Methylomirabilota	*Methylomirabilota* = 5797.752 − 58.409 × EC	0.908	69.196	0.000
Proteobacteria	*Proteobacteria* = 5845.745 + 21.631 × URE	0.761	22.343	0.002
Unclassified Bacteria	*Unclassified Bacteria* = −8218.610 + 1235.298 × pH + 84.372 × SWC	0.913	31.443	0.001
Verrucomicrobiota	*Verrucomicrobiota* = 5945.663 − 258.262 × SOC	0.575	9.469	0.018

Table A5. Redundancy analysis of soil bacteria and fungi, soil environmental variables.

Name	Explains (%)	F	P
Environment-Bacteria			
pH	24.5	13.6	0.002
TP	48.6	6.6	0.004
POD	11.2	3.6	0.022
EC	6.4	2.7	0.108
UE	3.8	2.0	0.146
SOC	2.3	1.4	0.312
SAK	1.8	1.3	0.434
SOP	1.4	<0.1	1
Environment-Fungi			
SAN	23.5	2.2	0.144
TK	11.6	1.1	0.330
BD	28.0	3.8	0.102
SWC	8.2	1.2	0.354
PPO	18.5	5.5	0.064
UE	9.4	26.5	0.032
POD	0.3	0.7	0.554
pH	0.4	<0.1	1

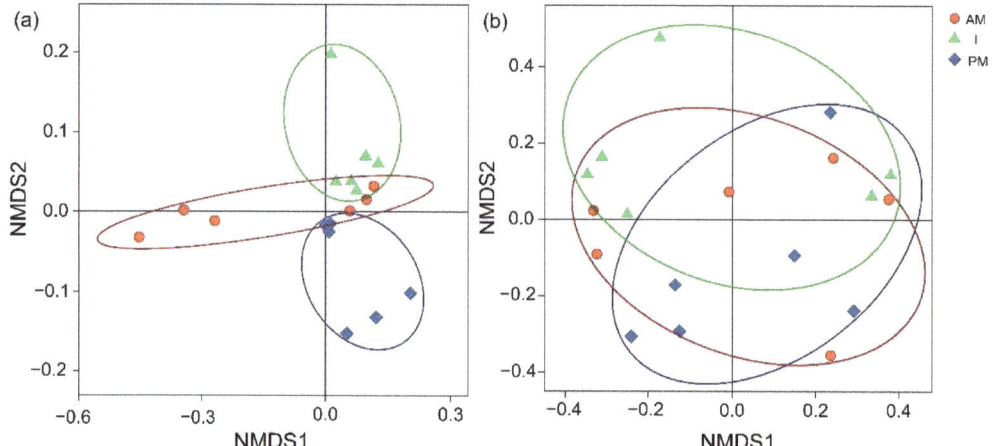

Figure A1. Effects of intercropping patterns on soil microbial (bacterial and fungal) beta diversity (NMDS) across the experimental period. AM represents areca nut monocropping; I represents areca nut intercropping with pandan; and PM represents pandan monocropping. ((**a**): $F = 1.9485$, $p = 0.09$; (**b**): $F = 1.969$, $p = 0.066$, calculated by PERMANOVA).

References

1. Duchene, O.; Vian, J.; Celette, F. Intercropping with legume for agroecological cropping systems: Complementarity and facilitation processes and the importance of soil microorganisms. A review. *Agric. Ecosyst. Environ.* **2017**, *240*, 148–161. [CrossRef]
2. Cuartero, J.; Pascual, J.A.; Vivo, J.; özbolat, O.; Sánchez-Navarro, V.; Egea-Cortines, M.; Zornoza, R.; Mena, M.M.; Garcia, E.; Ros, M. A first-year melon/cowpea intercropping system improves soil nutrients and changes the soil microbial community. *Agric. Ecosyst. Environ.* **2022**, *328*, 107856. [CrossRef]
3. Kang, Z.; Gong, M.; Li, Y.; Chen, W.; Yang, Y.; Qin, J.; Li, H. Low Cd-accumulating rice intercropping with *Sesbania cannabina* L. reduces grain Cd while promoting phytoremediation of Cd-contaminated soil. *Sci. Total Environ.* **2021**, *800*, 149600. [CrossRef] [PubMed]
4. Hong, Y.; Heerink, N.; Jin, S.; Berentsen, P.; Zhang, L.; van der Werf, W. Intercropping and agroforestry in China—Current state and trends. *Agric. Ecosyst. Environ.* **2017**, *244*, 52–61. [CrossRef]
5. Nyawade, S.O.; Karanja, N.N.; Gachene, C.K.K.; Gitari, H.I.; Schulte-Geldermann, E.; Parker, M.L. Short-term dynamics of soil organic matter fractions and microbial activity in smallholder potato-legume intercropping systems. *Appl. Soil Ecol.* **2019**, *142*, 123–135. [CrossRef]
6. Gu, C.; Bastiaans, L.; Anten, N.P.R.; Makowski, D.; van der Werf, W. Annual intercropping suppresses weeds: A meta-analysis. *Agric. Ecosyst. Environ.* **2021**, *322*, 107658. [CrossRef]
7. Wang, X.; Liu, J.; He, Z.; Xing, C.; Zhu, J.; Gu, X.; Lan, Y.; Wu, Z.; Liao, P.; Zhu, D. Forest gaps mediate the structure and function of the soil microbial community in a Castanopsis kawakamii forest. *Ecol. Indic.* **2021**, *122*, 107288. [CrossRef]
8. Bhat, R.; Sujatha, S.; Jose, C.T. Assessing soil fertility of a laterite soil in relation to yield of arecanut (*Areca catechu* L.) in humid tropics of India. *Geoderma* **2012**, *189-190*, 91–97. [CrossRef]
9. Li, J.; Liu, L.; Zhou, H.; Li, M. Improved Viability of Areca (*Areca catechu* L.) Seedlings under Drought Stress Using a Superabsorbent Polymer. *Hortscience* **2018**, *53*, 1872–1876. [CrossRef]
10. Santonja, M.; Rancon, A.; Fromin, N.; Baldy, V.; Hättenschwiler, S.; Fernandez, C.; Montès, N.; Mirleau, P. Plant litter diversity increases microbial abundance, fungal diversity, and carbon and nitrogen cycling in a Mediterranean shrubland. *Soil Biol. Biochem.* **2017**, *111*, 124–134. [CrossRef]
11. Ghasemzadeh, A.; Jaafar, H.Z. Profiling of phenolic compounds and their antioxidant and anticancer activities in pandan (*Pandanus amaryllifolius* Roxb.) extracts from different locations of Malaysia. *BMC Complement. Altern. Med.* **2013**, *13*, 341. [CrossRef] [PubMed]
12. Quyen, N.T.C.; Quyen, N.T.N.; Nhan, L.T.H.; Toan, T.Q. Antioxidant activity, total phenolics and flavonoids contents of *Pandanus amaryllifolius* (Roxb.). *IOP Conf. Ser. Mater. Sci. Eng.* **2020**, *991*, 12019. [CrossRef]
13. Routray, W.; Rayaguru, K. Chemical Constituents and Post-Harvest Prospects of *Pandanus amaryllifolius* Leaves: A Review. *Food Rev. Int.* **2010**, *26*, 230–245. [CrossRef]
14. Veres, Z.; Kotroczó, Z.; Fekete, I.; Tóth, J.A.; Lajtha, K.; Townsend, K.; Tóthmérész, B. Soil extracellular enzyme activities are sensitive indicators of detrital inputs and carbon availability. *Appl. Soil Ecol.* **2015**, *92*, 18–23. [CrossRef]

15. Gong, X.; Liu, C.; Li, J.; Luo, Y.; Yang, Q.; Zhang, W.; Yang, P.; Feng, B. Responses of rhizosphere soil properties, enzyme activities and microbial diversity to intercropping patterns on the Loess Plateau of China. *Soil Tillage Res.* **2019**, *195*, 104355. [CrossRef]
16. Cordero, I.; Snell, H.; Bardgett, R.D. High throughput method for measuring urease activity in soil. *Soil Biol. Biochem.* **2019**, *134*, 72–77. [CrossRef]
17. Lopes, É.M.G.; Reis, M.M.; Frazão, L.A.; Da Mata Terra, L.E.; Lopes, E.F.; Dos Santos, M.M.; Fernandes, L.A. Biochar increases enzyme activity and total microbial quality of soil grown with sugarcane. *Environ.Technol. Innov.* **2021**, *21*, 101270. [CrossRef]
18. Piotrowska-Długosz, A.; Długosz, J.; Frąc, M.; Gryta, A.; Breza-Boruta, B. Enzymatic activity and functional diversity of soil microorganisms along the soil profile—A matter of soil depth and soil-forming processes. *Geoderma* **2022**, *416*, 115779. [CrossRef]
19. Sinsabaugh, R.L. Phenol oxidase, peroxidase and organic matter dynamics of soil. *Soil Biol. Biochem.* **2010**, *42*, 391–404. [CrossRef]
20. Leptin, A.; Whitehead, D.; Anderson, C.R.; Cameron, K.C.; Lehto, N.J. Increased soil nitrogen supply enhances root-derived available soil carbon leading to reduced potential nitrification activity. *Appl. Soil Ecol.* **2021**, *159*, 103842. [CrossRef]
21. Vimal, S.R.; Singh, J.S.; Arora, N.K.; Singh, S. Soil-Plant-Microbe Interactions in Stressed Agriculture Management: A Review. *Pedosphere* **2017**, *27*, 177–192. [CrossRef]
22. Wang, Y.; Liu, L.; Yang, J.; Duan, Y.; Luo, Y.; Taherzadeh, M.J.; Li, Y.; Li, H.; Awasthi, M.K.; Zhao, Z. The diversity of microbial community and function varied in response to different agricultural residues composting. *Sci. Total Environ.* **2020**, *715*, 136983. [CrossRef] [PubMed]
23. Liu, Z.; Gu, H.; Yao, Q.; Jiao, F.; Liu, J.; Jin, J.; Liu, X.; Wang, G. Microbial communities in the diagnostic horizons of agricultural Isohumosols in northeast China reflect their soil classification. *Catena* **2022**, *216*, 106430. [CrossRef]
24. Zhou, X.; Yu, G.; Wu, F. Effects of intercropping cucumber with onion or garlic on soil enzyme activities, microbial communities and cucumber yield. *Eur. J. Soil Biol.* **2011**, *47*, 279–287. [CrossRef]
25. Jing, H.; Li, J.; Yan, B.; Wei, F.; Wang, G.; Liu, G. The effects of nitrogen addition on soil organic carbon decomposition and microbial C-degradation functional genes abundance in a Pinus tabulaeformis forest. *Forest Ecol. Manag.* **2021**, *489*, 119098. [CrossRef]
26. Huang, H.; Tian, D.; Zhou, L.; Su, H.; Ma, S.; Feng, Y.; Tang, Z.; Zhu, J.; Ji, C.; Fang, J. Effects of afforestation on soil microbial diversity and enzyme activity: A meta-analysis. *Geoderma* **2022**, *423*, 115961. [CrossRef]
27. Ren, J.; Liu, X.; Yang, W.; Yang, X.; Li, W.; Xia, Q.; Li, J.; Gao, Z.; Yang, Z. Rhizosphere soil properties, microbial community, and enzyme activities: Short-term responses to partial substitution of chemical fertilizer with organic manure. *J. Environ. Manag.* **2021**, *299*, 113650. [CrossRef]
28. Sekaran, U.; Loya, J.R.; Abagandura, G.O.; Subramanian, S.; Owens, V.; Kumar, S. Intercropping of kura clover (*Trifolium ambiguum* M. Bieb) with prairie cordgrass (*Spartina pectinata* link.) enhanced soil biochemical activities and microbial community structure. *Appl. Soil Ecol.* **2020**, *147*, 103427. [CrossRef]
29. Zhu, L.; He, J.; Tian, Y.; Li, X.; Li, Y.; Wang, F.; Qin, K.; Wang, J. Intercropping Wolfberry with Gramineae plants improves productivity and soil quality. *Sci. Hortic.* **2022**, *292*, 110632. [CrossRef]
30. Mathesius, U. Are legumes different? Origins and consequences of evolving nitrogen fixing symbioses. *J. Plant Physiol.* **2022**, *276*, 153765. [CrossRef]
31. Wang, L.; Deng, D.; Feng, Q.; Xu, Z.; Pan, H.; Li, H. Changes in litter input exert divergent effects on the soil microbial community and function in stands of different densities. *Sci. Total Environ.* **2022**, *845*, 157297. [CrossRef] [PubMed]
32. Fan, L.; Shao, G.; Pang, Y.; Dai, H.; Zhang, L.; Yan, P.; Zou, Z.; Zhang, Z.; Xu, J.; Zamanian, K.; et al. Enhanced soil quality after forest conversion to vegetable cropland and tea plantations has contrasting effects on soil microbial structure and functions. *Catena* **2022**, *211*, 106029. [CrossRef]
33. Gilmullina, A.; Rumpel, C.; Blagodatskaya, E.; Chabbi, A. Management of grasslands by mowing versus grazing – impacts on soil organic matter quality and microbial functioning. *Appl. Soil Ecol.* **2020**, *156*, 103701. [CrossRef]
34. van Agtmaal, M.; Straathof, A.L.; Termorshuizen, A.; Lievens, B.; Hoffland, E.; de Boer, W. Volatile-mediated suppression of plant pathogens is related to soil properties and microbial community composition. *Soil Biol. Biochem.* **2018**, *117*, 164–174. [CrossRef]
35. Ablimit, R.; Li, W.; Zhang, J.; Gao, H.; Zhao, Y.; Cheng, M.; Meng, X.; An, L.; Chen, Y. Altering microbial community for improving soil properties and agricultural sustainability during a 10-year maize-green manure intercropping in Northwest China. *J. Environ. Manag.* **2022**, *321*, 115859. [CrossRef]
36. Wu, J.; Zhang, Q.; Zhang, D.; Jia, W.; Chen, J.; Liu, G.; Cheng, X. The ratio of ligninase to cellulase increased with the reduction of plant detritus input in a coniferous forest in subtropical China. *Appl. Soil Ecol.* **2022**, *170*, 104269. [CrossRef]
37. Luo, G.; Sun, B.; Li, L.; Li, M.; Liu, M.; Zhu, Y.; Guo, S.; Ling, N.; Shen, Q. Understanding how long-term organic amendments increase soil phosphatase activities: Insight into phoD- and phoC-harboring functional microbial populations. *Soil Biol. Biochem.* **2019**, *139*, 107632. [CrossRef]
38. Zhou, Q.; Chen, J.; Xing, Y.; Xie, X.; Wang, L. Influence of intercropping Chinese milk vetch on the soil microbial community in rhizosphere of rape. *Plant Soil* **2019**, *440*, 85–96. [CrossRef]
39. Li, H.; Qiu, Y.; Yao, T.; Han, D.; Gao, Y.; Zhang, J.; Ma, Y.; Zhang, H.; Yang, X. Nutrients available in the soil regulate the changes of soil microbial community alongside degradation of alpine meadows in the northeast of the Qinghai-Tibet Plateau. *Sci. Total Environ.* **2021**, *792*, 148363. [CrossRef]
40. Wang, H.; Feng, D.; Zhang, A.; Zheng, C.; Li, K.; Ning, S.; Zhang, J.; Sun, C. Effects of saline water mulched drip irrigation on cotton yield and soil quality in the North China Plain. *Agric. Water Manag.* **2022**, *262*, 107405. [CrossRef]

41. Wang, Z.; Chen, L.; Liu, C.; Jin, Y.; Li, F.; Khan, S.; Liang, X. Reduced colloidal phosphorus loss potential and enhanced phosphorus availability by manure-derived biochar addition to paddy soils. *Geoderma* **2021**, *402*, 115348. [CrossRef]
42. Bray, R.H.; Kurtz, L.T. Determination of Total, Organic, and Available Forms of Phosphorus in Soils. *Soil Sci.* **1945**, *59*, 39–46. [CrossRef]
43. Chen, J.; Tao, W.; Wang, K.; Zheng, C.; Liu, W.; Li, X.; Ou, X.; Zhang, X. Highly efficient thermally activated delayed fluorescence emitters based on novel Indolo[2,3-b] acridine electron-donor. *Org. Electron.* **2018**, *57*, 327–334. [CrossRef]
44. Ren, C.; Zhao, F.; Kang, D.; Yang, G.; Han, X.; Tong, X.; Feng, Y.; Ren, G. Linkages of C:N:P stoichiometry and bacterial community in soil following afforestation of former farmland. *Forest Ecol. Manag.* **2016**, *376*, 59–66. [CrossRef]
45. Magoč, T.; Salzberg, S.L. FLASH: Fast length adjustment of short reads to improve genome assemblies. *Bioinformatics* **2011**, *27*, 2957–2963. [CrossRef] [PubMed]
46. Caporaso, J.G.; Kuczynski, J.; Stombaugh, J.; Bittinger, K.; Bushman, F.D.; Costello, E.K.; Fierer, N.; Peña, A.G.; Goodrich, J.K.; Gordon, J.I.; et al. QIIME allows analysis of high-throughput community sequencing data. *Nat. Methods* **2010**, *7*, 335–336. [CrossRef]
47. Schloss, P.D.; Westcott, S.L.; Ryabin, T.; Hall, J.R.; Hartmann, M.; Hollister, E.B.; Lesniewski, R.A.; Oakley, B.B.; Parks, D.H.; Robinson, C.J.; et al. Introducing mothur: Open-Source, Platform-Independent, Community-Supported Software for Describing and Comparing Microbial Communities. *Appl. Environ. Microb.* **2009**, *75*, 7537–7541. [CrossRef]
48. Yilmaz, P.; Parfrey, L.W.; Yarza, P.; Gerken, J.; Pruesse, E.; Quast, C.; Schweer, T.; Peplies, J.; Ludwig, W.; Glöckner, F.O. The SILVA and "All-species Living Tree Project (LTP)" taxonomic frameworks. *Nucleic Acids Res.* **2013**, *42*, D643–D648. [CrossRef]
49. Nilsson, R.H.; Larsson, K.; Taylor, A.F.S.; Bengtsson-Palme, J.; Jeppesen, T.S.; Schigel, D.; Kennedy, P.; Picard, K.; Glöckner, F.O.; Tedersoo, L.; et al. The UNITE database for molecular identification of fungi: Handling dark taxa and parallel taxonomic classifications. *Nucleic Acids Res.* **2018**, *47*, D259–D264. [CrossRef]
50. Louca, S.; Parfrey, L.W.; Doebeli, M. Decoupling function and taxonomy in the global ocean microbiome. *Science* **2016**, *353*, 1272–1277. [CrossRef]
51. Nguyen, N.H.; Song, Z.; Bates, S.T.; Branco, S.; Tedersoo, L.; Menke, J.; Schilling, J.S.; Kennedy, P.G. FUNGuild: An open annotation tool for parsing fungal community datasets by ecological guild. *Fungal Ecol.* **2016**, *20*, 241–248. [CrossRef]
52. Jing, Y.; Zhang, Y.; Han, I.; Wang, P.; Mei, Q.; Huang, Y. Effects of different straw biochars on soil organic carbon, nitrogen, available phosphorus, and enzyme activity in paddy soil. *Sci. Rep.* **2020**, *10*. [CrossRef] [PubMed]
53. Ma, Y.; Fu, S.; Zhang, X.; Zhao, K.; Chen, H.Y.H. Intercropping improves soil nutrient availability, soil enzyme activity and tea quantity and quality. *Appl. Soil Ecol.* **2017**, *119*, 171–178. [CrossRef]
54. Curtright, A.J.; Tiemann, L.K. Intercropping increases soil extracellular enzyme activity: A meta-analysis. *Agr. Ecosyst. Environ.* **2021**, *319*, 107489. [CrossRef]
55. Mouradi, M.; Farissi, M.; Makoudi, B.; Bouizgaren, A.; Ghoulam, C. Effect of faba bean (*Vicia faba* L.)–rhizobia symbiosis on barley's growth, phosphorus uptake and acid phosphatase activity in the intercropping system. *Ann. Agrar. Sci.* **2018**, *16*, 297–303. [CrossRef]
56. Chen, X.; Song, B.; Yao, Y.; Wu, H.; Hu, J.; Zhao, L. Aromatic plants play an important role in promoting soil biological activity related to nitrogen cycling in an orchard ecosystem. *Sci. Total Environ.* **2014**, *472*, 939–946. [CrossRef] [PubMed]
57. Wang, G.; Cao, F. Integrated evaluation of soil fertility in Ginkgo (*Ginkgo biloba* L.) agroforestry systems in Jiangsu, China. *Agroforest Syst.* **2011**, *83*, 89–100. [CrossRef]
58. Fan, L.; Tarin, M.W.K.; Zhang, Y.; Han, Y.; Rong, J.; Cai, X.; Chen, L.; Shi, C.; Zheng, Y. Patterns of soil microorganisms and enzymatic activities of various forest types in coastal sandy land. *Glob. Ecol. Conserv.* **2021**, *28*, e1625. [CrossRef]
59. Menezes, K.M.S.; Silva, D.K.A.; Gouveia, G.V.; Da Costa, M.M.; Queiroz, M.A.A.; Yano-Melo, A.M. Shading and intercropping with buffelgrass pasture affect soil biological properties in the Brazilian semi-arid region. *Catena* **2019**, *175*, 236–250. [CrossRef]
60. Li, N.; Gao, D.; Zhou, X.; Chen, S.; Li, C.; Wu, F. Intercropping with Potato-Onion Enhanced the Soil Microbial Diversity of Tomato. *Microorganisms* **2020**, *8*, 834. [CrossRef]
61. Li, S.; Wu, F. Diversity and Co-occurrence Patterns of Soil Bacterial and Fungal Communities in Seven Intercropping Systems. *Front. Microbiol.* **2018**, *9*, 1521. [CrossRef] [PubMed]
62. Jiang, Y.; Khan, M.U.; Lin, X.; Lin, Z.; Lin, S.; Lin, W. Evaluation of maize/peanut intercropping effects on microbial assembly, root exudates and peanut nitrogen uptake. *Plant Physiol. Bioch.* **2022**, *171*, 75–83. [CrossRef] [PubMed]
63. Wang, J.; Lu, X.; Zhang, J.; Wei, H.; Li, M.; Lan, N.; Luo, H. Intercropping perennial aquatic plants with rice improved paddy field soil microbial biomass, biomass carbon and biomass nitrogen to facilitate soil sustainability. *Soil Tillage Res.* **2021**, *208*, 104908. [CrossRef]
64. Singh, S.R.; Yadav, P.; Singh, D.; Shukla, S.K.; Tripathi, M.K.; Bahadur, L.; Mishra, A.; Kumar, S. Intercropping in Sugarcane Improves Functional Diversity, Soil Quality and Crop Productivity. *Sugar Tech* **2021**, 1–17. [CrossRef]
65. Cheng, H.; Yuan, M.; Tang, L.; Shen, Y.; Yu, Q.; Li, S. Integrated microbiology and metabolomics analysis reveal responses of soil microorganisms and metabolic functions to phosphorus fertilizer on semiarid farm. *Sci. Total Environ.* **2022**, *817*, 152878. [CrossRef] [PubMed]
66. Guo, F.; Wang, M.; Si, T.; Wang, Y.; Zhao, H.; Zhang, X.; Yu, X.; Wan, S.; Zou, X. Maize-peanut intercropping led to an optimization of soil from the perspective of soil microorganism. *Arch. Acker-Pflanzenbau Bodenkd.* **2021**, *67*, 1986–1999. [CrossRef]

67. Mouhamadou, B.; Puissant, J.; Personeni, E.; Desclos-Theveniau, M.; Kastl, E.M.; Schloter, M.; Zinger, L.; Roy, J.; Geremia, R.A.; Lavorel, S. Effects of two grass species on the composition of soil fungal communities. *Biol. Fert. Soils* **2013**, *49*, 1131–1139. [CrossRef]
68. Yang, J.; Duan, Y.; Liu, X.; Sun, M.; Wang, Y.; Liu, M.; Zhu, Z.; Shen, Z.; Gao, W.; Wang, B.; et al. Reduction of banana fusarium wilt associated with soil microbiome reconstruction through green manure intercropping. *Agric. Ecosyst. Environ.* **2022**, *337*, 108065. [CrossRef]
69. Yin, Y.; Yang, C.; Tang, J.; Gu, J.; Li, H.; Duan, M.; Wang, X.; Chen, R. Bamboo charcoal enhances cellulase and urease activities during chicken manure composting: Roles of the bacterial community and metabolic functions. *J. Environ. Sci.-China* **2021**, *108*, 84–95. [CrossRef]
70. Fan, F.; Yin, C.; Tang, Y.; Li, Z.; Song, A.; Wakelin, S.A.; Zou, J.; Liang, Y. Probing potential microbial coupling of carbon and nitrogen cycling during decomposition of maize residue by 13C-DNA-SIP. *Soil Biol. Biochem.* **2014**, *70*, 12–21. [CrossRef]
71. Alori, E.T.; Glick, B.R.; Babalola, O.O. Microbial Phosphorus Solubilization and Its Potential for Use in Sustainable Agriculture. *Front. Microbiol.* **2017**, *8*, 971. [CrossRef] [PubMed]
72. Bolo, P.; Kihara, J.; Mucheru-Muna, M.; Njeru, E.M.; Kinyua, M.; Sommer, R. Application of residue, inorganic fertilizer and lime affect phosphorus solubilizing microorganisms and microbial biomass under different tillage and cropping systems in a Ferralsol. *Geoderma* **2021**, *390*, 114962. [CrossRef]
73. Yang, Y.; Tong, Y.; Liang, L.; Li, H.; Han, W. Dynamics of soil bacteria and fungi communities of dry land for 8 years with soil conservation management. *J. Environ. Manag.* **2021**, *299*, 113544. [CrossRef] [PubMed]
74. Janvier, C.; Villeneuve, F.O.; Alabouvette, C.; Edel-Hermann, V.; Mateille, T.; Steinberg, C. Soil health through soil disease suppression: Which strategy from descriptors to indicators? *Soil Biol. Biochem.* **2007**, *39*, 1–23. [CrossRef]
75. Yu, L.; Luo, S.; Gou, Y.; Xu, X.; Wang, J. Structure of rhizospheric microbial community and N cycling functional gene shifts with reduced N input in sugarcane-soybean intercropping in South China. *Agric. Ecosyst. Environ.* **2021**, *314*, 107413. [CrossRef]
76. Zhang, X.; Gao, G.; Wu, Z.; Wen, X.; Zhong, H.; Zhong, Z.; Bian, F.; Gai, X. Agroforestry alters the rhizosphere soil bacterial and fungal communities of moso bamboo plantations in subtropical China. *Appl. Soil Ecol.* **2019**, *143*, 192–200. [CrossRef]
77. Wei, Z.; Liu, Y.; Feng, K.; Li, S.; Wang, S.; Jin, D.; Zhang, Y.; Chen, H.; Yin, H.; Xu, M.; et al. The divergence between fungal and bacterial communities in seasonal and spatial variations of wastewater treatment plants. *Sci. Total Environ.* **2018**, *628-629*, 969–978. [CrossRef]
78. Wang, Y.; Shi, X.; Huang, X.; Huang, C.; Wang, H.; Yin, H.; Shao, Y.; Li, P. Linking microbial community composition to farming pattern in selenium-enriched region: Potential role of microorganisms on Se geochemistry. *J. Environ. Sci.-China* **2022**, *112*, 269–279. [CrossRef]
79. Yu, Y.; Liu, L.; Wang, J.; Zhang, Y.; Xiao, C. Effects of warming on the bacterial community and its function in a temperate steppe. *Sci. Total Environ.* **2021**, *792*, 148409. [CrossRef]
80. Song, Z.; Kennedy, P.G.; Feng, J.L.; Schilling, J.S. Fungal endophytes as priority colonizers initiating wood decomposition. *Funct. Ecol.* **2016**, *31*, 407–418. [CrossRef]

Article

Effects of Intercropping between *Morus alba* and Nitrogen Fixing Species on Soil Microbial Community Structure and Diversity

Jiaying Liu [1,2], Yawei Wei [1,3], Haitao Du [1,3], Wenxu Zhu [1,3], Yongbin Zhou [2,4] and You Yin [1,3,*]

1. College of Forestry, Shenyang Agricultural University, Shenyang 110866, China
2. Institute of Modern Agricultural Research, Dalian University, Dalian 116622, China
3. Research Station of Liaohe-River Plain Forest Ecosystem, Chinese Forest Ecosystem Research Network (CFERN), Shenyang Agricultural University, Tieling 112000, China
4. Life Science and Technology College, Dalian University, Dalian 116622, China
* Correspondence: 1993500012@syau.edu.cn; Tel.: +86-15840436818

Citation: Liu, J.; Wei, Y.; Du, H.; Zhu, W.; Zhou, Y.; Yin, Y. Effects of Intercropping between *Morus alba* and Nitrogen Fixing Species on Soil Microbial Community Structure and Diversity. *Forests* 2022, 13, 1345. https://doi.org/10.3390/f13091345

Academic Editors: Chunjian Zhao, Zhi-Chao Xia, Chunying Li and Jingle Zhu

Received: 30 June 2022
Accepted: 22 August 2022
Published: 24 August 2022

Publisher's Note: MDPI stays neutral with regard to jurisdictional claims in published maps and institutional affiliations.

Copyright: © 2022 by the authors. Licensee MDPI, Basel, Switzerland. This article is an open access article distributed under the terms and conditions of the Creative Commons Attribution (CC BY) license (https://creativecommons.org/licenses/by/4.0/).

Abstract: The intercropping of nitrogen-fixing and non-nitrogen-fixing tree species changed the availability of soil nitrogen and soil microbial community structure and then affected the regulation process of soil carbon and nitrogen cycle by microorganisms in an artificial forest. However, there is no consensus on the effect of soil nitrogen on soil microorganisms. In this study, the intercropping of mulberry and twigs was completed through pot experiments. Total carbon, total nitrogen, and total phosphorus in the rhizosphere soil were determined, and the composition and structure of the soil microbial community were visualized by PCR amplification and 16S rRNA ITS sequencing. The analysis found that the intercropping of *Morus alba* L. and *Lespedeza bicolor* Turcz. had no significant effect on soil pH but significantly increased the contents of total carbon, total nitrogen, and total phosphorus in the soil. The effect on the alpha diversity of the bacterial community was not significant, but the effect on the evenness and diversity of the fungal community was significant ($p < 0.05$). It was also found that soil nutrients had no significant effect on bacterial community composition but had a significant effect on the diversity within the fungal community. This study added theoretical support for the effects of intercropping between non-nitrogen-fixing tree species and nitrogen-fixing tree species on soil nutrients and microbial community diversity.

Keywords: nitrogen-fixing plant; soil microorganism; nitrogen; *Lespedeza bicolor*; *Morus alba*

1. Introduction

Soil microorganisms are a kind of significant link between the aboveground vegetation community and the underground ecological process [1,2]. They are involved in soil organic matter decomposition, adjusting the forest ecosystem circulation process of material circulation, and energy flow in soil. It is the most momentous driver of ecosystem function and the "engine" of soil nutrient cycling [3,4]. Changes in the utilization pattern and utilization efficiency of soil organic carbon by microbial communities may affect forest ecosystems in organic matter decomposition, the emission of greenhouse gases (CO_2 and CH_4), and the carbon sequestration [5]. Soil phosphorus and nitrogen are mainly in the shape of an organic state, which cannot be directly absorbed and utilized by plants. The nutrients that must be transformed, absorbed, and "temporarily" preserved by soil microorganisms are the "effective reservoirs" for plants to absorb nutrients. Soil microbial community composition (bacteria and fungi) affects soil nutrient transformation and availability [6]. The intercropping between nitrogen-fixing plants and non-nitrogen-fixing plants changes the availability of soil nitrogen and microbial community structure and then affects the regulation process of soil carbon and nitrogen cycling by artificial forest microorganisms. [7].

However, there is still no consensus on the relationship between soil nitrogen and soil microorganisms. It has been found that increasing soil nitrogen content can increase microbial biomass [8]. Some studies have also found that an increase in soil nitrogen can sometimes harm soil microbial biomass, and sometimes this impact is negligible [9,10].

Nitrogen-fixing plants remarkably increased soil organic matter and nitrogen content [11,12]. The results of Wang et al. showed that the soil organic matter and nitrogen content in the surface layer of nitrogen fixation artificial forests were 40%–50% and 20%–50% higher than that of non-nitrogen fixation plantations, respectively [13]. The increase of soil nitrogen not only affects plant growth but also changes soil physicochemical properties and affects the structure and function of soil microorganisms, thereby affecting the activity of soil enzymes and their stoichiometric ratios. Wang et al. have shown that increases in soil nitrogen (such as nitrogen addition) can change the stoichiometric ratio of soil nutrients, such as carbon to nitrogen ratio (C/N), carbon to phosphorus ratio (C/P), and nitrogen to phosphorus ratio (N/P) [14]. Using five years of nitrogen addition experiments, Zeng et al. found that nitrogen application reduced the N requirement of soil microorganisms in *Phyllostachys pubescens* forest [15]. However, in a three-year nitrogen addition experiment in the mid-subtropical *Castanopsis carlesii* natural forest, it was found that nitrogen treatment could significantly accelerate the soil C cycle [16]. In addition to directly affecting the soil due to their nitrogen-fixing ability, nitrogen-fixing plants can also affect other plants through their interaction with the soil. Nitrogen-fixing plants can provide abundantly available nitrogen for other plants through root secretion or litter decomposition [17–19], which in turn promotes the growth of those [20], improves productivity [21], and promotes secondary growth of vegetation [22]. They have important ecological functions in the ecosystem. The roots of nitrogen-fixing plants can coexist with nitrogen-fixing bacteria, and the nitrogen-fixing effect of nitrogen-fixing bacteria can continuously increase the content and effectiveness of soil nitrogen, thereby improving plant productivity.

The intercropping compound system is a typical resource-efficient planting model, which can not only effectively utilize resources through the complementation of species vegetative niche and spatial niche but also increase species diversity and improve soil quality, thereby achieving high yield [23,24]. Intercropping can change soil physicochemical properties, further cause changes in soil microbial communities, and affect soil health and quality [25]. Legume crops have the function of biological nitrogen fixation, which can effectively supplement nitrogen in the soil, and their advantages in intercropping with non-legume crops are very prominent. Previous studies have shown that interbreeding with nitrogen-fixing plants has positive effects on increasing soil nitrogen content and nitrogen availability [26,27], thus increasing the organic carbon storage in ecosystems [28]. More importantly, the competition of intercropping species for soil nitrogen stimulates legume crops to rely more on symbiotic nitrogen fixation, broadening the nitrogen nutrient niche and further improving the nitrogen use efficiency of the intercropping system [29,30]. Soil nitrogen and atmospheric nitrogen fastened by legume crops are the main sources of nitrogen for low-input legume and non-legume intercropping. Legume crops show strong nitrogen dominance when participating in intercropping.

Lespedeza bicolor Turcz. belongs to the legume family and is an erect shrub of the genus Lespedeza, with good resistance and strong soil adaptability, and can grow on barren, newly cultivated land. It is a common species in the habitats of different desertification ecosystems and is suitable for constructing desertification control projects [31]. It is also a typical case of nitrogen fixation plants. *M. alba* L., which belongs to the Moraceae, is an excellent local plant in China. It has the characteristics of strong flexibility, wide geographical distribution, a high survival rate of afforestation, and a large crown, and can be used for ecological afforestation. At present, many scholars have found that *M. alba* and *L. bicolor* belong to the legume family and are erect shrubs of the genus Lespedeza that have the same strong resistance to various adverse site environments; that is, they have excellent salinity resistance, barren resistance, and drought resistance [32,33]. This lays the foundation for the possibility and extension of the intercropping of *M. alba* and *L. bicolor*. In this study, the native tree species *L. bicolor* in Zhangwu County area of Liaoning Province and "Shen Sang No.1" cultivated by the Forestry College of Shenyang Agricultural University were selected for intercropping experiments to investigate the effects of introducing nitrogen-fixing plants on soil nutrients and microbial community diversity in *M. alba* plantations and the effects of nitrogen-fixing plants on soil nutrients and microbial community diversity in *M. alba* plantation. The correlation between soil nutrients and microbial community diversity after the introduction of nitrogen-fixing plants was analyzed.

2. Materials and Methods

2.1. Tested Varieties and Planting Patterns

Planting experiments began in March 2021 in Zhangwu County, Fuxin City, Liaoning Province, China (122°29′52″ E, 42°21′24″ N) (Figure 1). It is located on the southern edge of the Horqin Sandy Land, which has a north temperate monsoon continental climate and is a typical sandy land in northwestern Liaoning. A total of three planting patterns were arranged in the pot experiment, namely *M. alba* pure planting (Ma), *L. bicolor* pure planting (Lb), and intercropping of *M. alba*-*L. bicolor* planting (MaLb). Pure forest samples were planted with four plants per pot, and MaLb was planted with two *M. alba* and two *L. bicolor* per pot. A total of four replicates were set up in the experiment, and each replicate was randomly placed.

Figure 1. The specific location of the pot experiment.

2.2. Gathered Soil Samples and Admeasurement of Physicochemical Properties

The soil around the roots of each potted plant was obtained using the root shaking method. Twelve groups of samples were put into sterile ziplock bags and numbered in August 2021. They were put in an ice box and returned to the lab for further manipulation. In the laboratory, plant residues, roots, stones, and other garbage were removed from the sample, and samples were sieved by a 100-mesh (the pore size of 0.015 mm) sieve. The sieved sample was divided into two parts. A part was air-dried and stored in a refrigerator at 4 °C for the determination of soil chemical properties (total soil carbon, total nitrogen, and total phosphorus). Another part of the samples was stored in centrifuge tubes according to the numbers and stored in a −80 °C refrigerator until they were used for the determination of soil microorganisms.

The soil pH value was extracted with CO_2-free water and determined by the potentiometric method (Mettler Toledo pH (FE20), and the water-soil ratio was 2.5:1. Soil total carbon and soil nitrogen contents were determined by the elemental analyzer (Elementar Vario EL III Germany). Total phosphorus in soil was determined by the molybdenum–antimony anti-colorimetric method.

2.3. DNA Extraction and Sequencing of Soil Microorganisms

The OMEGA Mag-bind Soil DNA Kit (Omega M5636-02) (Omega Bio-Tek, Norcross, GA, USA) was used to extract total DNA and based on the kit-specific extraction, procedures for each sample weighing 0.5 g sample. The quantity and quality of the extracted DNA were determined by a NanoDrop ND-1000 spectrophotometer (Thermo Fisher Scientific, Waltham, MA, USA). Primers 338F (5′-ACTCCTACGGGAGGCAGCA-3′) and 806R (5′-CGGACTACHVGGGTWTCTAAT-3′) amplified the V3-V4 region of bacterial 16S rRNA gene [34]; the fungal ITS region was amplified with primers ITS5 (5′-GGAAGTAAAAGTCGTAACAAGG-3′) and ITS2 (5′-GCTGCGTTCTTCATCGATGC-3′). 2 μL DNA template [35], 1 μL upstream and downstream primers (10 μmol/L), 5 μL buffer (×5), 5 μL Q5 high-fidelity buffer (×5), high-fidelity DNA polymerase (5 U/μL) 0.25 μL, 2 μL dNTP (2.5 mmol/L), 8.75 μL ultrapure water (ddH_2O) constitute the PCR reaction system. PCR amplification was first pre-denatured at 98 °C for 2 min, then repeated 25 times in a cycle of 98 °C for 15 s, 55 °C for 30 s, and 72 °C for 30 s, and finally extended at 72 °C for 5 min. The PCR-amplified products were checked by 2% agarose gel electrophoresis. Then the target fragments were recovered by gel cutting and then by AXYGEN's gel recovery kit with Axygen Axy Prep DNA Gel Extraction kit (AP-GX-500). The final concentrations obtained were 0.63–5.29 ng/μL for bacteria and 0.58–11 ng/μL for fungi. After the individual quantification step, amplicons were pooled in equal amounts, and pair-end 2 × 250 bp sequencing was performed using the Illumina NovaSeq platform with NovaSeq 6000 SP Reagent Kit (500 cycles) at Shanghai Personal Biotechnology Co., Ltd. (Shanghai, China). TruSeq Nano DNA LT Library Prep Kit (Illumina) was used in the sequencing library. In total, 1μL of the library was taken, and the Agilent High Sensitivity DNA Kit was used for 2100 quality inspections of the library on the Agilent Bioanalyzer machine. Qualified libraries were sequenced at 2 × 300 bp paired-end using the MiSeq Reagent Kit V3 (600 cycles) on the MiSeq machine.

2.4. Statistical Analysis

Raw sequence data were demultiplexed using the demux plugin, followed by primers cutting with the cutadapt plugin [36]. Sequences were then quality filtered, denoised, merged, and chimera removed using the DADA2 plugin [37]. Non-singleton amplicon sequence variants (ASVs) were aligned with mafft [38] and used to construct a phylogeny with fasttree2 [39].

Excel Office 2019 and IBM SPSS Statistics 26.0.0 (Chicago, USA) were used for data processing and statistical analysis. All the data in Table 1 were mean ± standard deviation of 4 replicates. The one-way ANOVA method (LSD) was used to compare the different soil physical and chemical properties and the significant differences of soil microbial communities. Differences in soil β-diversity were analyzed based on the Operational Taxonomic Units (OTU) table, and APE package in R (R v.3.4.4) (New Zealand). Common and unique OTUs of soil microbial communities in each sample were analyzed in R (R v.3.4.4). Venn diagrams were generated using the "Venn diagram" package [40]. Stacked histograms of species composition were the most commonly used means of characterizing the species composition of multiple samples. By statistical analysis of the feature table after removing Singleton, the visualization of the component distribution of each sample at different classification levels was realized, and the bar chart was drawn by QIIME2 (2019.4). Data were normalized during α-diversity analysis. The leveling rule adopts the qiime feature table refinement function, and the leveling depth is set to 95% of the minimum sample sequence size. Abundance is represented by the Chao1 index and observed_species index, while the Shannon and Simpson index represents diversity. The uniformity was characterized by the Pielou_e index, and the coverage was characterized by the Good_coverage index. Boxplots were produced using the ggplot2 package in R (R v.3.4.4) [41]. Non-metric multidimensional scaling analysis (NMDS) was done with R (R v.3.4.4)'s "vegan" package. It simplified the data structure by reducing the dimension of the sample distance matrix to trace the distribution characteristics of the sample under a specific distance scale. By rank ordering the sample distance, ordering the samples in the low-dimensional space was as close as possible to the similar distance relationship between each other (rather than the exact distance value). The smaller the stress value (Stress) of the NMDS results, the better. It is generally believed that when the value is less than 0.2, the NMDS analysis's results are more reliable [42]. Cluster analysis was performed to identify discontinuities in the data. Hierarchical clustering was often used in Beta diversity clustering analysis, which is in the form of a hierarchy tree according to the similarity between samples. Through the clustering tree branch, length measures the quality of the clustering effect. Using the "uclust" function of the R (R v.3.4.4)'s "stat" package, the Bray–Curtis distance matrix was clustered by the UPGMA algorithm (the clustering method was average) by default, and the "ggtree" package of R (R v.3.4.4) was used for visualization.

Table 1. Physical and chemical properties of soil under different planting methods.

Tree Species	pH Value	Total N/g kg^{-1}	Total C/g kg^{-1}	Total C/Total N	Total P/mg kg^{-1}
MaLb	7.94 ± 0.017 aA	0.53 ± 0.003 aA	7.60 ± 0.142 bB	14.43 ± 0.350 bA	4.93 ± 0.268 bB
Ma	7.88 ± 0.023 aA	0.46 ± 0.009 bB	4.38 ± 0.123 cC	9.80 ± 0.030 cB	2.23 ± 0.074 cC
Lb	7.89 ± 0.003 aA	0.54 ± 0.006 aA	8.30 ± 0.047 aA	15.38 ± 0.249 aA	7.12 ± 0.352 aA

Data was expressed as mean ± standard deviation (n = 4). Capital letters in the same row represent a significant difference ($p < 0.01$), and lower-case letters mean significant differences ($p < 0.05$). Ma: *Morus alba*; Lb: *Lespedeza bicolor*; MaLb: *Morus alba-Lespedeza bicolor*.

3. Results

3.1. Soil Physicochemical Properties of Different Planting Types

As can be seen from Table 1, there are significant differences in total carbon (total C), total nitrogen (total N), total phosphorus (total P), and total carbon/total nitrogen (total C/total N) of different tree species ($p < 0.01$). At the same time, the pH value differences between them were not significant ($p > 0.0$). The soil chemical indexes were the lowest in Ma and the highest in Lb, including total C, total N, total P, and total C/total N. Unlike soil chemical indicators, the pH of MaLb was higher than Lb (Table 1).

3.2. Soil Microbial Community Composition and Structural Diversity under Different Planting Methods

A total of 34,193 OTUs were detected in the presence of unique OTUs and shared OTUs in the three sample bacteria. OTUs of Ma, MaLb, and Lb were 13,515, 167,206, and 14,315, respectively. Among them, the number of OTUs in Ma, MaLb, and Lb was 2837, among which the unique OTUs in Ma, MaLb, and Lb respectively were 7941, 10,611, and 8115, (Figure 2a). 2035 OTUs were aggregated in the three sample fungi. Ma, MaLb, and Lb had 1411, 1227, and 1216 OTUs, respectively. Among them, 307 OTUs were shared by Ma, MaLb, and Lb, and only Ma, MaLb, and Lb shared 851, 642, and 630 OTUs, respectively (Figure 2b).

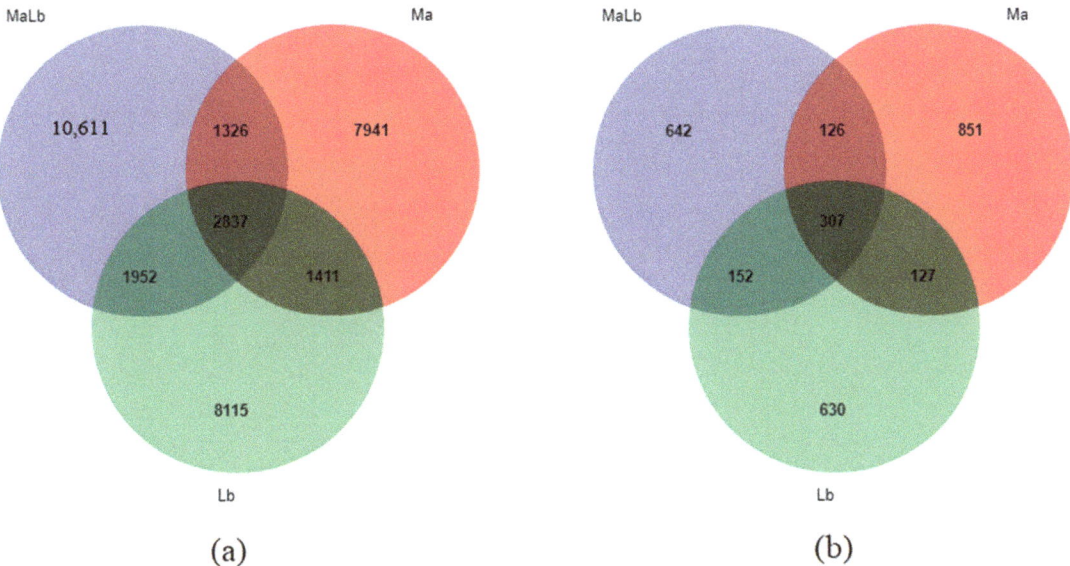

Figure 2. Venn diagram shows three different samples of unique and shared OTUs soil microorganisms. (**a**): Endemic and shared OTUs of soil bacteria from three different samples; (**b**): Endemic and shared OTUs of soil fungi from three different samples. Ma: *Morus alba*; Lb: *Lespedeza bicolor*; MaLb: *Morus alba-Lespedeza bicolor*.

The samples with different soil ratios were analyzed by the α–diversity index, and boxplots were drawn. The soil bacterial diversity indices, including the Chao 1 index (F = 0.751, p = 0.015), Pielou_e index (F = 1.802, p = 0.2), Goods_coverage (F = 0.849, p = 0.69), Shannon index (F = 0.677, p = 0.58), Simpson index (F = 0.174, p = 0.79), and Observed_species (F = 1.108, p = 0.37), showed no significant differences among Ma, MaLb, and Lb. MaLb had the highest Chao 1 index, Observed_species index, and Shannon index, which were 6942.935, 5682.775, and 11.205, respectively, followed by Lb, while Ma had the lowest. Ma had the highest abundance, diversity, and evenness (Figure 3a). MaLb had the lowest Goods_coverage index, Pielou_e index, and Simpson index, which were 0.949, 0.900, and 0.99841, followed by Lb, while Ma had the highest.

However, the results for fungi were different from those for bacteria. The soil bacterial diversity index, including Chao 1 index (F = 3.333, p = 0.018), Goods_coverage (F = 5.149, p = 0.038), Pielou_e index (F = 4.124, p = 0.024), Shannon index (F = 4.682, p = 0.024), Simpson index (F = 1.516, p = 0.087), and Observed_species (F = 4.237, p = 0.024), exhibited differences among Ma, MaLb as well as Lb. Ma had the highest Chao 1 index, Goods_coverage index, Observed_species index, Pielou_e index, Shannon index, and Simpson index, which were 556.214, 0.99947, 547.475, 0.715, 6.502, and 0.975, respectively. Lb had the lowest Chao 1 index (was 480.340) and Observed_species index (was 471.950). The Goods_coverage, Pielou_e index, Shannon index, and Simpson index of MaLb were the lowest, which were 0.99908, 0.624, 5.562, and 0.926, respectively (Figure 3b).

The relative abundance of microorganisms at the phylum level (others were shown) in the three soil samples was counted, as shown in Figure 4. At the bacterium phylum level, the top 10 relative abundances were Actinobacteria, Proteobacteria, Acidobacteria, Chloroflexi, Firmicutes, Gemmatimonadetes, Bacteroidetes, Rokubacteria, Nitrospirae, and Patescibacteria. Among them, Actinobacteria was the most abundant phylum in the three samples. Only Actinobacteria had the highest content in MaLb, which was 35.85%. Proteobacteria and Acidobacteria were Lb > Ma > MaLb (Figure 4a). At the level of fungal phylum, the top 10 contents were Ascomycota, Mortierellomycota, Basidiomycota,

Zoopagomycota, Basidiobolomycota, Blastocladiomycota, Glomeromycota, Olpidiomycota, Chytridiomycota, and Mucoromycota. Only Ascomycota, Mortierellomycota, Basidiomycota, and Zoopagomycota had a relative content of more than 1%. Ascomycota was the dominant phyla in the three soil samples, and Lb (82.35%) > MaLb (78.33%) > Ma (74.67%). Mortierellomycota had the highest content in MaLb at 10.48%, while Basidiomycota had it in Ma and Zoopagomycota in Lb (Figure 4b).

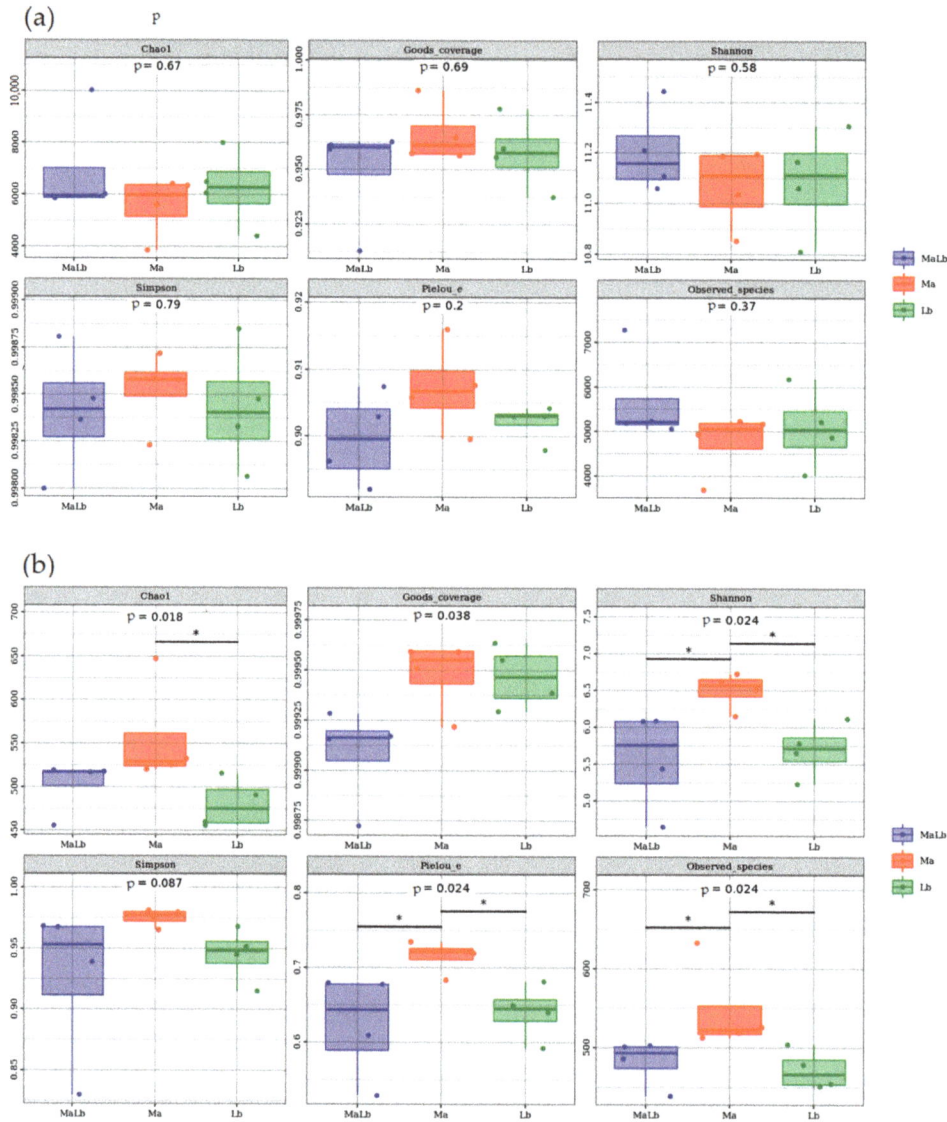

Figure 3. Soil microbial alpha diversity index in Ma, MaLb, and Lb. (**a**): Analysis of α diversity of soil bacterial community; (**b**): Analysis of α diversity of soil fungal community. Ma: *Morus alba*; Lb: *Lespedeza bicolor*; MaLb: *Morus alba-Lespedeza bicolor*. * indicates significant difference at 0.05 level.

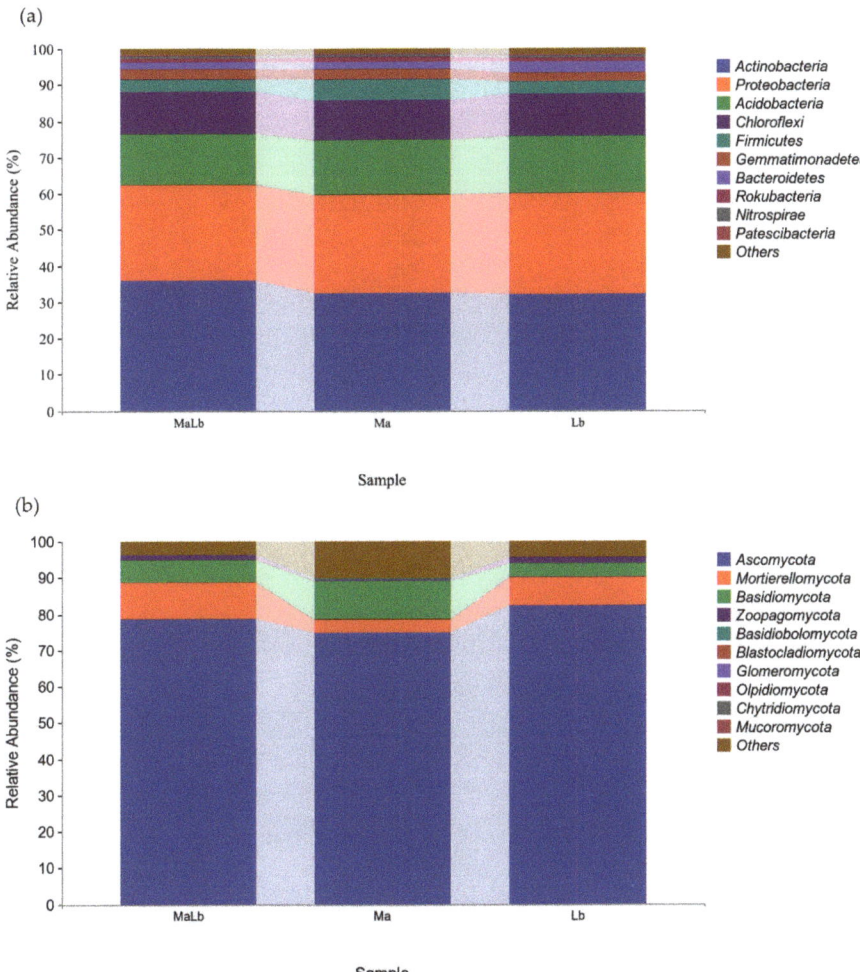

Figure 4. The relative content of species composition at the soil microbial phylum-level in different planting types (others were shown). (**a**): Relative content of bacterial phylum-level species composition; (**b**): Relative content of fungal phylum-level species composition. Ma: *Morus alba*; Lb: *Lespedeza bicolor*; MaLb: *Morus alba-Lespedeza bicolor*.

Hierarchical clustering analysis at the genus level of soil microbial community showed that bacterial and fungal communities exhibit the same regularity. Whether at the bacterial genus level or fungi, Ma clustered into one class, Ma and MaLb into the class. This indicated that the similarity between Ma and MaLb is high at the genus level, and the similarity to Ma was low. The inlay on the right showed that the genus-level abundance of the microbial community was not the same, and although the species composition of each treatment was similar, the abundance difference was obvious (Figure 5).

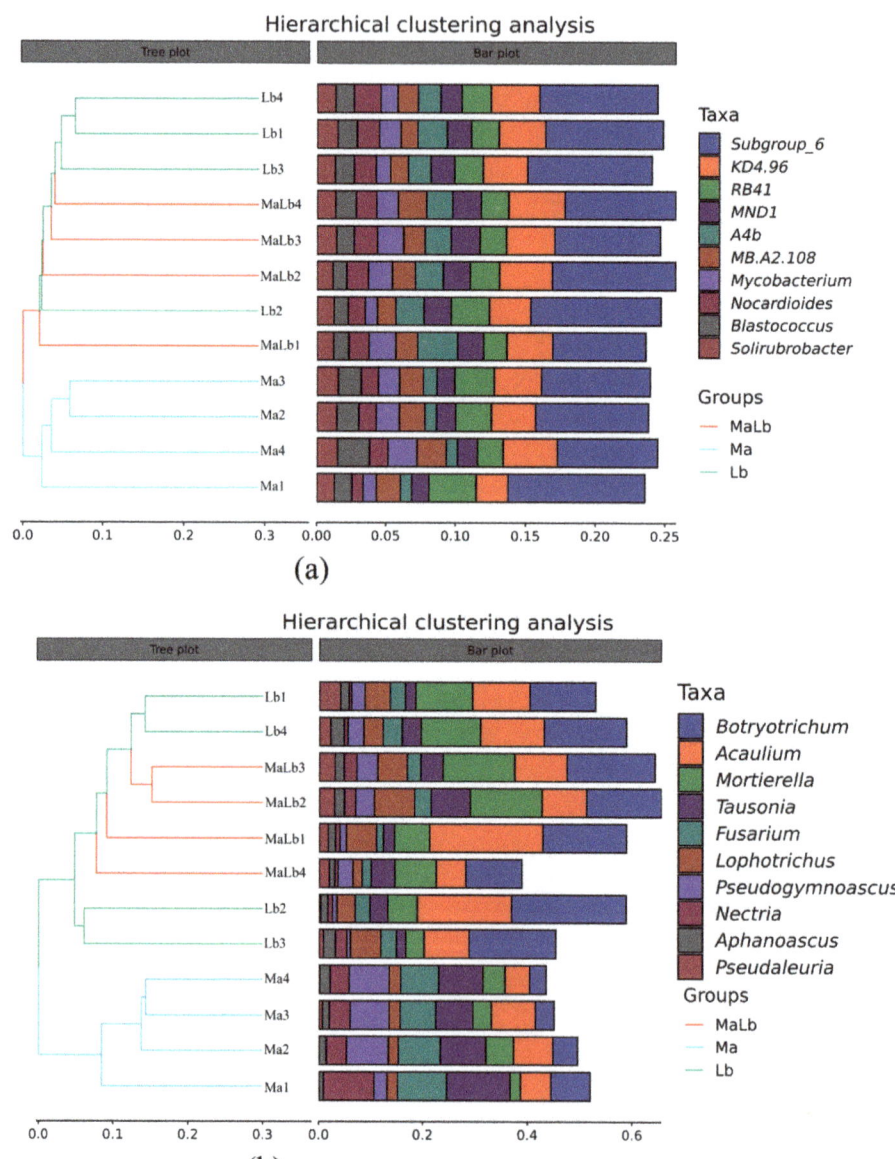

Figure 5. Hierarchical clustering analysis of soil microbial composition of different planting types at the genus level. (**a**): Hierarchical cluster analysis at genus level of bacteria; (**b**): Hierarchical cluster analysis at genus level of fungi. Ma: *Morus alba*; Lb: *Lespedeza bicolor*; MaLb: *Morus alba-Lespedeza bicolor*.

The relative abundances of microorganisms at the genus level (others were not shown) in the three soil samples were calculated, as shown in Figure 6. At the bacterial genus level, the relative contents of *Subgroup_6*, *KD4-96*, *RB41*, *MND1*, *A4b*, *MB-A2-108*, *Mycobacterium*, *Nocardioides*, *Blastococcus*, and *Solirubrobacter* were in the top 10, but their relative contents did not exceed 10%. *Subgroup_6* was the genus with the highest content in the three soil samples, with Lb of 8.79%, Ma of 8.22%, and MaLb of 7.75%, respectively. The *KD4-96* content of MaLb was higher than that of Ma and Lb, which was 3.65%, and the content of *RB41* was the highest in Ma (Figure 6a). At the fungal genus level, the top 10 relative contents are *Botryotrichum, Acaulium, Mortierella, Tausonia, Fusarium, Lophotrichus, Pseudogymnoascus, Nectria, Aphanoascus, Pseudaleuria*. The genus with the highest Ma content was *Tausonia*, accounting for 9.08%, while *Botryotrichum* had the highest Lb and MaLb content, accounting for 16.68% and 14.44%, respectively (Figure 6b).

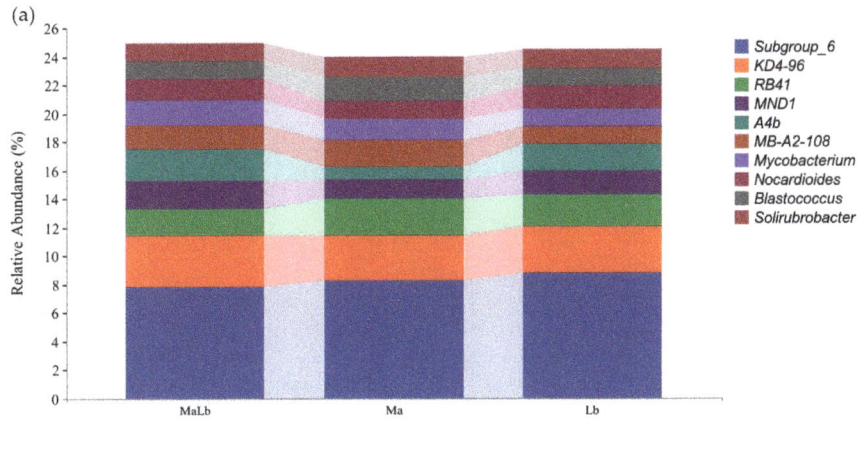

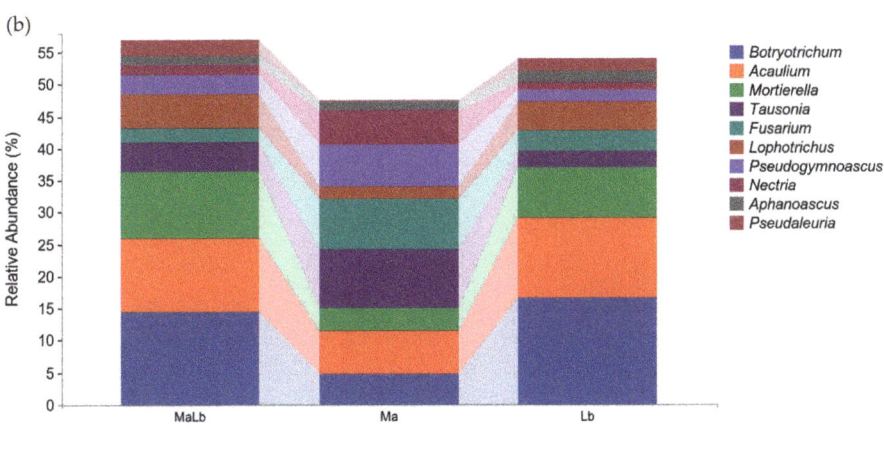

Figure 6. The relative content of species composition at the soil microbial genus level in different planting types (others were not shown). (**a**): Relative content of bacterial genus-level species composition; (**b**): Relative content of fungal genus-level species composition. Ma: *Morus alba*; Lb: *Lespedeza bicolor*; MaLb: *Morus alba-Lespedeza bicolor*.

As can be seen from the NMDS analysis chart, based on the Bray-Curtis algorithm, the population structures of three different soil microorganisms were significantly different. The degree of similarity in sample microbial population composition was indicated by the distance between samples in the figure. For soil bacterial communities, Ma was distributed in the first four quadrants, Lb was distributed in the second quadrant, and MaLb was distributed in the third quadrant (Figure 7a). Soil fungal communities differed from bacteria in that both MaLb and Lb were located in the second and third quadrants. Both Lb and MaLb were distributed on the negative semi-axis of the NMDS1 axis, while Ma was distributed on the positive NMDS1 axis. It could be seen that Lb and MaLb had great similarities in soil microbial structure and were quite different from Ma (Figure 7b).

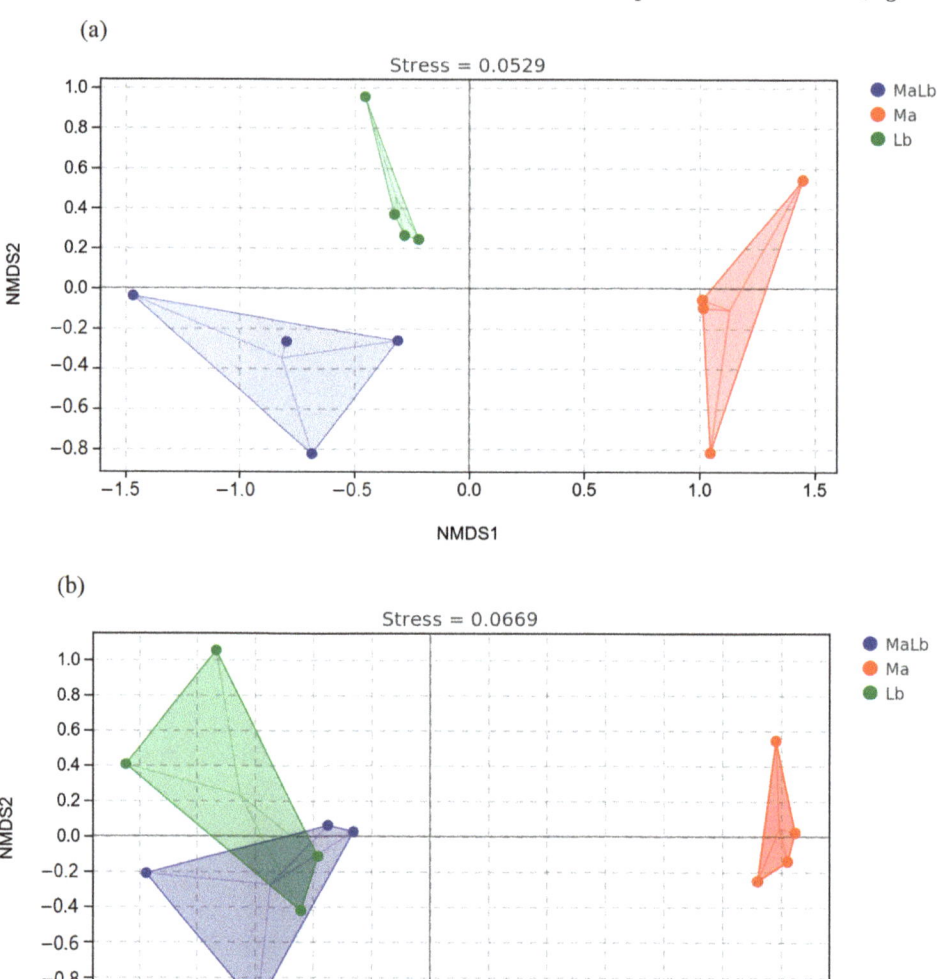

Figure 7. Nonmetric multidimensional scale analysis (NMDS) of microbial community structure in different soil samples. (**a**): soil bacterial communities; (**b**): soil fungal communities. Ma: *Morus alba*; Lb: *Lespedeza bicolor*; MaLb: *Morus alba-Lespedeza bicolor*.

3.3. Different Types of Planting Soil Physicochemical Properties and Microbial Community Diversity

As shown in Table 2, bacterial community alpha diversity had no significant relationship with the soil's physical and chemical properties. For the fungal community, soil pH was inversely related to the Goods_coverage index ($p < 0.01$), Pielou_e index, and Shannon index ($p < 0.05$). The soil total N, total C, and total C/total N were significantly negatively correlated with Chao 1 index, Observed_species index, Pielou_e index, and Shannon index ($p < 0.05$). Soil total P index was negatively correlated with Chao 1 index and Observed_species index ($p < 0.05$) (Table 3).

Table 2. The relationship between soil physicochemical properties and α-diversity of the bacterial community in different planting types.

	Chao1	Goods_Coverage	Observed_Species	Pielou_e	Shannon	Simpson
pH value	0.281	−0.305	0.308	−0.390	0.133	−0.200
Total N	0.253	−0.270	0.258	−0.494	0.146	−0.282
Total C	0.334	−0.350	0.326	−0.515	0.197	−0.190
Total C/total N	0.376	−0.392	0.365	−0.531	0.234	−0.158
Total P	0.132	−0.143	0.111	−0.351	0.006	−0.189

Table 3. The relationship between soil physicochemical properties and α-diversity of the fungal community in different planting types.

	Chao1	Goods_Coverage	Observed_Species	Pielou_e	Shannon	Simpson
pH value	−0.230	−0.760 **	−0.350	−0.614 *	−0.595 *	−0.444
Total N	−0.630 *	−0.253	−0.675 *	−0.616 *	−0.649 *	−0.426
Total C	−0.647 *	−0.286	−0.697 *	−0.656 *	−0.688 *	−0.430
Total C/total N	−0.638 *	−0.307	−0.691 *	−0.669 *	−0.698 *	−0.436
Total P	−0.613 *	−0.046	−0.624 *	−0.487	−0.528	−0.281

** indicates a significant difference at the 0.01 level. * indicates significant difference at 0.05 level.

At the phylum level, the bacterial phyla and soil physicochemical properties with the content of bacterial phyla in the top 10 and the fungal phylum content of >1% were compared for significance analysis. Firmicutes and Basidiomycota were significantly negatively correlated with soil total N, total C, total C/total N, and total P ($p < 0.01$). Nitrospirae was a notable negative correlation between total N ($p < 0.01$), total C, total C/total N, and total P ($p < 0.05$). Bacteroidetes was significantly positively correlated with soil total P ($p < 0.05$). Ascomycota was significantly positively correlated with total N, total C, total C/total N, and total P ($p < 0.05$). Zoopagomycota was significantly positively correlated with total C and total C/total N ($p < 0.05$). Other phyla had no correlation with the soil's biological and biological properties. (Table 4).

Table 4. Correlation between soil physicochemical properties and bacterial phyla levels in different planting types.

	pH Value	Total N	Total C	Total C/Total N	Total P
Actinobacteria	0.246	0.245	0.147	0.133	−0.025
Proteobacteria	−0.013	−0.078	0.057	0.075	0.172
Acidobacteria	−0.234	−0.128	−0.064	−0.064	0.116
Chloroflexi	0.093	0.575	0.493	0.491	0.452
Firmicutes	−0.525	−0.852 **	−0.921 **	−0.935 **	−0.855 **
Gemmatimonadetes	0.138	−0.452	−0.339	−0.313	−0.313
Bacteroidetes	−0.023	0.403	0.520	0.534	0.663 *
Rokubacteria	−0.450	−0.454	−0.521	−0.548	−0.367
Nitrospirae	−0.185	−0.788 **	−0.700 *	−0.658 *	−0.625 *
Patescibacteria	−0.288	0.195	0.202	0.202	0.273
Ascomycota	0.085	0.652 *	0.604 *	0.588 *	0.613 *
Mortierellomycota	0.272	0.633 *	0.586 *	0.565	0.451
Basidiomycota	−0.284	−0.848 **	−0.853 **	−0.853 **	−0.810 **
Zoopagomycota	0.364	0.499	0.602 *	0.634 *	0.572

The bacterial phylum level listed the top 10 relative content, and the fungal phylum level listed the relative content >1%. ** indicates a significant difference at the 0.01 level. * indicates significant difference at 0.05 level.

4. Discussion

Many early studies have shown that intercropping of nitrogen-fixing plants and other species can improve productivity [43,44]. Studies also found that intercropping with nitrogen-fixing plants can increase soil nitrogen content [45,46], thereby enhancing the accumulation of organic carbon and nitrogen in the ecosystem [28]. As a nitrogen fertilizer crop, the nitrogen content in Lb was significantly higher than in Ma. However, there were no significant changes between it and MaLb, which might be due to the apparent effect of *L. bicolor* in the intercropping of *M. alba* and *L. bicolor*. This study found that nitrogen-fixing and *M. alba* intercropping could significantly increase the soil total N, P, and K of *M. alba*, but had no significant impact on the soil pH. This indicated that nitrogen-fixing plants could not only enhance soil nutrients in pure *M. alba* forests but also have the potential to increase their soil carbon interception potential, similar to previous studies that concluded that those could significantly reduce soil carbon loss and increase soil carbon sequestration [47,48]. Nitrogen-fixing plants and *M. alba* were mixed to improve soil nitrogen availability and nutrients. The increase in available nitrogen content in soil was an important factor in significantly improving the net productivity of plants [49]. These results suggest that nitrogen-fixing plants enhance ground plant productivity by increasing the availability of soil nutrients, especially nitrogen, and provide essential basal metabolites for more microbial growth. In turn, the soil microbial biomass was increased, and to a certain extent, it was possible to increase the source of soil organic carbon.

Several recent studies have shown that soil nutrient quality (e.g., total soil nitrogen, total carbon, and total carbon/total nitrogen) is also a major factor driving changes in soil microbial communities [50,51], which is similar to the results of the present study. The introduction of nitrogen-fixing plants increases soil nitrogen content and reduces soil carbon-nitrogen ratio, which inhibits the growth of soil fungal communities to a certain extent [52]. The present study found that intercropping of *M. alba* and *L. bicolor* significantly increased soil nutrient content (total N, total C, total P). These changes significantly increased the biomass of soil bacterial communities but significantly decreased the proportion of fungal communities. This study was also supported by many studies that suggest that bacterial communities tend to be dominant in fertile soils [53,54].

Soil has a strong metabolic capacity due to the presence of soil microorganisms [55]. Soil microbes are the link between soil and plants, which drives plant growth [56,57]. Changes in pH have the greatest impact on soil bacteria [58,59], and higher pH increases the activity of soil microorganisms such as Nitrospira [60]. In this study, the pH value of MaLb was the highest. Still, there was no significant difference among the three treatments, so pH value was not the dominant factor affecting the soil microbial community in this experiment. After intercropping with *M. alba*, nitrogen fixed by rhizobia was available to mulberries, and this level of nitrogen addition increased the complexity of rhizosphere bacteria [61]. In this experiment, the top three bacteria in soil content of three treatments were Actinobacteria, Proteobacteria, and Acidobacteria. The abundance of actinomycetes is associated with fast-acting nutrients [62]. There are many reasons for the high abundance of actinomycetes in MaLb. One is because of nitrogen, and the other is because of the high content of organic matter, which can also be found from higher C/N [63]. In harsh environments such as salinity, Proteobacteria dominate the soil and have a certain resistance to such extreme environments [64]. In this experiment, the content of Proteobacteria showed Lb > Ma > MaLb, so it was speculated that the resistance of *M. alba* and *L. bicolor* to the harsh environment was higher, and the resistance was still retained after intercropping.

In addition, Fungi play an important role in soil, some of which promote crop growth and development, and some cause crop diseases [65]. Most soil fungi belong to the Ascomycetes or Basidiomycetes. In this study, Ascomycetes were significantly positively correlated with soil nutrients ($p < 0.05$), while Basidiomycetes were negatively correlated with those ($p < 0.01$). After the intercropping of *L. bicolor* and *M. alba*, the abundance of the bacterial community decreased. Basidiomycetes are abundant decomposing fungi in soil, and their abundance increases the decomposition of organic matter in the soil faster efficiently. Ascomycete fungi can be parasites, symbiotic, saprophytic, or facultative saprophytic, while Basidiomycete fungi are mostly saprophytic. Since plants have different life forms, their properties vary at the phylum level. The proportion of unclassified plants in the rhizosphere flora was low, indicating that more research is needed into the rhizosphere flora plants of *M. alba* and *L. bicolor*.

Nitrogen is an important nutrient in the soil and often acts as a soil factor limiting plant growth. Recent studies have shown that nitrogen-fixing plants are capable of increasing soil nitrogen content, increasing soil organic matter, and also enhancing soil [66–70]. Up to now, it has not been possible to establish a uniform conclusion on the mechanisms by which soil nitrogen content and its availability affect soil microbial biomass. Some studies have found that the increase of soil nitrogen content is beneficial to the reproduction and growth of soil microorganisms [71,72]. However, some studies have found that increasing soil nitrogen content sometimes inhibits their reproduction and growth, and sometimes this effect is minimal [73,74]. This also explained to a certain extent why the bacterial community structure and composition of different treatments in this study were not significantly different. In contrast, the fungal community structure and composition were significantly different. The past state of understanding the key abiotic and biotic factors that underlie the structural variations in soil microbial communities exists very uncertain [75,76].

5. Conclusions

In this study, the nutrient content, microbial community, and structural composition of the rhizosphere soil of different samples were measured in the pot experiments of *M. alba-L. bicolor* intercropping and pure breeding of *M. alba* and *L. bicolor*, respectively. We found that in terms of soil nutrients, the intercropping of *M. alba* and nitrogen-fixing tree species *L. bicolor* had no significant effect on soil pH but significantly increased the content of total C, total N, and total P in the soil and improved soil nutrients. Microbial community composition and structure were visualized by PCR amplification and 16S rRNA and ITS sequencing of corresponding primers for soil microbes in different samples. It was found that the intercropping of *M. alba* and *L. bicolor* had no significant effect on the alpha diversity of the bacterial community but did have a significant effect on the evenness and diversity of the fungal community ($p < 0.05$). The intercropping of *M. alba* and nitrogen-fixing species increased the relative abundance of Actinobacteria in the rhizosphere soil. Mortierellomycota was more abundant in MaLb than in the other two. According to NMDS analysis, the similarity between MaLb microbial community and Lb was higher than that of Ma. The study also found that soil nutrients had no significant effect on bacterial community composition ($p > 0.05$) but did have a significant effect on fungal community richness, diversity, and uniformity ($p < 0.05$). This study enriched our understanding of the effects of the introduction of nitrogen-fixing tree species on soil nutrients and microbial community diversity in *M. alba* plantations through the intercropping of mulberry and nitrogen-fixing tree species—*L. bicolor*. Add a theoretical basis for our understanding of the impact of soil nitrogen content on soil nutrients and microorganisms in the future.

Author Contributions: Conceptualization, Y.Y.; methodology, Y.W.; software, H.D.; validation, H.D., W.Z., and Y.Y.; formal analysis, J.L.; investigation, J.L., Y.W., W.Z., and Y.Z.; resources, H.D.; data curation, J.L.; writing—original draft preparation, J.L.; writing—review and editing, Y.Y.; visualization, Y.W.; supervision, J.L.; project administration, Y.Y.; funding acquisition, Y.Y. All authors have read and agreed to the published version of the manuscript.

Funding: This research was funded by Liaoning Province Scientific Research Funding Project, grant numbers LSNZD202002 and LSNQN202012, and the Science and Technology Program of Liaoning Province, grant number 2020020287-JH1/103-05-02.

Institutional Review Board Statement: Not applicable.

Informed Consent Statement: Not applicable.

Data Availability Statement: The data presented in this study are available on request from the corresponding author. The data are not publicly available due to the policy of the institute.

Conflicts of Interest: The authors declare no conflict of interest.

References

1. Hossain, M.; Okubo, A.; Sugiyama, S. Effects of grassland species on decomposition of litter and soil microbial communities. *Ecol. Res.* **2009**, *25*, 255–261. [CrossRef]
2. You, Y.M.; Wang, J.; Huang, X.M.; Tang, Z.X.; Liu, S.M.; Osbert, J.S. Relating microbial community structure to functioning in forest soil organic carbon transformation and turnover. *Ecol. Evol.* **2014**, *4*, 633–647. [CrossRef] [PubMed]
3. Xu, Z.W.; Yu, G.R.; Zhang, X.Y.; Ge, J.P.; He, N.P.; Wang, Q.F.; Wang, D. The variations in soil microbial communities, enzyme activities and their relationships with soil organic matter decomposition along the northern slope of changbai mountain. *Appl. Soil Ecol.* **2015**, *86*, 19–29. [CrossRef]
4. Maillard, F.; Leduc, V.; Bach, C.; Reichard, A.; Buée, M. Soil microbial functions are affected by organic matter removal in temperate deciduous forest. *Soil Biol. Biochem.* **2019**, *133*, 28–36. [CrossRef]
5. Heijden, M.G.A.V.D.; Bardgett, R.D.; Straalen, N.M.V. The unseen majority: Soil microbes as drivers of plant diversity and productivity in terrestrial ecosystems. *Ecol. Lett.* **2008**, *11*, 296–310. [CrossRef]
6. Urbanová, M.; Šnajdr, J.; Baldrian, P. Composition of fungal and bacterial communities in forest litter and soil is largely determined by dominant trees. *Soil Biol. Biochem.* **2015**, *84*, 53–64. [CrossRef]
7. Doran, R. Soil microbiology and biochemistry. *J. Range Manag.* **2014**, *51*, 254. [CrossRef]
8. Gutknecht, J.; Henry, H.; Balser, T.C. Inter-annual variation in soil extra-cellular enzyme activity in response to simulated global change and fire disturbance. *Pedobiologia* **2010**, *53*, 283–293. [CrossRef]
9. Wallenstein, M.D.; Mcnulty, S.; Fernandez, I.J.; Boggs, J.; Schlesinger, W.H. Nitrogen fertilization decreases forest soil fungal and bacterial biomass in three long-term experiments. *For. Ecol. Manag.* **2006**, *222*, 459–468. [CrossRef]
10. Treseder, K.K. Nitrogen additions and microbial biomass: A meta-analysis of ecosystem studies. *Ecol. Lett.* **2008**, *11*, 1111–1120. [CrossRef]
11. Hoogmoed, M.; Cunningham, S.C.; Baker, P.; Beringer, J.; Cavagnaro, T.R. N-fixing trees in restoration plantings: Effects on nitrogen supply and soil microbial communities. *Soil Biol. Biochem.* **2014**, *77*, 203–212. [CrossRef]
12. Curtis, P.S. A meta-analysis of leaf gas exchange and nitrogen in trees grown under elevated carbon dioxide. *Plant Cell Environ.* **2010**, *19*, 127–137. [CrossRef]
13. Wang, F.M.; Li, Z.A.; Xia, H.P.; Zou, B.; Li, N.Y.; Liu, J.; Zhu, W.X. Effects of nitrogen-fixing and non-nitrogen-fixing tree species on soil properties and nitrogen transformation during forest restoration in southern China. *J. Soil Sci. Plant Nutr.* **2010**, *56*, 297–306. [CrossRef]
14. Le, J.J.; Su, Y.; Luo, Y.; Geng, F.Z.; Liu, X.J. Effects of enclosure on leaves of four plants and soil stoichiometry in an alpine grassland of Tianshan mountains. *Acta Ecol. Sin.* **2020**, *40*, 1621–1628.
15. Zeng, Q.X.; Zhang, Q.F.; Lin, K.M.; Zhou, J.C.; Yuan, X.C.; Mei, K.C.; Wu, Y.; Cui, J.Y.; Xu, J.G.; Chen, Y.M. Enzyme stoichiometry evidence revealed that fiveyears nitrogen addition exacerbated the carbon and phosphorus limitation of soil microorganismsin a Phyllostachys pubescens forest. *Chin. J. Appl. Ecol.* **2021**, *32*, 521–528.
16. Zhou, J.C.; Liu, X.F.; Zheng, Y.; Ji, Y.; Li, X.F.; Xu, P.C.; Chen, Y.M.; Yang, Y.C. Effects of nitrogen deposition on soil microbial biomassand enzyme activities in *Castanopsis carlesii* natural forests in subtropical regions. *Acta Ecol. Sin.* **2017**, *37*, 127–135.
17. Bellingham, P.J.; Walker, L.R.; Wardle, D.A. Differential facilitation by a nitrogen-fixing shrub during primary succession influences relative performance of canopy tree species. *J. Ecol.* **2010**, *89*, 861–875. [CrossRef]
18. He, X.H.; Xu, M.G.; Qiu, G.Y.; Zhou, J.B. Use of 15N stable isotope to quantify nitrogen transfer between mycorrhizal plants. *J. Plant Ecol.* **2009**, *2*, 107–118. [CrossRef]
19. Schipanski, M.E.; Drinkwater, L.E. Nitrogen fixation in annual and perennial legume-grass mixtures across a fertility gradient. *Plant Soil* **2012**, *357*, 147–159. [CrossRef]
20. Png, G.K.; Lambers, H.; Kardol, P.; Turner, B.L.; Wardle, D.A.; Laliberté, E. Biotic and abiotic plant–soil feedback depends on nitrogen-acquisition strategy and shifts during long-term ecosystem development. *J. Ecol.* **2019**, *107*, 142–153. [CrossRef]
21. Olsen, S.L.; Sandvik, S.M.; Totland, O. Influence of two n-fixing legumes on plant community properties and soil nutrient levels in an alpine. *Arct. Antarct. Alp. Res.* **2013**, *45*, 363–371. [CrossRef]
22. Titus, J.H.; Bishop, J.G.; Moral, R.D. Propagule limitation and competition with nitrogen fixers limit conifer colonization during primary succession. *J. Veg. Sci.* **2014**, *25*, 990–1003. [CrossRef]
23. Tang, X.M.; Zhong, R.C.; Jiang, J.; He, L.Q.; Huang, Z.P.; Shi, G.Y.; Wu, H.N.; Liu, J.; Xiong, F.Q.; Han, Z.Q.; et al. Cassava/peanut intercropping improves soil quality via rhizospheric microbes increased available nitrogen contents. *BMC Biotechnol.* **2020**, *28*, 20. [CrossRef] [PubMed]
24. Li, H.; Zhu, N.; Wang, S.; Gao, M.; Wu, Y. Dual benefits of long-term ecological agricultural engineering: Mitigation of nutrient losses and improvement of soil quality. *Sci. Total Environ.* **2020**, *721*, 137848. [CrossRef] [PubMed]
25. Fu, Z.; Zhou, L.; Chen, P.; Du, Q.; Pang, T.; Song, C.; Wang, X.; Liu, W.; Yang, W.; Yong, T. Effects of maize-soybean relay intercropping on crop nutrient uptake and soil bacterial community. *J. Integr. Agr.* **2019**, *18*, 2006–2018. [CrossRef]

26. Forrester, D.I.; Bauhus, J.; Cowie, A.L. Growth dynamics in a mixed-species plantation of Eucalyptus globulus and Acacia mearnsii. *For. Ecol. Manag.* **2004**, *193*, 81–95. [CrossRef]
27. Kelty, M. The role of species mixture in plantation forestry. *For. Ecol. Manag.* **2006**, *233*, 195–204. [CrossRef]
28. Garcias-Bonet, N.; Arrieta, J.M.; Duarte, C.M.; Marbà, N. Nitrogen-fixing bacteria in mediterranean seagrass (*posidonia oceanica*) roots. *Aquat. Bot.* **2016**, *131*, 57–60. [CrossRef]
29. Zhan, X.; Clab, C.; Cza, C.; Yang, Y.; Wvdwb, C.; Fza, C. Intercropping maize and soybean increases efficiency of land and fertilizer nitrogen use; a meta-analysis. *Field Crop Res.* **2020**, *246*, 107661. [CrossRef]
30. Cong, W.F.; Hoffland, E.; Li, L.; Six, J.; Sun, J.H.; Bao, X.G.; Zhang, F.S.; Werf, V.D.W. Intercropping enhances soil carbon and nitrogen. *Glob. Chang. Biol.* **2015**, *21*, 1715–1726. [CrossRef]
31. Hu, Y.L.; Mgelwa, A.S.; Singh, A.N.; Zeng, D.H. Differential responses of the soil nutrient status, biomass production, and nutrient uptake for three plant species to organic amendments of placer gold mine-tailing soils. *Land Degrad. Dev.* **2018**, *29*, 2836–2845. [CrossRef]
32. Wang, L.; Wang, N.; Ji, G. Responses of biomass allocation and photosynthesis in mulberry to Pb-contaminated soil. *Acta Physiol. Plant* **2022**, *44*, 1–9. [CrossRef]
33. Xu, H.Y.; Gao, S.; Song, G.L.; Han, L.B. Effect of rocky slopes gradient on root growth and pull-out resistance of Lespedeza bicolor Turcz plants. In Proceedings of the Beijing International Symposium Land Reclamation & Ecological Restoration, Beijing, China, 16–19 October 2014.
34. Claesson, M.J.; O'Sullivan, O.; Wang, Q.; Nikkilä, J.; Marchesi, J.R.; Smidt, H.; De Vos, W.M.; Paul Ross, R.; O'Toole, P.W. Comparative Analysis of Pyrosequencing and a Phylogenetic Microarray for Exploring Microbial Community Structures in the Human Distal Intestine. *PLoS ONE* **2009**, *4*, e6669. [CrossRef] [PubMed]
35. White, T.J.; Bruns, T.; Lee, S.; Taylor, J. Amplification and direct sequencing of fungal ribosomal RNA genes for phylogenetics. In *PCR Protocols: A Guide to Methods and Applications*; Innis, M.A., Gelfand, D.H., Sninsky, J.J., White, T.J., Eds.; Academic Press: San Diego, CA, USA, 1994; pp. 315–322.
36. Martin, M. Cutadapt removes adapter sequences from high-throughput sequencing reads. *EMBnet* **2011**, *17*, 1. [CrossRef]
37. Callahan, B.J.; Mcmurdie, P.J.; Rosen, M.J.; Han, A.W.; Johnson, A.J.; Holmes, S.P. Dada2: High-resolution sample inference from illumina amplicon data. *Nat. Methods* **2016**, *13*, 581–583. [CrossRef]
38. Katoh, K. Mafft: A novel method for rapid multiple sequence alignment based on fast fourier transform. *Nucleic Acids Res.* **2002**, *30*, 3059–3066. [CrossRef]
39. Price, M.N.; Dehal, P.S.; Arkin, A.P. FastTree: Computing large minimum evolution trees with profiles instead of a distance matrix. *Mol. Biol. Evol.* **2009**, *26*, 1641–1650. [CrossRef]
40. Zaura, E.; Keijser, B.J.F.; Huse, S.M.; Crielaard, W. Defining the healthy "core microbiome" of oral microbial communities. *BMC Microbiol.* **2009**, *9*, 1–12. [CrossRef]
41. Liu, J.; Wei, Y.; Yin, Y.; Zhu, K.; Liu, Y.; Ding, H.; Lei, J.; Zhu, W.; Zhou, Y. Effects of Mixed Decomposition of Pinus sylvestris var. mongolica and Morus alba Litter on Microbial Diversity. *Microorganisms* **2022**, *10*, 1117. [CrossRef]
42. Erwin, D.H.; Laflamme, M.; Tweedt, S.M.; Sperling, E.A.; Pisani, D.; Peterson, K.J. The Cambrian conundrum: Early divergence and later ecological success in the early history of animals. *Science* **2011**, *334*, 1091–1097. [CrossRef]
43. Binkley, D.; Senock, R.; Bird, S.; Cole, T.G. Twenty years of stand development in pure and mixed stands of Eucalyptus saligna and nitrogen-fixing Facaltaria moluccana. *For. Ecol. Manag.* **2003**, *182*, 93–102. [CrossRef]
44. Forrester, D.I.; Bauhus, J.; Cowie, A.L.; Vanclay, J.K. Mixed-species plantations of eucalyptus with nitrogen-fixing trees: A review. *For. Ecol. Manag.* **2006**, *233*, 211–230. [CrossRef]
45. Soares, G.M.; Silva, L.D.; Higa, A.; Simon, A.A.; José, J.S. Artificial Neural Networks (Ann) For Height Estimation in A Mixed-Species Plantation of Eucalyptus Globulus Labill and Acacia Mearnsii De Wild. *Rev. Arvore* **2021**, *45*, e4512. [CrossRef]
46. Crowther, J.; Zimmer, H.; Thi, H.L.; Quang, T.L.; Nichols, J.D. Forestry in vietnam: The potential role for native timber species. *For. Policy Econ.* **2020**, *116*, 102182. [CrossRef]
47. Aosaar, J.; Varik, M.; Lõhmus, K.; Ostonen, I.; Becker, H.; Uri, V. Long-term study of above- and below-ground biomass production in relation to nitrogen and carbon accumulation dynamics in a grey alder (Alnus incana (L.) moench) plantation on former agricultural land. *Eur. J. For. Res.* **2013**, *132*, 737–749. [CrossRef]
48. Resh, S.C.; Binkley, D.; Parrotta, J.A. Greater soil carbon sequestration under nitrogen-fixing trees compared with eucalypt species. *Ecosystems* **2002**, *5*, 217–231. [CrossRef]
49. Vitousek, P. Ecosystem science and human-environment interactions in the Hawaiian archipelago. *J. Ecol.* **2006**, *94*, 510–521. [CrossRef]
50. Yokobe, T.; Hyodo, F.; Tokuchi, N. Volcanic deposits affect soil nitrogen dynamics and fungal–bacterial dominance in temperate forests. *Soil Biol. Biochem.* **2020**, *150*, 108011. [CrossRef]
51. Wan, X.H.; Huang, Z.Q.; He, Z.M.; Yu, Z.P.; Wang, M.H.; Davis, M.R.; Yang, Y.S. Soil C:N ratio is the major determinant of soil microbial community structure in subtropical coniferous and broadleaf forest plantations. *Plant Soil* **2015**, *387*, 103–116. [CrossRef]
52. Huang, X.M.; Liu, S.R.; Wang, Z.D.; You, Y.M. Changes of soil microbial biomass carbon and community composition through mixing nitrogen-fixing species with Eucalyptus urophylla in subtropical China. *Soil Biol. Biochem.* **2014**, *73*, 42–48. [CrossRef]

53. Wu, J.P.; Liu, Z.F.; Wang, X.L.; Sun, Y.X.; Zhou, L.X.; Lin, Y.B.; Fu, S.L. Effects of understory removal and tree girdling on soil microbial community composition and litter decomposition in two Eucalyptus plantations in South China. *Funct. Ecol.* **2011**, *25*, 921–931. [CrossRef]
54. Caracciolo, A.B.; Bustamante, M.A.; Nogues, I.; Lenola, M.D.; Luprano, M.L.; Grenni, P. Changes in microbial community structure and functioning of a semiarid soil due to the use of anaerobic digestate derived composts and rosemary plants. *Geoderma* **2015**, *245–246*, 89–97. [CrossRef]
55. Zou, Y.; Liang, N.; Zhang, X.; Han, C.; Nan, X. Functional differentiation related to decomposing complex carbohydrates of intestinal microbes between two wild zokor species based on 16srrna sequences. *BMC Vet. Res.* **2021**, *17*, 2–12. [CrossRef] [PubMed]
56. Bardgett, R.D.; Van, D.P. Belowground biodiversity and ecosystem functioning. *Nature* **2014**, *515*, 505–511. [CrossRef]
57. Fierer, N. Embracing the unknown: Disentangling the complexities of the soil microbiome. *Nat. Rev. Microbiol.* **2017**, *15*, 579–590. [CrossRef]
58. Zhao, C.; Zhang, Y.; Liu, X.; Ma, X.; Wang, H. Comparing the effects of biochar and straw amendment on soil carbon pools and bacterial community structure in degraded soil. *J. Soil Sci. Plant Nutr.* **2019**, *20*, 751–760. [CrossRef]
59. Xun, W.B.; Huang, T.; Zhao, J.; Ran, W.; Wang, B.R.; Shen, Q.R.; Zhang, R.F. Environmental conditions rather than microbial inoculum composition determine the bacterial composition, microbial biomass and enzymatic activity of reconstructed soil microbial communities. *Soil Biol. Biochem.* **2015**, *90*, 10–18. [CrossRef]
60. Cao, Y.; Yan, X.; Luo, H.; Jia, Z.; Jiang, X. Nitrification activity and microbial community structure in purple soils with different pH. *Acta Pedol. Sin.* **2018**, *1*, 194–202.
61. Wang, J.; Liao, L.R.; Ye, Z.C.; Liu, H.F.; Zhang, C.; Zhang, L.; Liu, G.B.; Wang, G.L. Different bacterial co-occurrence patterns and community assembly between rhizosphere and bulk soils under N addition in the plant-soil system. *Plant Soil* **2020**, *471*, 697–713. [CrossRef]
62. Zeng, J.; Liu, X.; Song, L.; Lin, X.; Chu, H. Nitrogen fertilization directly affects soil bacterial diversity and indirectly affects bacterial community composition. *Soil Biol. Biochem.* **2016**, *92*, 41–49. [CrossRef]
63. Zeng, J.; Liu, X.J.; Song, L.; Lin, X.G.; Zhang, H.Y.; Shen, C.C.; Chu, H.Y. Effects of changes in straw chemical properties and alkaline soils on bacterial communities engaged in straw decomposition at different temperatures. *Sci. Rep.* **2016**, *6*, 1–12.
64. Yang, Y.L.; Xu, M.; Zou, X.; Chen, J.; Zhang, J. Effects of different vegetation types on soil bacterial community characteristics in the hilly area of Qianzhong Mountai. *J. Ecol. Rural. Environ.* **2021**, *37*, 518–525.
65. Kazerooni, E.A.; Rethinasamy, V.; Al-Sadi, A.M. Talaromyces pinophilus inhibits Pythium and Rhizoctonia-induced damping-off of cucumber. *J. Plant Pathol.* **2018**, *101*, 377–383. [CrossRef]
66. Chen, X.P.; Yang, J.N.; Zhu, X.; Liang, X.; Lei, Y.R.; He, C.Q. N-fixing trees in wetland restoration plantings: Effects on nitrogensupply and soil microbial communities. *Environ. Sci. Pollut. R* **2016**, *23*, 24749–24757. [CrossRef]
67. Zuazo, V.; Pleguezuelo, C.; Tavira, S.C.; Martínez, J.R.F. Linking soil organic carbon stocks to land-use types in a Mediterranean agroforestry landscape. *J. Agr. Sci. Tech.* **2014**, *16*, 667–679.
68. Varma, V.; Iyengar, S.B.; Sankaran, M. Effects of nutrient addition and soil drainage on germination of n-fixing and non-n-fixing tropical dry forest tree species. *Plant Ecol.* **2016**, *217*, 1–12. [CrossRef]
69. Cech, P.G.; Venterink, H.O.; Edwards, P.J. N and P cycling in tanzanian humid savanna: Influence of herbivores, fire, and N_2-fixation. *Ecosystems* **2010**, *13*, 1079–1096. [CrossRef]
70. Macedo, M.O.; Resende, A.S.; Garcia, P.C.; Boddey, R.M.; Jantalia, C.P.; Urquiaga, S.; Campello, E.F.C.; Franco, A.A. Changes in soil C and N stocks and nutrient dynamics 13 years after recovery of degraded land using leguminous nitrogen-fixing trees. *For. Ecol. Manag.* **2008**, *255*, 1516–1524. [CrossRef]
71. Waldrop, M.P.; Zak, D.R.; Sinsabaugh, R.L.; Gallo, M.; Lauber, C. Nitrogen deposition modifies soil carbon storage through changes in microbial enzymatic activity. *Ecol. Appl.* **2004**, *14*, 1172–1177. [CrossRef]
72. Zeglin, L.H.; Stursova, M.; Sinsabaugh, R.L.; Collins, S.L. Microbial responses to nitrogen addition in three contrasting grassland ecosystems. *Oecologia* **2007**, *154*, 349–359. [CrossRef]
73. Deforest, J.L.; Zak, D.R.; Pregitzer, K.S.; Burton, A.J. Atmospheric nitrate deposition, microbial community composition, and enzyme activity in northern hardwood forests. *Soil Sci. Soc. Am. J.* **2004**, *68*, 132–138. [CrossRef]
74. Gao, S.; Deluca, T.H.; Cleveland, C.C. Biochar additions alter phosphorus and nitrogen availability in agricultural ecosystems: A meta-analysis. *Sci. Total Environ.* **2019**, *654*, 463–472. [CrossRef] [PubMed]
75. Brockett, B.F.T.; Prescott, C.E.; Grayston, S.J. Soil moisture is the major factor influencing microbial community structure and enzyme activities across seven biogeoclimatic zones in western Canada. *Soil Biol. Biochem.* **2012**, *44*, 9–20. [CrossRef]
76. Li, Y.; Zhou, C.; Qiu, Y.; Tigabu, M.; Ma, X. Effects of biochar and litter on carbon and nitrogen mineralization and soil microbial community structure in a china fir plantation. *J. For. Res.* **2019**, *30*, 1913–1923. [CrossRef]

MDPI AG
Grosspeteranlage 5
4052 Basel
Switzerland
Tel.: +41 61 683 77 34

Forests Editorial Office
E-mail: forests@mdpi.com
www.mdpi.com/journal/forests

Disclaimer/Publisher's Note: The statements, opinions and data contained in all publications are solely those of the individual author(s) and contributor(s) and not of MDPI and/or the editor(s). MDPI and/or the editor(s) disclaim responsibility for any injury to people or property resulting from any ideas, methods, instructions or products referred to in the content.

www.ingramcontent.com/pod-product-compliance
Lightning Source LLC
LaVergne TN
LVHW070637100526
838202LV00012B/826